Twenty-Ninth World Conference on Boundary Elements and Other Mesh Reduction Methods

BEM/MRM XXIX

Conference Chairmen

C.A. Brebbia
Wessex Institute of Technology, UK

V. Popov
Wessex Institute of Technology, UK

International Scientific Advisory Committee

Organised by
Wessex Institute of Technology, UK

Sponsored by
International Journal of Engineering Analysis with Boundary Elements (EABE)

WIT Transactions on Modelling and Simulation

Transactions Editor

Editorial Board

Boundary Elements
and
Other Mesh Reduction Methods
XXIX

Editors

C.A. Brebbia
Wessex Institute of Technology, UK

D. Poljak
University of Split, Croatia

V. Popov
Wessex Institute of Technology, UK

WIT*PRESS* **Southampton, Boston**

Editors:

C.A. Brebbia
Wessex Institute of Technology, UK

D. Poljak
University of Split, Croatia

V. Popov
Wessex Institute of Technology, UK

Published by

WIT Press
Ashurst Lodge, Ashurst, Southampton, SO40 7AA, UK
Tel: 44 (0) 238 029 3223; Fax: 44 (0) 238 029 2853
E-Mail: witpress@witpress.com
http://www.witpress.com

For USA, Canada and Mexico

Computational Mechanics Inc
25 Bridge Street, Billerica, MA 01821, USA
Tel: 978 667 5841; Fax: 978 667 7582
E-Mail: infousa@witpress.com
http://www.witpress.com

British Library Cataloguing-in-Publication Data

A Catalogue record for this book is available
from the British Library

ISBN: 978-1-84564-076-7
ISSN: (print) 1746-4064
ISSN: (on-line) 1743-355X

The texts of the papers in this volume were set individually by the authors or under their supervision. Only minor corrections to the text may have been carried out by the publisher.

Printed in Great Britain by Athenaeum Press Ltd.

Preface

The Boundary Element Conference – or to give it its full name – the International Conference on Boundary Element and other Mesh Reduction Methods – is now in its 29th year. The continued success of this important meeting is due to the strength of the work carried out by the community since the original conference took place in Southampton in 1978.

At that time Southampton University was a well know international centre for the application of boundary integral techniques in engineering, work initiated by Professor Hugh Tottenham and continued by the senior editor. The research taking place at that moment resulted in a series of theses, written by colleagues now recognised as the pioneers of BEM, and produced the first book on the technique ("The Boundary Element Method for Engineers" by C A Brebbia, Pentech Press, London, Halstead Press New York, 1978).

It also has two other important by products, the first was the beginning of this conference series and the second the launching of the International Journal of Engineering Analysis with Boundary Elements, which took place in the 1980's. These initiatives produced two focuses for the work of the international scientific community and resulted in a wealth of papers that are frequently referenced in the literature.

Following the creation of the Wessex Institute of Technology in 1986, the bulk of the BEM research activities took place at WIT, where they still are one of the stronger areas of research.

WIT also continues to produce outstanding researchers, similar to those who in the past have now established their own groups worldwide and are directly associated with this conference.

The current meeting continues with the tradition of innovation and quality which have always been the trademarks of these conferences. The book contains a series of important sections:

- Meshless techniques
- Advanced formulations
- Dual reciprocity method

- Computational issues
- Fluid mechanics applications
- Heat and mass transfer
- Plates and shells
- Wave propagation
- Damage mechanics and fracture
- Electrical engineering & electromagnetics
- Inverse problems

BEM/07 was held on WIT's Campus in the New Forest which gave the participants the opportunity of appreciating the substantial growth of the Institute in the last few years. They were also able to discuss with WIT staff the application of BEM in a diversity of new areas.

The Editors appreciate the continuous support of the international community over a long period which resulted in the outstanding contributions published in this volume and the other 50 or so in the Series. They are also indebted to the members of the International Scientific Advisory Committee and other colleagues who helped to review the papers contained in this book.

The Editors
New Forest, UK
2007

Contents

Section 3: Dual reciprocity method

Section 4: Computational issues

Section 5: Fluid mechanics applications

Section 6: Heat and mass transfer

Section 7: Plates and shells

Section 8: Wave propagation

Section 9: Damage mechanics and fracture

Section 10: Electrical engineering and electromagnetics

Section 11: Inverse problems

Section 1
Meshless techniques

Stress analysis by local integral equations

V. Sladek[1], J. Sladek[1] & Ch. Zhang[2]
[1]*Institute of Construction and Architecture, Slovak Academy of Sciences, Bratislava, Slovakia*
[2]*Department of Civil Engineering, University of Siegen, Germany*

Abstract

This paper is a comparative study for various numerical implementations of local integral equations developed for stress analysis in plane elasticity of solids with functionally graded material coefficients. Besides two meshless implementations by the point interpolation method and the moving least squares approximation, the element based approximation is also utilized. The numerical stability, accuracy, convergence of accuracy and cost efficiency (assessed by CPU-times) are investigated in numerous test examples with exact benchmark solutions.
Keywords: elasticity, functionally graded materials, boundary value problems, force equilibrium, meshless implementations.

1 Introduction

A rapid progress can be observed in the development of various meshless techniques especially in fluid problems. Simultaneously, a considerable expansion of such techniques can be found also in various applications to engineering and science problems. This can be explained by the fact that there are known certain limitations of standard discretization techniques especially when applied to some classes of problems (e.g. problems in separable media, problems with free or moving boundaries; crack problems; problems with large distortions, etc.). Although the standard discretization techniques are applicable to the numerical solution of boundary value problems in continuously non-homogeneous elastic media, the formulations developed for homogeneous media are not applicable directly, since the governing equations are now given by partial differential equations with variable coefficients.

There has not been a unique classification of meshless techniques up to now. Mostly they are classified according to the employed approximation. Some of

WIT Transactions on Modelling and Simulation, Vol 44, © 2007 WIT Press
www.witpress.com, ISSN 1743-355X (on-line)
doi:10.2495/BE070011

the techniques utilize meshless approximation of field variables but a background mesh is still required for numerical integration especially in approaches based on global formulations. On the other hand, the local formulations bring a possibility to avoid the mesh completely with using nodes alone for approximation. Then, the physical principles can be formulated in integral forms on local sub-domains. A large group of meshless techniques are denoted as meshless local Petrov–Galerkin methods [1, 2] with bearing in mind that the Petrov–Galerkin weak form idea is applied in a local sense with selecting the trial and test functions independently and approximating the field variables in a meshless way.

Some comparative studies might be desired in this stage of rapid increase of literature devoted to various meshless techniques as well as their applications to various problems.

2 Local integral equations

Under assumption of static loading conditions, the demand of the force equilibrium in an arbitrary but small part of the elastic body results in the strong formulation of the governing equations given by the partial differential equations

$$\sigma_{ij,j}(\mathbf{x}) + X_i(\mathbf{x}) = 0 \ \text{ in } \Omega, \tag{1}$$

supplemented by the generalized Hooke's law

$$\sigma_{ij}(\mathbf{x}) = c_{ijkl}(\mathbf{x}) u_{k,l}(\mathbf{x}). \tag{2}$$

In the case of isotropic FGM, the spatial variation of the tensor of material coefficients is usually given via the variable Young's modulus as

$$c_{ijkl}(\mathbf{x}) = E(\mathbf{x}) c^{o}_{ijkl}, \quad c^{o}_{ijkl} = \frac{1}{2(1+\nu)}\left(\delta_{ik}\delta_{jl} + \delta_{il}\delta_{jk} + \frac{2\bar{\nu}}{1-2\bar{\nu}}\delta_{ij}\delta_{kl}\right), \tag{3}$$

with the material parameter $\bar{\nu}$ being expressed in terms of the constant Poisson ratio ν by

$$\bar{\nu} = \begin{cases} \nu/(1+\nu), & \text{for plane stress conditions} \\ \nu, & \text{otherwise} \end{cases}.$$

Inserting (3) into (1), one obtains the governing PDE for displacements

$$E(\mathbf{x})\, c^{o}_{ijkl}\, u_{k,lj}(\mathbf{x}) + E_{,j}(\mathbf{x})\, c^{o}_{ijkl}\, u_{k,l}(\mathbf{x}) = -X_i(\mathbf{x}). \tag{4}$$

The standard boundary conditions prescribe a half of the boundary quantities $\{\tilde{u}_i(\boldsymbol{\eta}), \tilde{t}_i(\boldsymbol{\eta})\}$ for ($i = 1,...,d$) at each boundary point $\boldsymbol{\eta} \in \partial\Omega$, with the traction vector being given by

$$t_i(\boldsymbol{\eta}) = n_j(\boldsymbol{\eta}) c_{ijkl}(\boldsymbol{\eta}) u_{k,l}(\boldsymbol{\eta}), \tag{5}$$

where $n_j(\mathbf{\eta})$ denotes the Cartesian components of the unit outward normal vector on the boundary $\partial\Omega$.

In numerical formulations for solution of b.v.p., weak formulations are frequently utilized instead of the strong formulation. The governing equation is satisfied in a weak sense if the weighted integral of the governing equation is fulfilled

$$\int_{\Omega}\left(\sigma_{ij,j}(\mathbf{x})+X_i(\mathbf{x})\right)w_{ik}(\mathbf{x})d\Omega(\mathbf{x})=0\,. \tag{6}$$

Since the test (or weight) functions can be arbitrary, the weak formulation might have no physical interpretation. In order to apply the formulation with clear physical interpretation, we shall use the test functions given by the Heaviside functions with support on local sub-domains Ω^c of the whole analysed domain Ω

$$w_{ik}^{c}(\mathbf{x})=\begin{cases}\delta_{ik}\,,\mathbf{x}\in\Omega^{c}\\ 0,\quad \mathbf{x}\notin\Omega^{c}\end{cases}.$$

Then, the weak formulation (6) after using the Gauss divergence theorem yields the well known force equilibrium on local sub-domains Ω^c

$$\int_{\partial\Omega^c} n_j(\mathbf{\eta})c_{ijkl}(\mathbf{\eta})u_{k,l}(\mathbf{\eta})d\Gamma(\mathbf{\eta})=-\int_{\Omega^c} X_i(\mathbf{x})d\Omega(\mathbf{x})\,, \tag{7}$$

which is the weak formulation with the clear physical interpretation.

Recall that the local integral equations (7) are non-singular, since there are no singular fundamental solutions involved in contrast to the singular integral equations employed in the boundary integral equation method. Moreover, the integration of unknown (approximated) field variables is constraint to the boundary of local sub-domains even in the case of non-homogeneous media. This can be effectively utilized by decreasing the amount of integration points as compared with the formulations involving the domain integrals.

3 Numerical implementations

In numerical solving, in general, the amount of degrees of freedom is decreased from infinity to a finite number by approximating the field variable in terms of certain shape functions and nodal unknowns. The nodal unknowns are determined by the set of equations obtained by collocating the prescribed boundary conditions at boundary nodes and force equilibrium equations at interior nodal points.

We shall consider domain-type approximations $u_i^h(\mathbf{x})$ for the displacements $u_i(\mathbf{x})$ within a sub-domain $\Omega_s\subset(\Omega\cup\partial\Omega)$. Then, it is possible to get also the approximations for displacement gradients by differentiating the approximated displacements. Thus, the discretized equations take the form

$$u_i^h(\zeta) = \tilde{u}_i(\zeta) \quad \text{at} \quad \zeta \in \partial\Omega \ \text{where} \ \tilde{u}_i(\zeta) \ \text{is prescribed}, \tag{8a}$$

$$n_j(\zeta) c_{ijkl}(\zeta) u_{k,l}^h(\zeta) = \tilde{t}_i(\zeta) \quad \text{at} \quad \zeta \in \partial\Omega \ \text{where} \ \tilde{t}_i(\zeta) \ \text{is prescribed}, \tag{8b}$$

$$\int_{\partial\Omega^c} n_j(\boldsymbol{\eta}) c_{ijkl}(\boldsymbol{\eta}) u_{k,l}^h(\boldsymbol{\eta}) d\Gamma(\boldsymbol{\eta}) = -\int_{\Omega^c} X_i(\mathbf{x}) d\Omega(\mathbf{x}) \tag{8c}$$

on sub-domains Ω^c around interior nodes $\mathbf{y}^c$.

3.1 Quadrilateral quadratic elements

A 2-d plane domain Ω is assumed to be subdivided into m conforming quadrilateral serendipity elements S_e [3] with quadratic polynomial interpolation for the approximation of both the geometry and displacements. Then,

$$\Omega = \bigcup_{e=1}^{m} S_e, \quad x_i\big|_{S_e} = \sum_{a=1}^{8} x_i^{ae} N^a(\xi_1, \xi_2), \quad u_i^h(\mathbf{x})\Big|_{S_e} = \sum_{a=1}^{8} u_i(\mathbf{x}^{ae}) N^a(\xi_1, \xi_2), \tag{9}$$

where x_i^{ae} are the Cartesian coordinates of the a-th nodal point on S_e and N^a represent the shape functions. Since the interpolation polynomials are expressed as functions of intrinsic coordinates, the expressions for displacement gradients are not trivial [4] and integrations are to be carried out in the transformed intrinsic space. The local sub-domain Ω^c is specified as union of elements adjacent to the interior node $\mathbf{y}^c$.

3.2 Point interpolation method (PIM)

As in all meshless approximation techniques, the shape functions derived for the approximation of the field variable $u_i(\mathbf{x})$ within a sub-domain Ω_s utilize only nodes scattered arbitrarily in the analyzed domain without any predefined mesh to provide a connectivity of the nodes.

Without going into details [5, 6], we present the interpolation formula for displacements in surroundings of the nodal point $\mathbf{x}^q$ in terms of the shape functions and nodal values as

$$u_i^h(\mathbf{x})\Big|_{\Omega^q} = \sum_{\alpha=1}^{N^q} u_i(\mathbf{x}^{n(q,\alpha)}) \varphi^{(q,\alpha)}(\mathbf{x}), \tag{10}$$

where $n(q,\alpha)$ stands for the global number of nodes from the interpolation domain Ω_i^q. If Ω_i^q is defined as a circle with the radius r^q the number of nodes involved in Ω_i^q is given as $N^q = \sum_{a=1}^{N_t} H\left(r^q - \left|\mathbf{x}^a - \mathbf{x}^q\right|\right)$, where $H(z)$ is the Heaviside unit step function and N_t is the total number of nodes. The numerically stable development of the shape functions can be achieved by combining the polynomials and RBFs as basis functions in a PIM(P+RBF) approach [5, 6]. The explicit expression for the shape functions being given elsewhere [6]. Recall that the Kronecker-delta property is satisfied

$$\varphi^{(q,\alpha)}(\mathbf{x}^{n(q,\beta)}) = \delta_{\alpha\beta} .$$

Finally, the displacement gradients are approximated as gradients of approximated displacements

$$u_{i,j}^h(\mathbf{x})\Big|_{\Omega^q} = \sum_{\alpha=1}^{N^q} u_i(\mathbf{x}^{n(q,\alpha)})\varphi_{,j}^{(q,\alpha)}(\mathbf{x}) , \tag{11}$$

i.e., in terms of the nodal values of displacements and the derivatives of the shape functions.

Note that the shape functions and their derivatives are not available in closed form. Thus, certain computational algorithm is to be repeated at each evaluation point. Nevertheless, in the present formulation, some of the inverse matrices can be pre-computed and stored in the memory for each nodal point in order to save CPU-time.

3.3 Moving least squares (MLS) approximation

The MLS-approximation is widely used in meshless methods. The displacements are approximated in terms of certain shape functions and nodal unknowns as

$$u_i^h(\mathbf{x}) = \sum_{a=1}^{N_t} \phi^a(\mathbf{x})\hat{u}_i^a . \tag{12}$$

The shape functions are expressed in terms monomial basis functions and weights associated with each nodal point. They have to be computed according to certain algorithm at each evaluation point. Since the shape functions do not possess the Kronecker delta property, $\phi^a(\mathbf{x}^b) \neq \delta_{ab}$, in general, the nodal unknowns are expansion coefficients (fictitious nodal displacements) which are different from the actual nodal values of displacements.

Since the number of nodal points which contribute to the sum in Eq. (12) is controlled by the weights, one has to consider all the nodes in the summation. To decrease the amount of the considered nodes, the Central Approximation Node (CAN) concept can be used. Then, the number of considered nodes in each evaluation at $\mathbf{x}$ is reduced from N_t to N^q, where $N^q < N_t$ is the number of nodes supporting the approximation at the central approximation node $\mathbf{x}^q$. The nodes supporting the approximation with the CAN located at $\mathbf{x}^q$ lie in the Ω_{CAN}^q specified by the radius r^q. Then, instead of the approximation given by Eq. (12), one can use

$$u_i^h(\mathbf{x}) = \sum_{a=1}^{N^q} \hat{u}_i^{n(q,a)} \phi^{n(q,a)}(\mathbf{x}). \tag{13}$$

The gradients of displacements are approximated as gradients of approximated displacements given by Eqs. (12) and (13).

In contrast to the implementation based on finite elements, the integration in meshless approaches is carried out in the global Cartesian coordinate system. The choice of simple shape for sub-domains yields simple integration. The Gaussian quadrature proved to be more convenient than the trapezoidal rule for integration over the boundary of circular sub-domains, since the later exhibits very slow improvement of accuracy with fining the subdivision of the integration interval, what results in enormous increasing the computational time needed for evaluation of shape functions.

4 Numerical tests

In order to test the proposed numerical methods, we consider examples for which analytical solutions are available. The body forces are vanishing in Ω, the Poisson ratio is constant $\nu = 0.25$, plane stress conditions are assumed and for conciseness, we present only the numerical results for exponential gradation $E(\mathbf{x}) = E_o \exp(2\delta x_2 / L)$ with $\delta = 3$. The considered domain is a square $L \times L$ with applied tension load on the top, fixed bottom in vertical direction and tractions on the lateral sides are given by the analytical solution [7].

In the study of the convergence and accuracy of the numerical results with respect to the increasing density of nodal points, we use the displacement norm % error defined as

$$\text{displ. norm error (\%)} = 100 \left\{ \sum_{a=1}^{N_t} \Delta u_i^a \Delta u_i^a \right\}^{1/2} \Big/ \left\{ \sum_{a=1}^{N_t} u_i^{ex}(\mathbf{x}^a) u_i^{ex}(\mathbf{x}^a) \right\}^{1/2}$$

$$\Delta u_i^a = u_i^{num}(\mathbf{x}^a) - u_i^{ex}(\mathbf{x}^a), \tag{14}$$

where N_t is the total number of nodes on the closed domain $\Omega \cup \partial\Omega$.

In most of the presented computations, we shall use a homogeneous distribution of nodes with $h^a = \min_{\forall b} \left\{ \left\| \mathbf{x}^a - \mathbf{x}^b \right\| \right\} = \text{const} = h$.

In the PIM, we have used combination of polynomial functions (given by six monomial basis) with RBFs for which we considered multiquadrics, Gaussian RBFs, and the 8-order spline. Similar in the case of MLS-approximation, we have used three different kinds of weights given by Gaussian, exponential, and 8-order spline weights. Although the shape and size of sub-domains can be chosen arbitrarily, the results of numerical computations may depend on these aspects and similarly on the shape parameters involved in both the RBFs and weights of the MLS-approximation. Therefore, firstly we have investigated the stability of numerical computations with respect to those three indicators. The use of square shape for sub-domains yields better results as compared with the circular shape. In the next computations, we used optimal values for the shape parameter and the size of sub-domains which guarantee the numerical stability. The numerical instability with respect to acceptable accuracy was observed in case of exponential weights used in MLS-approximation. The CAN concept with the nearest node to the evaluation point proved to give the best results in both the meshless techniques. The radius of the interpolation domain is taken as

$$r^a = 3.001\, h^a .$$

Fig. 1 shows the convergence of the numerical solutions by various PIM(P+RBF) approaches. The increasing density of nodes is represented by the decreasing parameter h/L. Fig.2 illustrates the variation of the displacement field $u_2(L/2, x_2)$ along the vertical line $(L/2, x_2)$ with $x_2 \in [0, L/2]$. It can be seen that excellent accuracy is achieved even in the case of strong gradation of Young's modulus when the variation of displacements differs dramatically from the case of homogeneous medium. The accuracy of numerically computed interior stresses is also reasonable (the results will be summarized in Tab.1). Thus, the PIM based on the combination of polynomials and the multiquadrics with $m = 5/2$ seems to be appropriate even for strong non-homogeneity $\delta = 3$, when the Young modulus on the top of the square domain is 403 times higher than on the bottom.

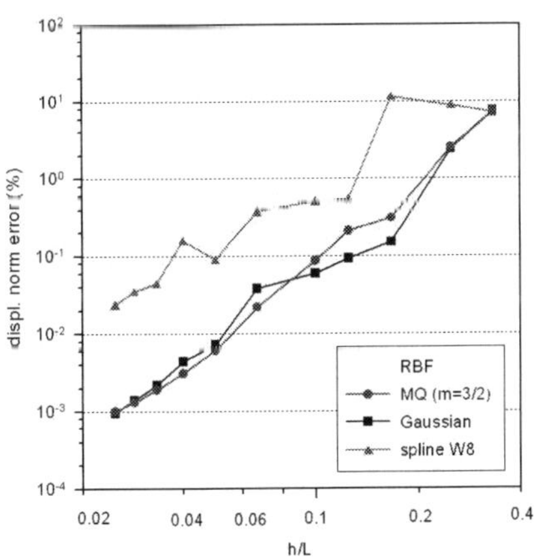

Figure 1: Convergence study.

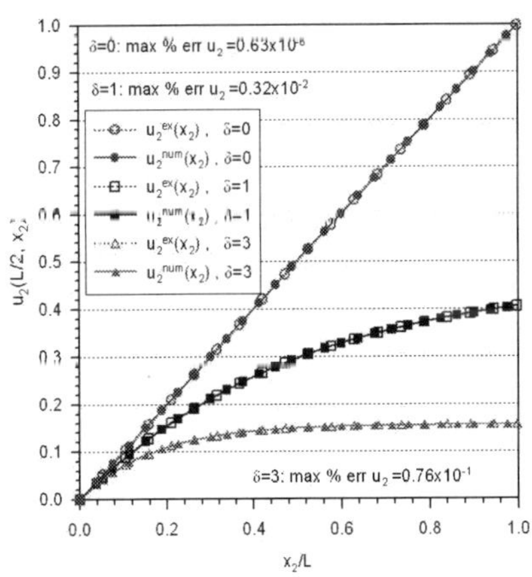

Figure 2: Displacement results.

It is interesting to compare the results by two variants of the MLS-approximation: standard formulation vs CAN-nearest node concept. Fig. 3 shows the comparison of the convergence of numerical results by using these two different approaches with Gaussian weights. It can be seen that the accuracy is almost invariant with respect to the predefinition of supporting nodes. On the other hand, the influence on the CPU-times is much more significant (Fig. 4). Although we can see negligible differences between the CPU-times resulting from both the formulations provided that the densities of nodes are low, the rates for the CPU-time increase are significantly different for each of the employed approaches. The difference between the CPU-times by the CAN-nearest node approach and standard MLS approach is increasing remarkably with increasing the density of nodes.

Finally, we present some comparisons of the best formulations based on the use of three different kinds of domain-type approximations. The best meshless formulations utilize the CAN-nearest node concept and are characterized by selection of square shapes for sub-domains, and optimal values of the shape parameter (involved in RBFs and/or weights) as well as the sub-domain size parameter.

The QE-approach exhibits reliable convergence of accuracy with increasing the density of nodes, but lower accuracy is achieved in the FGM sample with strong gradation of the Young modulus ($\delta = 3$) as compared with the meshless PIM and/or MLS results (Fig. 5).

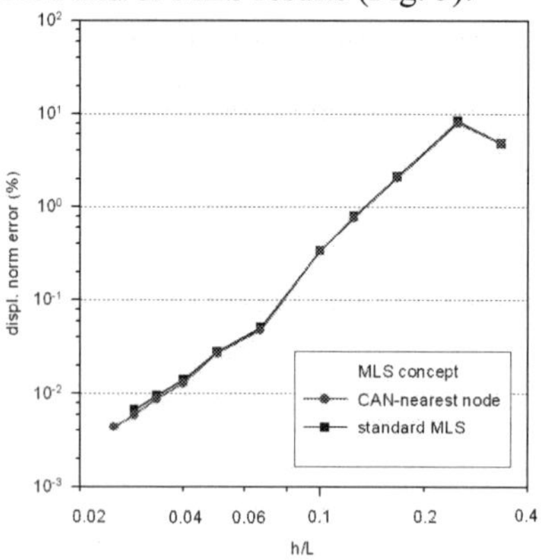

Figure 3: Accuracies by two MLS concepts.

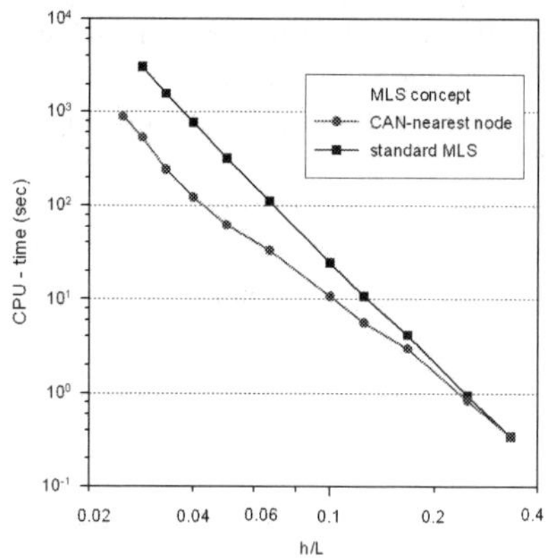

Figure 4: CPU by two MLS concepts.

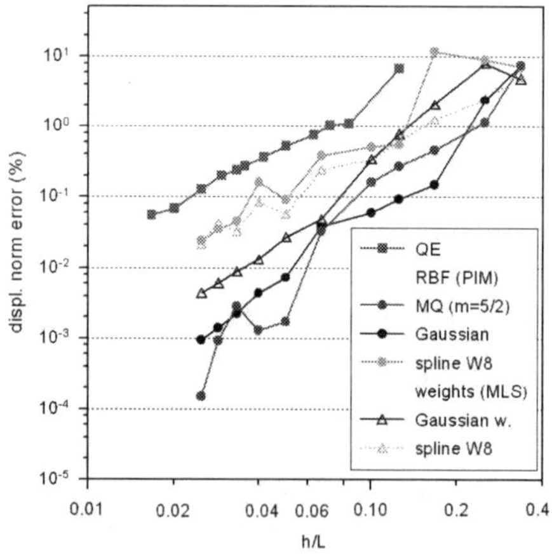

Figure 5: Comparison of accuracies by various numerical techniques.

A comparison of the CPU-times by the QE-approach with various meshless approaches is given in Fig. 6. The CPU-times by meshless approaches converge to each other by increasing the density of nodes and the differences between the QE and meshless approaches are diminished. This can be explained by the fact that with increasing the amount of nodes the time needed for solution of the system of discretized equations is becoming dominant in comparison with the time needed for evaluation of the system matrix.

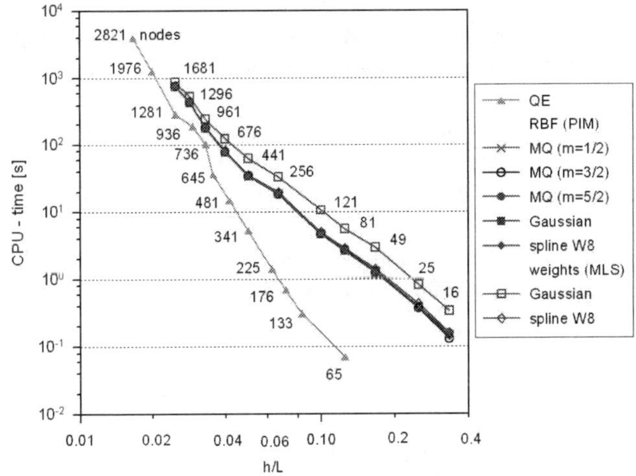

Figure 6: Comparison of CPU-times by various numerical techniques.

Table 1: Maximal % errors for displacements and stresses computed at interior points along the vertical line ($L/2.\ x_2$) in square sample.

computational method	grade parameter	max % error			CPU [sec]
		u_2	σ_{22}	σ_{11}	
LIE-QE (400 elem.) 1281 nodes	$\delta = 0$	0.84×10^{10}	0.39×10^{-9}		290
	$\delta = 1$	0.099	0.18	0.31	
	$\delta = 3$	0.93	1.93	5.83	
LIE-PIM(P+MQ) (441 nodes)	$\delta = 0$	0.63×10^{-6}	0.88×10^{-6}	-	35
	$\delta = 1$	0.0032	0.0083	0.032	
	$\delta = 3$	0.076	0.22	3.77	
LIE-MLS (441 nodes)	$\delta = 0$	0.10×10^{-9}	0.57×10^{-9}	-	65
	$\delta = 1$	0.11	0.16	0.20	
	$\delta = 3$	0.78	1.54	1.68	

The slightly higher relative error for the stresses σ_{11} is due to small value of this component in the considered b.v.p.

Similar results have been obtained also for stress analysis in both the transversal and axial cross-section of the thick-wall tube.

5 Conclusions

Both the meshless techniques proved to be useful for numerical implementation of the LIEs applied to stress analysis problems even in the case of strong gradation of the Young modulus.

- Acceptable accuracy, convergence of accuracy and numerical stability are guaranteed by using the proposed techniques.
- Great savings in the CPU-time are achieved by using the CAN-nearest node concept.
- The accuracy by the QE-approach is slightly worse than by meshless approaches, but reliable convergence is achieved with increasing the density of nodes.

Acknowledgements

The research has been supported by the Slovak Grant Agencies VEGA, APVV and German Research Foundation (DFG), which are gratefully acknowledged.

References

[1] Atluri S.N., Shen S., *The meshless local Petrov-Galerkin (MLPG) method*, Tech Science Press: Encino, 2002.

[2] Atluri S.N., *The meshless method (MLPG) for domain & BIE discretizations*, Tech Science Press: Forsyth, 2004.

[3] Hughes T.J.R., *The Finite Element Method. Linear Static and Dynamic Finite Element Analysis*. Prentice-Hall, Inc.: Englewood Cliffs, 1987.

[4] Sladek V., Sladek J., Zhang Ch., Local integro-differential equations with domain elements for numerical solution of PDE with variable coefficients. *J. Eng. Mathematics* **51**, pp. 261-282, 2005.

[5] Liu G.R., *Mesh Free Methods, Moving beyond the Finite Element Method.* CRC Press: Boca Raton, 2003.

[6] Sladek V., Sladek J., Tanaka M., Local integral equations and two meshless polynomial interpolations with application to potential problems in non-homogeneous media. *Computer Modeling in Engineering & Sciences* 7, pp. 69-83, 2005.

[7] Sladek V., Sladek J., Zhang Ch., A meshless Point Interpolation Method for Local Integral Equations in elasticity of non-homogeneous media. *Advances in the Meshless Method,* eds. J. Sladek, V. Sladek, Tech Science Press: Forsyth, 2006.

Non-overlapping domain decomposition scheme for the symmetric radial basis function meshless approach with double collocation at the sub-domain interfaces

H. Power, A. Hernandez & A. La Rocca
The University of Nottingham,
School of Mechanical, Material and Manufacturing Engineering, UK

Abstract

In the particular case of solving large-scale boundary value problems, the computational cost derived as a result of the application of any numerical scheme represents a determinant factor in the determination of its computational efficiency. The present work studies the influence of the non-overlapping domain decomposition on the symmetric radial basis collocation method, as a way to improve its efficiency under high demanding numerical conditions. Due to the Hermitian character of the symmetric scheme at each of the collocations points of the sub-domain interfaces it is possible to impose simultaneously all the corresponding matching conditions. A multi-zone problem is considered as a test example, comparison between the numerical result and the analytical solution for two set of different physical parameters are presented.
Keywords: symmetric RBF meshless approach, domain decomposition and double collocation.

1 Introduction

The use of a mesh is a basic characteristic of traditional numerical approaches for the solution of partial differential equations, as is the case of the finite difference, element, volume and the boundary element methods. In the first cases, assumptions are made for the local approximation, which require internal mesh to support them. On the other hand, in the case on boundary methods, a

WIT Transactions on Modelling and Simulation, Vol 44, © 2007 WIT Press
www.witpress.com, ISSN 1743-355X (on-line)
doi:10.2495/BE070021

boundary mesh is required to obtain a numerical approximation of the resulting boundary integrals.

During recent years, considerable effort has been given to the development of the so-called free-mesh methods (meshless approach). The aim of this type of approach is to eliminate at least the structure of the mesh and approximate the solution entirely using nodes values inside and/or in the boundary quasi random distributed in the domain.

Recently, some significant developments in meshless methods for solving boundary value problems of partial differential equations have been reported in the literature. Kansa [1, 2] introduced the concept of solving PDEs using radial basis functions (RBFs) (Unsymmetric scheme). This type of approach, which approximates the whole solution of the PDE directly using RBFs, is very attractive due to the fact that this is truly a mesh free technique.

The Kansa's method has been applied successfully in several cases (see for example [3–5]). However, no existence of solution and convergence analysis is available in the literature and for some cases, it has been reported that the resulting matrix was extremely ill-conditioned and even singular for some distribution of the nodal points (see [6]).

Several techniques have been proposed to improve the conditioning of the coefficient matrix and the solution accuracy, as are: the use of high order interpolation functions, replacement of global solvers by block partitioning, LU decomposition schemes, matrix preconditioners, overlapping and non-overlapping domain decomposition etc (see [7]). Fedoseyev et al. [8] proposed the use of a set of additional nodes at the boundary and beyond the boundary (at the exterior) where the governing equation is required to be satisfied. It was found that the suggested approach yields to more accurate results than only imposing the governing equation at internal nodes.

Fasshauer [9] suggested an alternative approach to the Unsymmetric scheme based on the Hermite interpolation property of the radial basis functions, which states that the RBFs not only are able to interpolate a given function but also its derivatives. The convergence proof for RBF Hermite-Brikhoff interpolation was given by Wu [10] who also proved the convergence of this approach when solving PDEs (see Wu [11] and Schaback and Franke [12]). Another advantage of the Hermite based approach is that the matrix resulting from the scheme is symmetric, as opposed to the completely unstructured matrix of the same size resulting from Unsymmetric schemes.

The main objective of this work is to study and test some of the above mentioned techniques previously used to improve the efficiency of the Unsymmetric approach in order to increase the computational efficiency of the radial basis function symmetric approach. In particular, we will study the non-overlapping domain decomposition with a double collocation at the sub-domain interfaces. The domain decomposition approach is itself a very powerful and popular scheme in numerical analysis, which have recently increased its popularity due to its use in parallel computing algorithms.

2 Symmetric radial basis function meshless approach

Let us consider a boundary value problem defined by:

$$L[C](x) = f(x) \tag{1-a}$$
$$B[C](x) = g(x) \tag{1-b}$$

where the operators L and B are linear partial differential operators on the domain Ω and at the contour Γ respectively.

A symmetric RBF collocation method (Fasshauer [9]), represents the solution of the above boundary value problem by the interpolation function:

$$C(x) = \sum_{k=1}^{n} \lambda_k B_\xi \Psi(\|x-\xi_k\|) + \sum_{k=n+1}^{N} \lambda_k L_\xi \Psi(\|x-\xi_k\|) + P_{m-1}(x) \tag{2}$$

with n as the number of nodes on the boundary of Ω and $N-n$ the number of internal nodes. Here, $\Psi(\|x-x_j\|)$ is a conditionally positive definite RBF of order m and P a polynomial term of order $m-1$.

In the above expression L_ξ and B_ξ are the differential operators used in (1-a,b), but acting on Ψ viewed as a function of the second argument ξ (see Fasshauer [9]). This expansion for C leads to a collocation matrix A, which is of the form

$$A = \begin{pmatrix} B_x B_\xi[\Psi] & B_x L_\xi[\Psi] & B_x P_{m-1} \\ L_x B_\xi[\Psi] & L_x L_\xi[\Psi] & L_x P_{m-1} \\ B_x P_{m-1}^T & L_x P_{m-1}^T & 0 \end{pmatrix} \tag{3}$$

where the following ortogonality conditions is required to complete the system:

$$\sum_{k=1}^{n} \lambda_k B_x P_{m-1}^T + \sum_{k=n+1}^{N} \lambda_k P_{m-1}^T = 0 \tag{4}$$

The matrix (3) is of the same type as the scattered Hermite interpolation matrices and thus non-singular as long as Ψ is chosen appropriately (see Wu [6]). A major point in favour of the Hermite based approach is that the matrix resulting from the scheme is symmetric, as opposed to the completely unstructured matrix of the same size resulting from Unsymmetric schemes.

3 Convection-diffusion problem

The steady state differential equation considered in this work is of the form:

$$D\frac{d^2C}{dx_j^2} - u_j \frac{dC}{dx_j} + kC = 0 \tag{5}$$

The partial differential operators on the matrix representation (3) of the symmetric collocation numerical solution of equation (5), when satisfying boundary conditions of the first and second kind (Dirichlet and Neumann), are defined by the following expressions:

$$L_x = D\frac{d^2}{dx_j^2} - u_j\frac{d}{dx_j} + k,\ L_\xi = D\frac{d^2}{d\xi_j^2} - u_j\frac{d}{d\xi_j} + k,$$
$$B_x^D = 1,\ B_x^N = \frac{d}{dx_j}n_j(x),\ B_\xi^D = 1,\ B_\xi^D = \frac{d}{d\xi_j}n_j(\xi) \tag{6}$$

In the above relations the super index D and N in the operator B represent the type of boundary conditions implemented, i.e. Dirichlet and Neumann.

In this work we will use the generalized TPS. Furthermore to avoid singularity at $r = 0$ on the resulting differential operators of the matrix A, we use in the representation formula (2) the generalized TPS

$$\psi = r^6 \log r \tag{7}$$

together with the corresponding cubic polynomial.

$$P_3(x) = \lambda_{N+1}x_1^3 + \lambda_{N+2}x_2^3 + \lambda_{N+3}x_1^2x_2 + \lambda_{N+4}x_1x_2^2 + \lambda_{N+5}x_1^2 +$$
$$\lambda_{N+6}x_2^2 + \lambda_{N+7}x_1x_2 + \lambda_{N+8}x_1 + \lambda_{N+9}x_2 + \lambda_{N+10} \tag{8}$$

4 Domain decomposition approach

Domain decomposition methods are frequently used in two contexts. First: the division of problems into smaller problems usually through artificial subdivisions of the domain, as a way to improve the performance of a numerical technique. Second, many problems involve more than one mathematical model each posed on a different domain, so that domain decomposition occurs naturally.

When dealing with the numerical simulation of large problems, it is usual to use the method of domain decomposition, in which the original domain is divided into sub-regions, and on each of them the original governing equations are imposed.

The main objective of the domain decomposition method is to decompose one large global problem into smaller sub-domain problems. In the implementation of the domain decomposition approach, two different alternatives are possible to use: overlapping and non-overlapping schemes.

In the non-overlapping technique, the domain is divided into non-overlapping sub-domains having common interface surfaces. In each sub-domain the original numerical scheme is implemented. Owing to the lack of the boundary condition on the interface between sub-domains, additional surface unknowns need to be determined, i.e. in the present case the value of the concentration and the surface flux. For each interface boundary point, the number of unknowns is more than the number of the equations and therefore the resulting system is underdetermined. However, once the matching conditions are imposed and the sub-domain assembled, then is possible to obtain a close system.

In the overlapping approach, the problem can be solved by an iterative scheme in terms of one of the Schwarz methods, i.e. by solving recursively each of the sub-domains, or instead the complete close system can be solved directly, after imposing the interface matching conditions between subdomains, without the used of an iterative scheme.

It is important to observe that the non-overlapping domain decomposition approach is naturally suited for the numerical solution of multi-zone problems, where the governing equations have different values of the problem parameters at different regions of the problem domain.

Several application of the domain decomposition approach has been reported in the literature when solving partial differential equations with the use of the Unsymmetric radial basis function collocation approach. In the work by Kansa and Carlson [13] they conclude that one of the most efficient technique when solving dense system of linear equation is to use preconditioning and to make use of domain decomposition techniques.

In this work will be implement the non-overlapping non-iterative domain decomposition approach for the numerical solution of boundary values problems based on the symmetric radial basis function collocation approach, with application to multi-zone problems.

5 Multi-zone problems

Considering a problem that contains different regions, in which the coefficients of the governing equation are constant but different in each of them. In the implementation of the non-overlapping domain decomposition approach for the solution of multi-zone problems, the problem's domain is divided into a finite number of non-overlapping zones according with the behaviour of the governing equation.

In order to implement the symmetric approach to solve this type of multi-zone domain problem, the solution at each zone is represented by its corresponding symmetric interpolation using a set of collocation points within each of the zones and the points at the interface between them. At the interfaces points that coincide with the physical boundary of the problem, the corresponding boundary conditions are imposed, while at the internal points, it is required that the governing equation, with corresponding value of the parameters at each zone, should be satisfy.

To solve this type of problem it is necessary to impose the continuity of flux at the interfaces between the zones, i.e. the flux leaving one sub-zone has to be equal to the flux entering the other. Therefore, it is necessary that the following flux matching conditions hold at the mth interface of the sub-zones i and i+1:

$$\left[D^i \frac{\partial C^i}{\partial n} - C^i u_j^i n_j \right]_m = \left[D^{i+1} \frac{\partial C^{i+1}}{\partial n} - C^{i+1} u_j^{i+1} n_j \right]_m \qquad (9)$$

Besides the above conditions, the concentration at each interface needs to be continuous, i.e.:

$$\left[C^i\right]_m = \left[C^{i+1}\right]_m \tag{10}$$

The Hermitian interpolation property of the symmetric approach (which takes into account the function and it's derivate) makes this method a natural technique to deal with the above matching conditions. There are two different alternatives to impose the interface matching conditions. First two different set of points at each sub-domains interface are defined. In each set of points a different matching condition is imposed, i.e. of the $n_i = n_{i1} + n_{i2}$ interface points n_{i1} are required to satisfy the continuity of the concentration and on the remaining n_{i2} points the flux matching condition is satisfied. On the other hand, due to the dependence of the Hermite interpolation on the partial differential operators, it is possible to impose simultaneously both conditions at each interface point. In this last case, as we are using a Hermite interpolation scheme, the resulting matrix system is non-singular as long as the partial differential operators applied to each point are linearly independent, even if in a single node we impose two different differential conditions (see Wu [11]). In this case, at each interface point both matching conditions, i.e. concentration and flux, are required to be satisfied. Therefore, the Hermite interpolation function is represented by:

$$C(x) = \sum_{k=1}^{n_b} \lambda_k B_\xi \Psi\left(\left\|x - \xi_k\right\|\right) + \sum_{k=n_b+1}^{n_i} \lambda_k \Psi\left(\left\|x - \xi_k\right\|\right) + \sum_{k=n_b+n_i+1}^{n_b+2n_i} \lambda_k \left(D^i \frac{\partial}{\partial n_\xi} \Psi\left(\left\|x - \xi_k\right\|\right) \right) + \sum_{k=n_b+2n_i}^{N} \lambda_k L_\xi^i \Psi\left(\left\|x - \xi_k\right\|\right) + P_{m-1}(x) \tag{11}$$

with n_b as the number of nodes on the boundary of a subdomain that coincide with the physical boundary, i.e. at an internal subdomain $n_b \equiv 0$, n_i the number of nodes on the interfaces common with other subdomains and $N-(n_b+n_i)$ the number of internal nodes at the subdomain. As before, in the above expression L_ξ^i and B_ξ are the differential operators corresponding to the partial differential equation at the subdomain and the boundary differential operator.

In the above interpolation function, the flux matching condition (9) at the interface *m* is reduced to

$$\left[D^i \frac{\partial C^i}{\partial n}\right]_m = \left[D^{i+1} \frac{\partial C^{i+1}}{\partial n}\right]_m \tag{12}$$

since we are imposing simultaneously at each interface point the continuity of concentration and flux, besides the convective velocity field needs to be continuous across the sub-domains.

It is important to point out that the above double collocation strategy at the interface points can also be used in the standard Unsymmetric approach (Kansa method). However due to lack of dependence on the differential operators of the corresponding interpolation function, this alternative will results in an over-determined system of algebraic equations.

5.1 Numerical examples

Let us consider the steady state heat transfer problem in a circular cylinder with a circular hollow. At the origin a constant value of temperature is given as well as at the exterior wall. The cylinder consists of three rings of constant but different parameters (see figure 1), under these conditions, in cylindrical co-ordinates, the problem is described by the following equation:

$$\frac{K^i}{r}\frac{d}{dr}\left(r\frac{dT^i}{dr}\right)+\alpha^i=0 \qquad i=1,2,3 \tag{13}$$

where K^i and α^i are the thermal conductivity and heat generation rate at the zone (ring) i, respectively, and r is the radial distance. The matching conditions at the contact region between rings, i.e. continuity of temperature and flux are given by:

$$T^i=T^{i+1} \text{ and } \frac{K^i}{r}\frac{d}{dr}\left(rT^i\right)=\frac{K^{i+1}}{r}\frac{d}{dr}\left(rT^{i+1}\right) \text{ at } r=r_m,\ m=1,2,3 \tag{14}$$

The analytical solution of the problem is given by Carslaw and Jaeger [13].
By expanding the cylindrical Laplacian operator in equation (21), we obtain the following expression:

$$K^i\frac{d^2T^i}{dr^2}+\frac{K^i}{r}\frac{dT^i}{dr}+\alpha^i=0 \tag{15}$$

which can be interpreted as a one-dimensional convection diffusion equation with variable negative convective velocity field, $u_r^i=-K^i/r$.

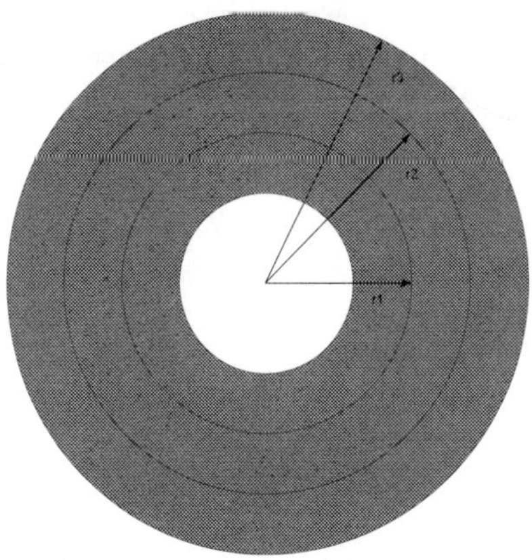

Figure 1: Cylindrical domain consisting of three rings with different constant coefficients at each ring.

The one-dimensional problem defined by equation (15), the matching conditions given by equation (14) and the corresponding boundary conditions at $r = r_0$ and $r = r_3$, will be solve here as two dimensional convection diffusion problem in the rectangular domain $1 \leq x \leq r_3 = 7\, dm$ and $0 \leq y \leq 2\, dm$ with zero lateral flux, were at each zone (ring) the following governing equation is satisfied:

$$K^i \frac{\partial^2 T^i}{\partial x_j^2} - u_1^i \frac{\partial T^i}{\partial x_1} + \alpha^i = 0 \text{ where } u_1^i = -K^i / x_1 \tag{16}$$

Two cases are considered, with different parameters in each zone and the same boundary conditions; $T(0,y) = 1$, $T(7,y) = 2$ and $\partial T / \partial n = 0$ at $y = 0$ and $y = 2$.

In figures 2 and 3, it is possible to appreciate the excellent agreement found between the numerical results using the above symmetric meshless collocation method and the analytical solution. In the first example (figure 2), a total of 841 collocation points uniformly distributed were used in order to achieve the obtained accuracy. The second case is more computational demanding due to the drastic changes in the heat production term, α^i $(i = 1,2,3)$, between the different zones, given by $\alpha^1 = 5, \alpha^2 = -3, \alpha^3 = 10$ (temperature/sec). In this case, we compare how the numerical result is affect by increasing the total number of collocation points. In figure 3, the results for two different set of uniform distribution of collocation points (841 and 2987) are presented, showing the convergence of the numerical scheme.

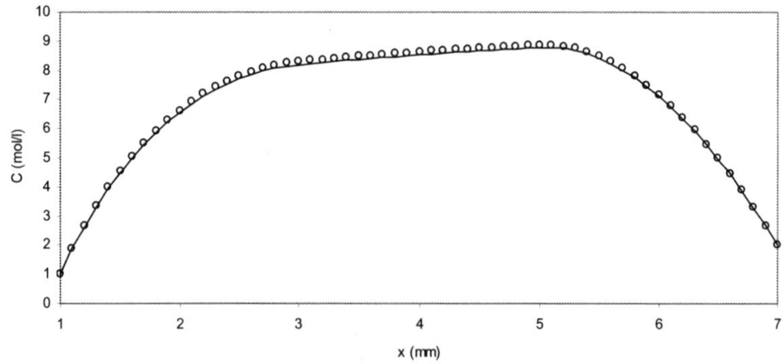

Figure 2: Comparison between the analytical solution — and the numerical results ○ for the following parameter: $\alpha^1 = 2, \alpha^2 = 0, \alpha^3 = 4$ (temperature/sec) and $K^1 = K^2 = K^3 = 1$ $(dm^2 / \sec)$.

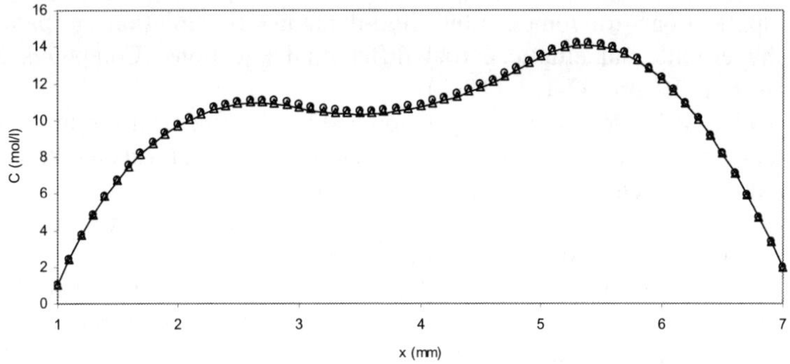

Figure 3: Comparison between the analytical solution — and the numerical results obtained with a total of 841, (∘), and 2987, (Δ), collocation points, for the following parameters: $\alpha^1 = 5, \alpha^2 = -3, \alpha^3 = 10$ (temperature/sec) and $K^1 = K^2 = K^3 = 1$ (dm^2 / sec).

6 Conclusions

The use of symmetric radial basis function collocation method to solve partial differential equations provides a simply accurate and truly meshes free technique. It is important to point out that in the case when this scheme is applied to solve large scale problems with a large number of data points, the conditional number of the resulting collocation matrix could be very large and the computational performance poor. As a way to overcome these problems, a domain decomposition scheme with double collocation at the interfaces joining neighbouring sub-domains is proposed. The proposed domain decomposition technique makes possible to improve the ill-conditioning problem through the reduction of the size of the full coefficient matrix to be solved in a global manner.

Acknowledgements

This research was been partially sponsored by the GABARDINE project (Contract number 518118) – part of the FP6-2006-TTC-TU European Commission Programme.

References

[1] E.J.Kansa, Multiquadrics- A scattered data approximation scheme with applications to computation fluid-dynamics-I: Surface approximations and partial derivatives estimates; Computers Math. Applic. 19, pp 127-145, (1990)

[2] E.J.Kansa, Multiquadrics- A scattered data approximation scheme with applications to computation fluid-dynamics-II: Solution to parabolic, hyperbolic and elliptic partial differential equations; Computers Math. Applic. 19, pp 147-161, (1990)

[3] Dubal M.R. Domain decomposition and local refinement for multiquadric approximations. I: second-order equations in one-dimension, Journal of Applied Science, 1, No.1, 146-171 (1994).

[4] Y. C. Hon and X. Z. Mao An efficient numerical scheme for Burgers' equations, Appl. Math. and Comp. 95, 37-50 (1998).

[5] Z Zerroukat M., Power H. and Chen C.S., A numerical method for heat transfer problems using collocation and radial basis functions, Int. J. Numer. Meth. Engng, 42, 1263-1279, (1998).

[6] Dubal M.R., Olivera S.R. and Matzner R.A. In Approaches to Numerical Relativity, Editors: R.d Inverno, Cambridge University Press, Cambridge UK, (1993).

[7] Kansa E.J. and Hon Y.C., Circumventing the ill conditioning problem with multiquadric radial basis functions: applications to elliptic partial differential equations, 39, 123-137, (2000).

[8] Fedoseyev AI, Friedmann MJ, Kansa EJ. Improved multiquadratic method for elliptic partial differential equation via PDE collocation on the boundary. Comput. Math. Appl. 2002, 43, 439-455

[9] Fasshauer G.E. Solving Partial Differential Equations by Collocation with Radial Basis functions, Proceedings of Chamonix, Editors: A. Le Méchauté, C. Rabut and L.L. Schumaker, 1-8, Vanderbilt University Press, Nashville, TN (1996).

[10] Wu Z., Hermite-Birkhoff interpolation of scattered data by radial basis functions; Approx. Theory, 8:2, 1-11 (1992).

[11] Wu Z., Solving PDE with radial basis function and the error estimation; Advances in Computational Mathematics, Lecture Notes on Pure and Applied Mathematics, 202, Editors: Z. Chen, Y. Li, C.A. Micchelli, Y. Xu and M. Dekker, GuangZhou (1998).

[12] Schaback R and Franke C., `Covergence order estimates of meshless collocation methods using radial basis functions', Advances Computational Mathematics, 8, Issue 4, 381-399, (1998).

[13] Kansa E.J. & Carlson 1992, 'Improved accuracy of multiquadric interpolation using variable shape parameters', Computers & Mathematic with Application, vol. 24, 99-120.

[14] Carslaw H.S. and Jaeger J.C., Conduction of heat in solids, Oxford at the Clarendon press, Oxford, (1959).

Initial stress formulation for three-dimensional elastoplastic analysis by the triple-reciprocity boundary element method

Y. Ochiai
Department of Mechanical Engineering, Kinki University, Japan

Abstract

In general, internal cells are required to solve elastoplastic problems using a conventional boundary element method (BEM). However, in this case, the merit of BEM, which is ease of data preparation, is lost. Triple-reciprocity BEM can be used to solve two-dimensional elastoplasticity problems with a small plastic deformation. It has been shown that three-dimensional elastoplastic problems can be solved, without the use of internal cells, by the triple-reciprocity BEM and initial strain method. In this study, an initial stress formulation is adopted and the initial stress distribution is interpolated using boundary integral equations. A new computer program was developed and applied to solving several problems.
Keywords: elastoplastic problem, initial stress method, BEM.

1 Introduction

Elastoplastic problems can be solved by a conventional boundary element method (BEM) using internal cells for domain integrals [1, 2]. In this case, however, the merit of BEM, which is ease of data preparation, is lost. On the other hand, several countermeasures have been considered. Ochiai and Kobayashi proposed the triple-reciprocity BEM (improved multiple-reciprocity BEM) without the use of internal cells for two-dimensional elastoplastic problems using an initial stress and strain formulations [3]. By this method, a highly accurate solution can be obtained using only fundamental solutions of a low order. It has been shown by Ochiai that three-dimensional elastoplastic problems can be solved, without the use of internal cells, by the triple-reciprocity BEM and initial strain method.

WIT Transactions on Modelling and Simulation, Vol 44, © 2007 WIT Press
www.witpress.com, ISSN 1743-355X (on-line)
doi:10.2495/BE070031

In this study, the initial stress formulation and triple-reciprocity BEM are adapted to three-dimensional elastoplastic problems, and new fundamental solutions for this method are shown. In this method, only boundary elements and internal points are used. The arbitrary distributions of the initial stress for elastoplastic analysis are interpolated using boundary integral equations and internal points. In this method, strong singularities in the calculation of stresses at internal points become weak. A new computer program was developed and applied to several elastoplastic problems to clearly understand the theory.

2 Theory

2.1 Initial stress formulation

To analyze the elastoplastic problems using the initial strain formulation, the following boundary integral equation must be solved [1, 2].

$$c_{ij}(P)\dot{u}_j(P) = \int_\Gamma [u_{ij}^{[1]}(P,Q)\dot{p}_j(Q) - p_{ij}(P,Q)\dot{u}_j(Q)]d\Gamma + \int_\Omega \varepsilon_{jki}^{[1]}(P,q)\dot{\sigma}_{I\,jk}^{[1]}(q)d\Omega \tag{1}$$

Here, $\dot{\sigma}_{I\,jk}^{[1]}$ is the initial stress rate and c_{ij} is the free coefficient. Moreover, $\dot{u}_i$ and $\dot{p}_i$ are the j-th components of the displacement rate and the surface traction rate, respectively. On the other hand, Γ and Ω are the boundary and the domain, respectively. As shown in Eq. (1), when there is an arbitrary initial stress rate, a domain integral becomes necessary. Denoting the distance between the observation point and the loading point by r, Kelvin's solution $u_{ij}^{[1]}$ and p_{ij} are given by

$$u_{ij}^{[1]} = \frac{1}{16\pi(1-\nu)Gr}\{(3-4\nu)\delta_{ij} + r,_i r,_j\} \tag{2}$$

$$p_{ij} = \frac{1}{8\pi(1-\nu)Gr^2}\{[(1-2\nu)\delta_{ij} + 3r,_i r,_j]\frac{\partial r}{\partial n} - (1-2\nu)(r,_i n_j - r,_j n_i)\}\,, \tag{3}$$

where ν is Poisson's ratio and G is the shear modulus. The i-th component of a unit normal vector is denoted by n_i. Moreover, let us set $r,_i=\partial r/\partial x_i$. The function $\varepsilon_{ijk}^{[1]}$ in Eq. (1) is given by [1]

$$\varepsilon^{[1]}{}_{jki} = \frac{-1}{16\pi(1-\nu)Gr^2}\{(1-2\nu)(\delta_{ji}r,_k + \delta_{ki}r,_j) - \delta_{jk}r,_i + 3r,_i r,_j r,_k\}\,. \tag{4}$$

2.2 Interpolation of initial stress

Interpolation using boundary integrals is introduced to avoid the domain integral in Eq. (1). The distribution of the initial stress $\dot{\sigma}_{I\,jk}^{[1]}$ in the case of a three-dimensional problem is interpolated using the integral equation to transform the domain integral into a boundary integral. The following equations are used for interpolation [5-8]:

$$\nabla^2 \dot{\sigma}_{I\,jk}^{[1]S} = -\dot{\sigma}_{I\,jk}^{[2]S}, \tag{5}$$

$$\nabla^2 \dot{\sigma}_{I\,jk}^{[2]S} = -\sum_{m=1}^{M} \dot{\sigma}_{I\,jk(m)}^{[3]PA}, \tag{6}$$

where $\nabla^2 = \partial^2/\partial x^2 + \partial^2/\partial y^2 + \partial^2/\partial z^2$. From Eqs. (6) and (7), we obtain

$$\nabla^4 \dot{\sigma}_{I\,jk}^{[1]S} = \sum_{m=1}^{M} \dot{\sigma}_{I\,jk(m)}^{[3]PA}, \tag{7}$$

where the function $\dot{\sigma}_{I\,jk}^{[3]PA}$ expresses a state of a uniformly distributed polyharmonic function in a spherical region with radius A. We must emphasize that Eqs. (6) and (7) can be used for interpolating the complicated distribution of the initial stress $\dot{\sigma}_{I\,jk}^{[1]}$. These equations are the same as those used to generate a free-form surface using an integral equation [6]. In this method, each component of initial stress $\dot{\sigma}_{I\,jk}^{[1]}$ (j, k=1,2,3) is interpolated.

2.3 Representation of initial stress by integral equation

The distribution of the initial stress is represented by an integral equation. The polyharmonic function $T^{[f]}$ and its normal derivatives are given by

$$T^{[f]} = \frac{r^{2f-3}}{4\pi(2f-2)!} \tag{8}$$

$$\frac{\partial T^{[f]}}{\partial n} = \frac{(2f-3)r^{2f-4}}{4\pi(2f-2)!}\frac{\partial r}{\partial n} \tag{9}$$

Figure 1 shows the shape of polyharmonic functions; the biharmonic function $T^{[2]}$ is not smooth at r=0. In the three-dimensional case, a smooth interpolation cannot be obtained using solely the biharmonic function $T^{[2]}$. In order to obtain a smooth interpolation, the polyharmonic function with volume distribution $T^{[2]A}$ is introduced. A polyharmonic function with volume distribution $T^{[f]A}$, as shown in Fig.1, is defined as [5]

$$T^{[f]A} = \int_0^A [\int_0^{2\pi} \{\int_0^{\pi} T^{[f]} a^2 \sin\theta \, d\theta\} d\phi] da \ . \tag{10}$$

The function $T^{[f]A}$ can be easily obtained using the relationships $r^2 = R^2 + a^2 - 2aR\cos\theta$ and $dr = aR\sin\theta d\theta$, as shown in Fig.1. This function is written using r instead of R, similarly to Eqs. (8) and (9), though the function in Eq. (10) is a function of R. The newly defined function $T^{[f]A}$ can be explicitly shown as

$$T^{[f]A} = \frac{1}{2r(2f+1)!}\{(2fA-r)(r+A)^{2f} + (2fA+r)(r-A)^{2f}\} \quad r > A \tag{11}$$

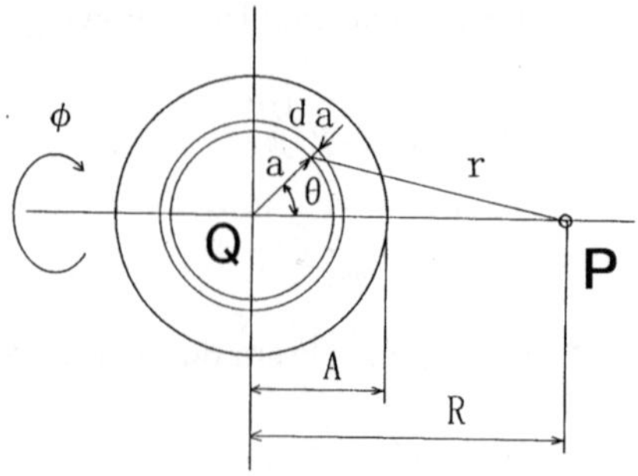

Figure 1: Notations for polyharmonic function with volume distribution.

$$T^{[f]A} = \frac{1}{2r(2f+1)!}\{(2fA-r)(A+r)^{2f} - (2fA+r)(A-r)^{2f}\} \quad r \le A. \tag{12}$$

Denoting the number of points $\dot{\sigma}_{I\,jk}^{[3]P}$ as *M*, the curvature of the initial stress rate $\dot{\sigma}_{I\,jk}^{[2]S}$ is given by Green's second identity and Eq. (6) as [4–6]

$$c\dot{\sigma}_{I\,jk}^{[2]S}(P) = \int_\Gamma \{T^{[1]}(P,Q)\frac{\partial \dot{\sigma}_{I\,jk}^{[2]S}(Q)}{\partial n} - \frac{\partial T^{[1]}(P,Q)}{\partial n}\dot{\sigma}_{I\,jk}^{[2]S}(Q)\}d\Gamma$$
$$+ \sum_{m=1}^{M} T^{[1]A}(P,q)\dot{\sigma}_{I\,jk(m)}^{[3]PA}(q)\,. \tag{13}$$

The initial stress rate $\dot{\sigma}_{I\,jk}^{[1]}$ is given by Green's theorem and Eqs. (5) and (6) as [4–6]

$$c\dot{\sigma}_{I\,jk}^{[1]S}(P) = -\sum_{f=1}^{2}(-1)^f \int_\Gamma \{T^{[f]}(P,Q)\frac{\partial \dot{\sigma}_{I\,jk}^{[f]S}(Q)}{\partial n} - \frac{\partial T^{[f]}(P,Q)}{\partial n}\dot{\sigma}_{I\,jk}^{[f]S}(Q)\}d\Gamma$$
$$- \sum_{m=1}^{M} T^{[2]A}(P,q)\dot{\sigma}_{I\,jk(m)}^{[3]PA}(q)\,, \tag{14}$$

where c=0.5 on the smooth boundary and *c*=1 in the domain. It is assumed that $\dot{\sigma}_{I\,jk}^{[2]S}$ (Q) is zero. For internal points, the next equation is obtained similarly to Eq. (14).

$$c\dot{\varepsilon}_{I\,jk}^{[1]S}(p) = -\sum_{f=1}^{2}(-1)^f \int_\Gamma \{T^{[f]}(p,Q)\frac{\partial \dot{\varepsilon}_{I\,jk}^{[f]S}(Q)}{\partial n} - \frac{\partial T^{[f]}(p,Q)}{\partial n}\dot{\varepsilon}_{I\,jk}^{[f]S}(Q)\}d\Gamma$$

$$- \sum_{m=1}^{M} T^{[2]A}(p,q)\dot{\varepsilon}_{I\,jk(m)}^{[3]PA}(q) \tag{15}$$

If the boundary is divided into N_0 constant elements, and N_1 internal points are used, the simultaneous linear algebraic equations with $(2N_0+N_I)$ as unknowns must be solved.

2.4 Triple-reciprocity boundary element method for the representation of initial stress

The function $\varepsilon_{jki}^{[f]}$ is defined as

$$\nabla^2 \varepsilon_{jki}^{[f+1]} = \varepsilon_{jki}^{[f]} . \tag{16}$$

Using eqs (5), (6) and (16) and Green's second identity, eq (1) becomes

$$c_{ij}(P)\dot{u}_j(P) = \int_\Gamma [u_{ij}^{[1]}(P,Q)\dot{p}_j(Q) - p_{ij}(P,Q)\dot{u}_j(Q)]d\Gamma - \sum_{f=1}^{2}(-1)^f \int_\Gamma \{\frac{\partial \varepsilon_{jki}^{[f+1]}(P,Q)}{\partial n}\dot{\sigma}_{I\,jk}^{[f]S}(Q)$$
$$- \varepsilon_{jki}^{[f+1]}(P,Q)\frac{\partial \dot{\sigma}_{I\,jk}^{[f]S}(Q)}{\partial n}\}d\Gamma + \sum_{m=1}^{M}\varepsilon_{jki}^{[3]A}(P,q)\dot{\sigma}_{I\,jk(m)}^{[3]PA}(q) \tag{17}$$

$\varepsilon_{ijk}^{[f]}$ for the representation of initial stress is obtained as [4]

$$\varepsilon_{ijk}^{[f]} = \frac{(2f-1)(2f-3)r^{2f-4}}{8\pi(1-\nu)(2f)!G}$$
$$\{(2f-1-2f\nu)(\delta_{jk}r,_i+\delta_{ik}r,_j) - \delta_{ij}r,_k - (2f-5)r,_i r,_j r,_k\} . \tag{18}$$

Moreover, the normal derivative $\partial\varepsilon_{ijk}^{[f]}/\partial n$ and $\varepsilon_{ijk}^{[3]A}$ are given by

$$\frac{\partial \varepsilon^{[f]}{}_{ijk}}{\partial n} = \frac{(2f-1)(2f-3)r^{2f-5}}{8\pi(1-\nu)(2f)!G}[(2f-5)\{(2f-1-2f\nu)(\delta_{jk}r,_i+\delta_{ik}r,_j)$$
$$-\delta_{ij}r,_k - (2f-7)r,_i r,_j r,_k\}\frac{\partial r}{\partial n} - (2f-5)(r,_j r,_k n_i + r,_i r,_k n_j + r,_i r,_j n_k)$$
$$+(2f-1-2f\nu)(\delta_{jk}n_i + \delta_{ik}n_j) - \delta_{ij}n_k] \tag{19}$$

$$\varepsilon_{ijk}^{[3]A} = \frac{A^3}{15120(1-\nu)r^4 G}\{-(\delta_{jk}r,_i+\delta_{ik}r,_j+\delta_{ij}r,_k)(105r^6+63A^2r^4-9A^4r^2+A^6)$$
$$-r,_i r,_j r,_k(105r^6-63A^2r^4+27A^4r^2-5A^6)$$
$$+18(1-\nu)(\delta_{jk}r,_i+\delta_{ik}r,_j)r^2(35r^4+14A^2r^2-A^4)\} \qquad r > A \tag{20}$$

$$\varepsilon_{ijk}^{[3]A} = \frac{r}{7560(1-\nu)G}\{-(\delta_{jk}r,_i+\delta_{ik}r,_j+\delta_{ij}r,_k)(-r^4+18A^2r^2+63A^4)$$
$$-4r,_i r,_j r,_k r^2(-r^2+9A^2) + 9(1-\nu)(\delta_{jk}r,_i+\delta_{ik}r,_j)(-r^4+14A^2r^2+35A^4)\}$$
$$r \le A . \tag{21}$$

2.5 Internal stresses

The internal stress is given by

$$\dot{\sigma}_{ij}(p) = \int_\Gamma [-\sigma_{kij}^{[1]}(p,Q)\dot{p}_k(Q) - S_{kij}(p,Q)\dot{u}_k(Q)]d\Gamma$$
$$+\int_\Omega \varepsilon_{ijkl}^{[1]}(p,q)\dot{\sigma}_{I\,kl}^{[1]}(q)d\Omega - \dot{\sigma}_{I\,ij}^{[1]}(q) , \tag{22}$$

where $\dot{\sigma}_{Iij}^{[1]}$ is the initial stress obtained. The functions S_{kij} and $\sigma_{ijks}^{[1]}$ in Eq. (22) are given by

$$S_{kij} = \frac{G}{2\pi(1-\nu)r^2}[2\frac{\partial r}{\partial n}\{(1-2\nu)\delta_{ij}r_{,k} + \nu(\delta_{ik}r_{,j} + \delta_{jk}r_{,i}) - 4r_{,i}r_{,j}r_{,k}\} \\ + 2\nu(n_i r_{,j}r_{,k} + n_j r_{,i}r_{,k}) + (1-2\nu)(2n_k r_{,i}r_{,j} + n_j\delta_{ik} + n_i\delta_{jk}) - (1-4\nu)n_k\delta_{ij}] \tag{23}$$

$$\sigma_{ijkl}^{[1]} = \frac{1}{2\pi(1-\nu)r^2}[2(1-2\nu)(\delta_{ij}r_{,k}r_{,l} + \delta_{kl}r_{,i}r_{,j}) \\ +2\nu(\delta_{il}r_{,j}r_{,k} + \delta_{jk}r_{,i}r_{,l} + \delta_{ik}r_{,j}r_{,s} + \delta_{jl}r_{,i}r_{,k}) \\ +(1-2\nu)(\delta_{ik}\delta_{lj} + \delta_{jk}\delta_{li}) - (1-4\nu)\delta_{ij}\delta_{kl} - 8r_{,i}r_{,j}r_{,k}r_{,l}]. \tag{24}$$

The function $\varepsilon_{ijkl}^{[f]}$ is defined as

$$\nabla^2\varepsilon_{ijkl}^{[f+1]} = \varepsilon_{ijkl}^{[f]}. \tag{25}$$

Using Eq. (25) and Green's theory, Eq. (22) becomes

$$\dot{\sigma}_{ij}(p) = \int_\Gamma[-\sigma_{kij}^{[1]}(p,Q)\dot{p}_k(Q) - S_{kij}(p,Q)\dot{u}_k(Q)]d\Gamma \\ - \sum_{f=1}^{2}(-1)^f\int_\Gamma[\frac{\partial\sigma_{ijkl}^{[f+1]}(p,Q)}{\partial n}\dot{\varepsilon}_{Ikl}^{[f]S}(Q) - \sigma_{ijkl}^{[f+1]}(p,Q)\frac{\partial\dot{\varepsilon}_{Ikl}^{[f]S}(Q)}{\partial n}]d\Gamma \\ + \sum_{m=1}^{M}\varepsilon_{ijkl}^{[3]A}(p,q)\dot{\sigma}_{Ikl(m)}^{[3]PA}(q) - \dot{\sigma}_{Iij}^{[1]}(q) \tag{26}$$

Using Eq. (22) and the relationship between displacement and strain, $\varepsilon_{ijkl}^{[f]}$ is obtained as

$$\varepsilon_{ijkl}^{[f]} = \frac{(2f-1)(2f-3)r^{2f-5}}{4\pi(1-\nu)(2f)!}\langle\ [(2f-1-2f\nu)(\delta_{ik}\delta_{jl} + \delta_{il}\delta_{jk}) \\ +(2f-5)(f-1-f\nu)(\delta_{jl}r_{,i}r_{,k} + \delta_{jk}r_{,i}r_{,l} + \delta_{il}r_{,j}r_{,k} + \delta_{ik}r_{,j}r_{,l}) \\ -(1-2f\nu)\{(2f-5)\delta_{kl}r_{,i}r_{,j} + \delta_{ij}\delta_{kl}\} + (2f-5) + \delta_{ij}r_{,k}r_{,l} \\ -(2f-5)(2f-7)r_{,i}r_{,j}r_{,k}r_{,l}]>. \tag{27}$$

Similarly, $\partial\varepsilon_{ijkl}^{[f]}/\partial n$ and $\varepsilon_{ijkl}^{[3]A}$ are obtained as

$$\frac{\partial\varepsilon^{[f]}{}_{ijkl}}{\partial n} = \frac{(2f-1)(2f-3)(2f-5)r^{2f-6}}{2\pi(1-\nu)(2f)!}\langle\ [(2f-1-2f\nu)(\delta_{ik}\delta_{jl} + \delta_{il}\delta_{jk}) \\ +(2f-7)(f-1-f\nu)(\delta_{jl}r_{,i}r_{,k} + \delta_{jk}r_{,i}r_{,l} + \delta_{il}r_{,j}r_{,k} + \delta_{ik}r_{,j}r_{,l}) \\ -(1-2f\nu)\{(2f-7)(\delta_{kl}r_{,i}r_{,j} + \delta_{ij}r_{,k}r_{,l}) + \delta_{ij}\delta_{kl}\} - (2f-7)(2f-9)r_{,i}r_{,j}r_{,k}r_{,l}]\frac{\partial r}{\partial n} \\ + (f-1-f\nu)\{(\delta_{jl}n_k + \delta_{jk}n_l)r_{,i} + (\delta_{il}n_k + \delta_{ik}n_l)r_{,j} \\ +(\delta_{jl}n_i + \delta_{il}n_j)r_{,k} + (\delta_{jk}n_i + \delta_{ik}n_j)r_{,l}\}$$

$$-(1-2f\nu)\{\delta_{ij}(r_{,l}n_k+r_{,k}n_l)+\delta_{kl}(r_{,j}n_i+r_{,i}n_j)\}$$
$$-(2f-7)\{(r_{,l}n_k+r_{,k}n_l)r_{,i}r_{,j}+(r_{,j}n_i+r_{,i}n_j)r_{,k}r_{,l}\}\rangle \quad (28)$$

$$\varepsilon^{[3]A}_{ijkl}=\frac{A^3}{7560(1-\nu)r^5}[18\nu r^2\delta_{ij}\{\delta_{kl}(35r^4+14A^2r^2-A^4)+r_{,k}r_{,l}(35r^4-14A^2r^2$$
$$+3A^4)\}-(\delta_{ij}\delta_{kl}+\delta_{kj}\delta_{li}+\delta_{ki}\delta_{li})(105r^6+63A^2r^4-9A^4r^2+A^6)$$
$$-(\delta_{ij}r_{,k}r_{,l}+\delta_{kj}r_{,i}r_{,l}+\delta_{ki}r_{,j}r_{,l}+\delta_{kl}r_{,i}r_{,j}+\delta_{il}r_{,k}r_{,j}+\delta_{jl}r_{,k}r_{,i})(105r^6$$
$$-63A^2r^4+27A^4r^2-5A^6)-r_{,j}r_{,i}r_{,k}r_{,l}(-105r^6+189A^2r^4-135A^4r^2+35A^6)$$
$$+9(1-\nu)r^2\{2(\delta_{ki}\delta_{jl}+\delta_{kj}\delta_{il})(35r^4+14A^2r^2-A^4)$$
$$+(\delta_{ki}r_{,j}r_l+\delta_{kj}r_{,i}r_l+\delta_{li}r_{,j}r_{,k}+\delta_{lj}r_{,i}r_{,k})(35r^4-14A^2r^2+3A^4)\}] \quad r>A \quad (29)$$

$$\varepsilon^{[3]A}_{ijkl}=\frac{1}{3780(1-\nu)}[9\nu\delta_{ij}\{\delta_{kl}(-r^4+14A^2r^2+35A^4)+4r_{,k}r_{,l}r^2(-r^2+7A^2)\}$$
$$-(\delta_{ij}\delta_{kl}+\delta_{ik}\delta_{jl}+\delta_{il}\delta_{jk})(-r^4+18A^2r^2+63A^4)$$
$$-4(\delta_{kl}r_{,i}r_{,j}+\delta_{jl}r_{,i}r_{,k}+\delta_{jk}r_{,i}r_{,l}+\delta_{il}r_{,j}r_{,k}+\delta_{ik}r_{,j}r_{,l}+\delta_{ij}r_{,k}r_{,l})$$
$$r^2(-r^2+9A^2)+8r_{,i}r_{,j}r_{,k}r_{,l}r^4+9(1-\nu)\{(\delta_{ik}\delta_{jl}+\delta_{il}\delta_{jk})(-r^4+14A^2r^2+35A^4)$$
$$+2(\delta_{jl}r_{,i}r_{,k}+\delta_{jk}r_{,i}r_{,l}+\delta_{il}r_{,j}r_{,k}+\delta_{ik}r_{,j}r_{,l})r^2(-r^2+7A^2)\}] \quad r\le A \quad (30)$$

3 Numerical examples

In order to ensure the accuracy of the present method, the stress in a thick cylinder, which is made of an elastoplastic material, subjected to internal pressure is obtained. It is assumed that the inner and outer radii are 10 and 30 mm. The von Mises yield criterion is used, and the cylinder is free in the z direction. Young's modulus E =210 GPa and Poisson's ratio $\nu=0.30$ are assumed. Internal pressure p_O = 1.2 GPa, yield stress σ_Y = 1.2 GPa and strain hardening $H=0.1E$ are assumed. The numbers of discretized boundary elements and internal points are 680 and 315, as shown in Fig. 2. Internal points are used to interpolate the distribution of initial stress. Figure 3 shows the circumferential and radial stress distributions. Boundary element results are shown with FEM solutions in Fig. 2. The stress distributions agree well with the FEM solutions.

Next elastoplastic problem is a notched tensile specimen which is shown in Fig.4. This example is one of the very early plasticity problems solved using the finite-element technique, and boundary element solutions were presented by Telles [1]. Von Mises yield criterion and no-strain hardening is assumed. The number of discretized boundary elements is 856, and the number of internal points for interpolation is 715, as shown in Fig. 5. Young's modulus E=70.0 GPa, Poisson's ratio ν=0.2 and yield stress σ_o =243 MPa are assumed. The spread of plastic zones at different load levels presented in Fig.6 exhibits good agreement with the conventional boundary element solution for the same problem.

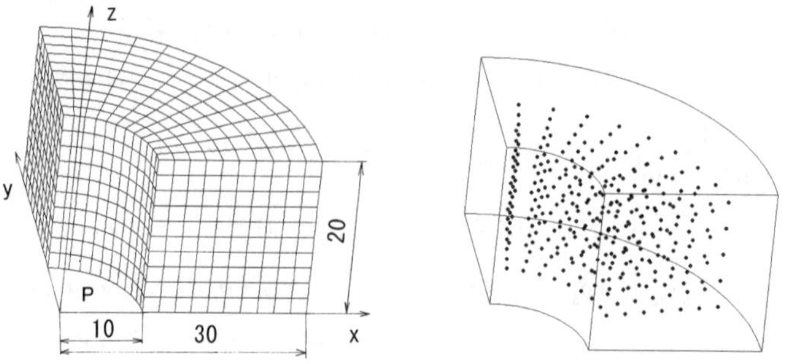

(a) (b)

Figure 2: Boundary elements and internal points in quarter-region (Number of boundary elements: 680). (a) Boundary elements, (b) internal points.

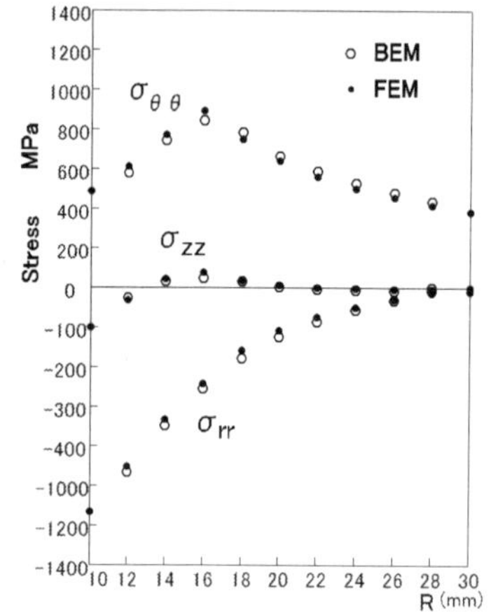

Figure 3: Stress distribution in hollow cylinder with internal pressure.

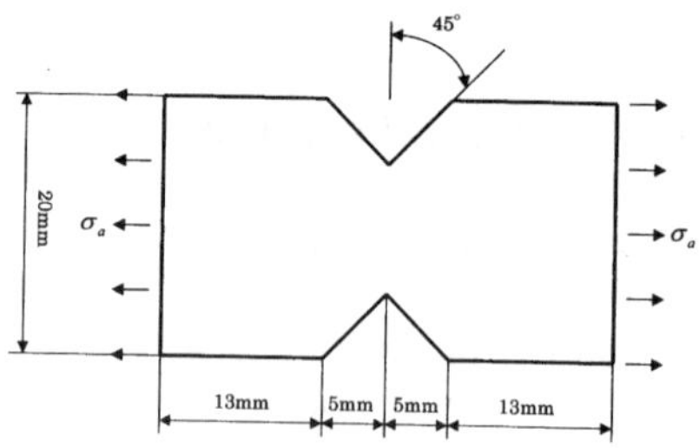

Figure 4: Notched tensile specimen.

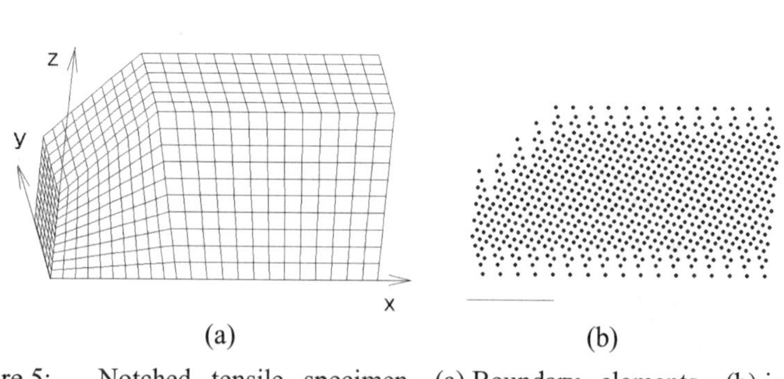

(a) (b)

Figure 5: Notched tensile specimen. (a) Boundary elements, (b) internal points.

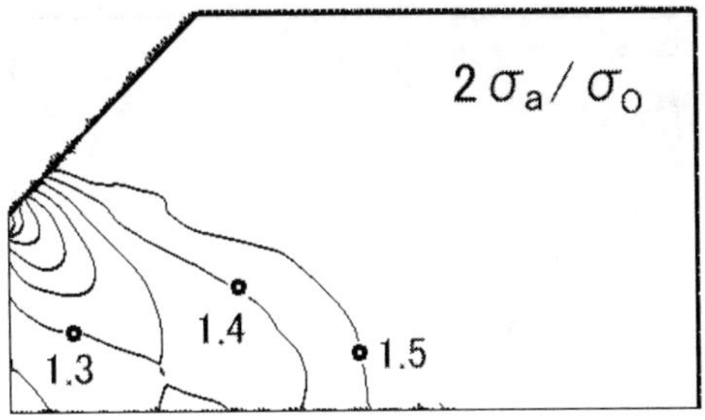

Figure 6: Plastic zone obtained for various values of $2\sigma_a/\sigma_o$.

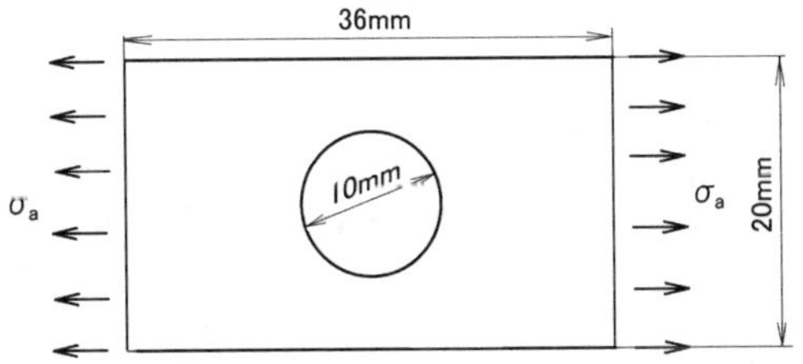

Figure 7: Perforated tension strip.

The elastoplastic problem of a plate with a circle hole, as shown in Fig. 7, is solved by using Mises yield criterion. Uniform traction is σ_a, and the thickness of the plate is 1mm. The number of discretized boundary elements is 856, and the number of internal points for interpolation is 660 as shown in Fig.8. Young's modulus E=70 GPa, Poisson's ratio ν=0.2 and the yield stress σ_Y=243 MPa are assumed. Fig.9 shows the plastic zone obtained by this method for various values $2\sigma_a/\sigma_Y$ in a quarter region. This result is in good agreement with the FEM analysis.

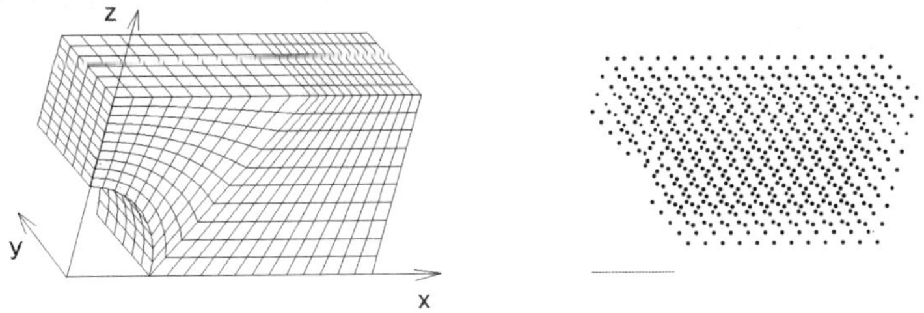

(a) (b)

Figure 8: Boundary element in quarter region. (a) Boundary elements, (b) internal points.

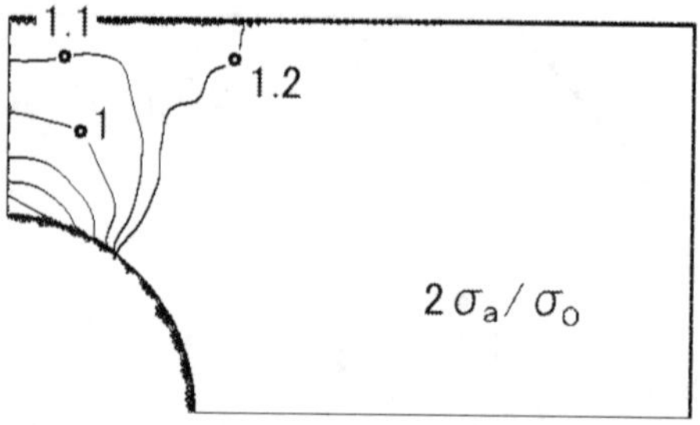

Figure 9: Plastic zone obtained for various values of $2\sigma_a/\sigma_o$.

4 Conclusion

It was shown that three-dimensional elastoplastic analysis can be carried out, without the use of internal cells, using the triple-reciprocity boundary element method and initial stress method. The fundamental solutions for initial stress method were shown. In this method, the strong singularity that appears in the calculation of internal stress by the conventional boundary element method becomes weak. Using numerical examples, the effectiveness and accuracy of this method were demonstrated. In this method, the merit of BEM, which is ease of data preparation, is not lost because internal cells are not necessary.

References

[1] Telles, J. C. F., The Boundary Element Method Applied to Inelastic Problems, Springer-Verlag, Berlin, 1983.
[2] Ochiai, Y. and Kobayashi, T., Initial Stress Formulation for Elastoplastic Analysis by Improved Multiple-Reciprocity Boundary Element Method, Engineering Analysis with Boundary Elements, Vol. 23, pp. 167-173, 1999.
[3] Ochiai, Y. and Kobayashi, T., Initial Strain Formulation without Internal Cells for Elastoplastic Analysis by Triple-Reciprocity BEM, International Journal for Numerical Methods in Engineering, Vol. 50, pp. 1877-1892, 2001.
[4] Ochiai, Y. and Sladek, V. Numerical Treatment of Domain Integrals without Internal Cells in Three-Dimensional BIEM Formulations, CMES (Computer Modeling in Engineering & Sciences), Vol. 6, No. 6, pp. 525-536, 2004.
[5] Ochiai, Y., Nishitani, H. and Sekiya, T., Stress Analysis with Arbitrary Body Force by Boundary Element Method, Engineering Analysis with Boundary Elements, Vol. 17, pp. 295-302, 1996.
[6] Ochiai, Y., Multidimensional Numerical Integration for Meshless BEM, Engineering Analysis with Boundary Elements, Vol. 27, No. 3, pp. 241-249, 2003.

A meshless solution for potential equations using a continuous-valued circular line source

P. Mitic[1] & Y. F. Rashed[2]
[1]*Positive Corporation Ltd., Hampshire, UK*
[2]*Department of Structural Engineering, University of Cairo, Egypt*

Abstract

We find solutions to $\nabla^2 U = 0$ in a simply-connected 2-D domain D, using a continuous line source associated with a concentration function comprising n undetermined parameters. This choice reduces ill-conditioning effects by reducing the number of parameters involved. The choice of a continuous circular line source C around D follows from previous results indicating that, when solving the same problem with discrete point sources, the result is independent of precise placement of sources. The circle is associated with a concentration function that is constrained to satisfy the problem's boundary conditions. Accuracy is achievable using a number of parameters which, had discrete sources been used, would be insufficient to represent the geometry of D, thus giving inaccurate results. Empirical investigations with various forms of concentration function show that with some domains, the error in calculated values of U can be less than 0.1%: an order of magnitude improvement over discrete methods. More complex domains yield less accuracy, and, after testing on a range of domains, we formulate an empirical rule for an appropriate form for the concentration function for a generic domain. Code requiring high-precision arithmetic was developed in *Mathematica*, which also simplifies routine tasks of solving linear systems and integrations.

1 Introduction

Previous research has shown that when using meshless discrete sources in the MFS, the configuration of sources relative to the domain D is extremely flexible. In [1] we showed that, within certain limits, the source distribution can be random, and in [2] we showed that sources "at infinity" (i.e. a large distance

WIT Transactions on Modelling and Simulation, Vol 44,
www.witpress.com, ISSN 1743-355X (on-line)
doi:10.2495/BE070041

from D compared to the size of D) can produce very accurate results for the simple domain discussed in this paper. Fam's study [4] confirmed our findings on source distributions near D, using extensions of the MFS. Studies [1, 2, 4] show that previous attempts to analyse meshless source distributions are incorrect. For example, Alves [3], in an analysis of Poisson equations, considers a 'natural' radius for a circular distribution of discrete sources of "5~10 times the diameter of D" (without any precise definition for the term 'diameter' for a non-circular domain). We claim, using [1, 2], that a 'natural' radius is infinite, and reiterate that view from the results of this paper. Alves says that ill-conditioning problems preclude the use of very large radii, but we use Mathematica to analyse potential problems with sources at arbitrary distances from D without significant ill-conditioning.

2 The continuous line source method

For discrete point sources S_j exterior to a domain D, the potential $U(m)$ at a point m in D or on the boundary of D is given by

$$U(m) = \sum_j U^*(S_j)\, c_j \tag{1}$$

where U^* is the fundamental solution at S_j with respect to m and the c_j are undetermined coefficients. Figure 1 shows such a domain D (with boundary ∂D and interior point $m(x,y)$) in which the point sources have been replaced by a continuous circular line source with constant radius R, of which AB is an arc. This circular source has an associated concentration function $c(\theta,\mathbf{c})$ (abbreviated to $c(\theta)$), where the parameter θ is the angle shown in Figure 1 and $\mathbf{c}$ is a vector of n undetermined coefficients.

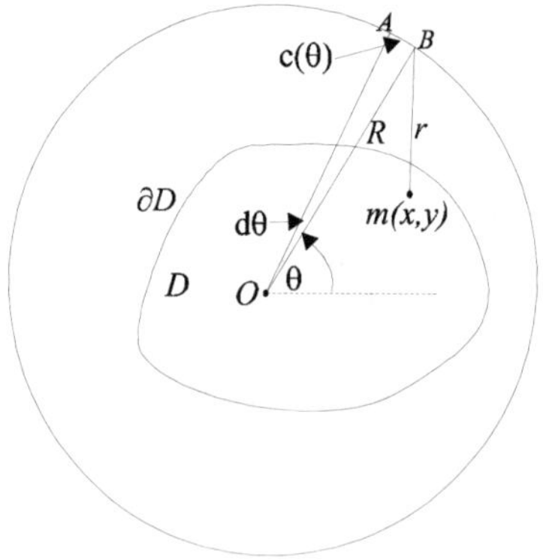

Figure 1: Circular source and domain.

The continuous equivalent of eqn. (1) is derived by considering the potential dU due to the infinitesimal line source S_θ on the arc AB, on which it is assumed that $c(\theta)$ is approximately constant. Assuming throughout that the Origin O is in the interior of D.and that that m is not at O,

$$dU(m,\theta) = U^*(S_\theta)c(\theta)d\theta \tag{2}$$

Hence the total potential at m due to the whole circular source is

$$U(m) = \int_0^{2\pi} U^*(S_\theta)\, c(\theta)d\theta = \int_0^{2\pi} \frac{\ln(r)}{2\pi} c(\theta)d\theta, \tag{3}$$

where $$r^2 = (R\cos(\theta) - x)^2 + (R\sin(\theta) - y)^2 \tag{4}$$

Similarly the flux $Q(m)$ at m due the entire circular source is given by

$$Q(m) = \int_0^{2\pi} Q^*(S_\theta)\, c(\theta)d\theta = \int_0^{2\pi} \frac{1}{2\pi r} \frac{n_x r_x + n_y r_y}{r} c(\theta)d\theta \tag{5}$$

where $r_x = R\cos(\theta) - x$, $r_y = R\sin(\theta) - y$ and (n_x, n_y) are direction cosines at m.

In the case where m is at O, eqn (3) simplifies to

$$U(m) = \frac{\ln(R)}{2\pi} \int_0^{2\pi} c(\theta)\, d\theta \tag{6}$$

and eqn (5) simplifies to

$$Q(m) = \frac{1}{2\pi R} \int_0^{2\pi} (n_x \cos(\theta) + n_y \sin(\theta))\, c(\theta)d\theta. \tag{7}$$

Our method then proceeds by discretising ∂D and setting up a set of linear equations based on known potentials on ∂D. Let $\mathbf{m} = (m_0, m_1, \cdots, m_{N-1})$ be a vector of the midpoints of N boundary elements on ∂D with corresponding known boundary values (either potential or flux) $\mathbf{b} = (b_0, b_1, \cdots, b_{N-1})$. For each element of $\mathbf{m}$ in turn we use one of eqns (3, 5) (depending on whether the relevant boundary condition is a known potential or flux) to obtain a system of linear equations in the undetermined parameters $\mathbf{c}$ for $c(\theta)$. This gives a matrix equation

$$\mathbf{Mc} = \mathbf{b} \tag{8}$$

where the coefficients of $\mathbf{M}$ come from one of eqns (3,5). *Mathematica* provides a convenient inversion for $\mathbf{M}$ using its pseudo-inverse, and this accounts for over-determined systems and can be done to arbitrary precision (within the limits of computer memory). Thus we can obtain a relatively accurate approximation $\bar{\mathbf{c}}$ for $\mathbf{c}$:

$$\bar{\mathbf{c}} = \text{PseudoInverse}[\mathbf{M}]\ \mathbf{b}. \tag{9}$$

The potential and flux at an interior or boundary point **p** can then be calculated from any one of eqns (3,5,6,7), depending on whether a flux or potential is required and whether **p** is at the Origin or not, using the known elements of $\overline{\mathbf{c}}$ in the integrals involving $c(\theta)$.

2.1 Choice of concentration function

Choosing a concentration function involves finding suitable linearly independent basis functions, and a subjective judgement about the number of undetermined parameters involved. On balance we have found that there is rarely much to be gained by choosing over-complicated basis functions, since simple polynomials suffice in all the cases we have tried. The numbers of undetermined parameters is more of a problem. Choosing too few cannot reflect the domain geometry accurately, and often gives completely wrong solutions. Choosing too many often has little effect but produces a progressive loss of accuracy due to ill-conditioning. In some cases this loss of accuracy is significant.

The only other constraint on the concentration function is a continuity condition $c(0)=c(2\pi)$. Collocation polynomials satisfying this continuity condition and evaluating to the undetermined parameter c_i of **c** at ordinates i initially produced promising results in simple rectangular domains, even with high values of n (>30). But for more complex domains, either ill-conditioning or a rapidly oscillating function (or both) gave less accurate results. Hence we concentrated on piecewise linear functions only. The n-parameter continuous piecewise linear function *pwlin*, is defined in eqn (10).

$$pwlin(\theta,n)=\begin{cases}\alpha_r\theta+\beta_r \;:\theta\in\left[\dfrac{2\pi r}{n},\dfrac{2\pi(r+1)}{n}\right), & r=0..n-1,\\[2ex] 0 \qquad\quad :\theta\notin\left[\dfrac{2\pi r}{n},\dfrac{2\pi(r+1)}{n}\right), & r=0..n-1,\quad \theta\in[0,2\pi]\end{cases} \qquad (10)$$

$$\alpha_r=\frac{n}{2\pi}(c_{r+1}-c_r),\quad \beta_r=(r+1)c_r-rc_{r+1}$$

This set of concentration functions is useful because the integrals in matrix **M** in eqn (8) are simple for all values of n (although it's technically harder to define their domains correctly). No assumption are made about placement of the c_r. This fact, and the results using piecewise linear concentration functions, effectively counter Poullikkis's assertion [7], that point sources must be placed uniformly at a fixed distance from the boundary.

2.2 Conjectures on parameters of the concentration function

In the examples that follow, it is apparent that in some cases the choice of parameters of the concentration function matters very little, whereas in other cases it matters a great deal. We aim to determine whether or not it is possible to find general results for the number and position of the parameters of the concentration function which are independent of the domain. If this can be done, our method can become a generic method to solve any potential problem $\nabla^2 U = 0$ in any D subject to boundary conditions on ∂D. Trials from this study and results from [1] and [9] suggest that, given the definition for Nominal Radius of the domain, *NR*[*D*], in section 2.2.1, we should consider placing the enclosing circle either 'at infinity' or between 10 and 20 times *NR*[*D*]. Furthermore, the number of concentration function parameters should be a simple function of the number of boundary elements on the convex and concave parts of the discretised boundary. .

2.2.1 Definition: *Nominal Radius*

For a simply-connected domain *D*, the *Nominal Radius* of *D*, denoted by *NR*[*D*] is half of the maximum (straight line) distance between pairs of points on ∂D.

$$NR[D] = \frac{1}{2} \sup_{x,y \in \partial D} \left(|x - y| \right) \tag{11}$$

This definition is intended to be no more than a general guideline in choosing the radius of the enclosing circle. It is used to relate the circle radius to a single spatial characteristic of *D*. *NR*[*D*] is "loosely" the radius of the smallest circle that can be drawn around *D* without intersecting with the boundary of *D*.

3 Convex domain example: torsion of an elliptical bar

This is an example of torsion of a bar with an elliptical cross-section, and is taken from [10]. Figure 2 shows the configuration of a quarter of the domain. Dirichlet boundary conditions *U*=0 apply on the straight sides and a Neumann boundary condition $Q1(x,y) = \dfrac{75xy}{\sqrt{25x^2 + 10000y^2}}$ applies on the curved side.

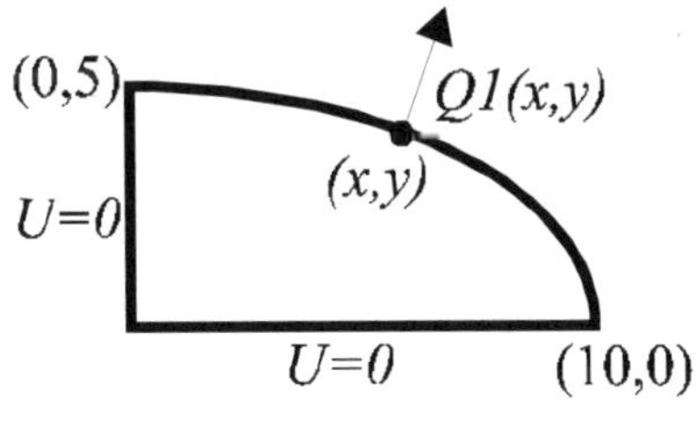

Figure 2: Torsion of an elliptical bar.

Brebbia and Dominguez [10] use reference points (2,2) and (4,3.5) for their calculations. At these points their quadratic BEM gives *U* as –2.431 and –8.472 respectively, with analytical results of –2.400 and –8.400 respectively. Table 1 shows our results at these reference points. In this case the nominal radius for the domain is $NR(D)=\frac{1}{2}\sqrt{5^2+10^2}\sim 5.6$. Rounded up to 6, this is the minimal radius that yields reasonable results The discretisation used 12 boundary elements, so we consider among others, 12 concentration function parameters (CFP). With high numbers of parameters such as these, using the collocation concentration function is not practical, as failure to converge is frequent, and the results in Table 1 are for the piecewise linear concentration function.

Table 1: Results for torsion of an elliptical bar.

CFP	10	10	12	12	15	15	20	20
Radius	U(2,2)	U(4,3.5)	U(2,2)	U(4,3.5)	U(2,2)	U(4,3.5)	U(2,2)	U(4,3.5)
6	-2.0862	-6.6759	-2.0862	-6.6759	-2.2935	-8.258	-2.4444	-8.61
50	-2.4116	-8.4406	-2.4116	-8.4406	-2.4116	-8.44053	-2.4116	-8.4406
100	-2.4116	-8.4406	-2.4116	-8.4406	-2.4116	-8.44064	-2.4116	-8.4406
200	-2.4116	-8.4407	-2.4116	-8.4407	-2.4117	-8.44068	-2.4116	-8.4406
400	-2.4115	-8.4406	-2.4115	-8.4406	-2.4116	-8.44066	-2.4116	-8.4407
1000	-2.4116	-8.4406	-2.4116	-8.4406	-2.4116	-8.44063	-2.4116	-8.4406
5000	-2.4116	-8.4407	-2.4116	-8.4407	-2.4121	-8.44109	-2.4116	-8.4406

The results in Table 1 show an improvement on the quadratic BEM calculations and those obtained in [1] and [9], but there appears to be a limit on the maximum accuracy achievable (about 0.5% in this case). In all cases, the results are largely independent of the circle radius, except at a radius which is marginally larger than *NR*(*D*)). There is also very little dependence of the results on the number of parameters in the concentration function. Other purely convex domains yielded results of similar accuracy, with maximum errors in *U* and *Q* of ~1%.

4 Convex domain example: flow past a circular cylinder

Zhang [6] provides an example of a concave domain, for which our meshless method needs more careful investigation. In Figure 3, fluid flows past the circular arc *AB*. *U* is the stream function and $Q=\nabla U$. There are 'natural' boundary conditions on the boundary segments *OA*, *AB* and *BC*, and the boundary conditions on *CD* and *DO* are calculated from the analytical solution, eqn (13) with y=2 and x=0 respectively.

$$U(x,y)=1-\frac{1}{y^2+(x-4)^2} \tag{13}$$

In contrast to the number of nodes used by Zhang [6] to discretise the boundary (ranging from 26 to 104), we obtained similar accuracy using a crude discretisation with at most 20 parameters. We used 18 boundary elements: 3 on *OA*, 8 on *AB*, 1 on *BC*, 4 on *CD* and 2 on *DO*. For the following results we calculated the potential at a selection of the same 19 interior points as Zhang: $\left\{\left(\frac{n}{5},\frac{n}{10}\right), n=1\ldots19\right\}$ (equally spaced on diagonal OC in Figure 3). Table 2 gives summary results for these 19 reference points. The Nominal Radius is $\sqrt{5}$: half the diagonal OC in Figure 3, and the column marked 'R/*NR*' shows the radius of the enclosing circle R as a multiple of the Nominal Radius. CFP is the number of Concentration function parameters and "% error" is the mean absolute percentage error for all 19 reference point.

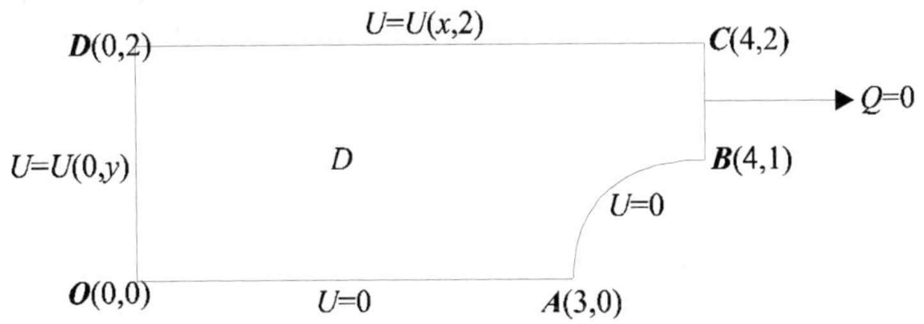

Figure 3: Fluid flow.

Table 2: Fluid flow.

CFP		10	13	15	20	25
R	R/*NR*	% error	% error	% error	% error	% error
2.5	1.1	1.2	1.2	22.2	50.3	254.3
10	4.5	9.5	3.6	56.6	84.0	72.3
20	8.9	10.0	3.2	27.2	33.1	33.8
30	13.4	10.1	3.1	4.8	31.1	28.8
40	17.9	10.1	3.0	3.5	5.2	4.7
50	22.4	10.1	2.9	3.4	5.2	3.5
100	44.7	10.1	2.8	5.2	2.7	2.6
200	89.4	10.1	6.6	6.6	6.6	6.6
500	223.6	9.9	3.5	9.9	9.9	9.9
1000	447.2	9.9	9.9	7.4	9.9	9.9

There is clearly a much wider variation in results than there was for the convex domain example. With 13 concentration function parameters (the number of boundary elements on the convex sides), there is some consistency of results for a wide range of radius. For cases where the radius is between 10 and 20 times the nominal radius (actual radius is roughly between 20 and 50), the mean relative % error is about 3%. For an "infinite" radius ($R > 100$), the mean relative % error is more than 5%, which is unacceptable. Using a "minimal" radius only gives good accuracy with a careful choice for the number of concentration function parameters. Solving other problems with concave boundaries leads us to believe that this meshless method results in ill-conditioned with respect to the number of concentration function parameters

The main source of error in the overall error measure in Table 2 is nearly always due to the reference point (0.2, 0.1), which is the reference point nearest O in Figure 3. This point typically accounts for between 35% and 65% of the total error. The next reference point (0.4, 0.2) usually accounts for a further 10% of the total error.

Table 3 shows the results of calculations of U for five reference points using the optimal 13 concentration function parameters. Most results are within 2% of the exact values. In this case the major source of error is the point (3.8, 1.9).

Table 3: Fluid flow - variation of U with radius at five reference points.

Radius	U(0.2, 0.1)	U(1.0, 0.5)	U(2.0, 1.0)	U(3, 1.5)	U(3.8, 1.9)
2.5	0.100555	0.439841	0.802767	1.03351	1.42095
10	0.116463	0.431713	0.802797	1.05629	1.27069
20	0.112426	0.434135	0.801896	1.05633	1.27231
30	0.110742	0.435021	0.801614	1.0563	1.27302
40	0.109861	0.43547	0.801478	1.05627	1.27344
50	0.109323	0.43574	0.801396	1.05625	1.27372
100	0.108171	0.436266	0.80118	1.05615	1.2742
200	0.136966	0.415435	0.819446	1.05093	1.24058
500	0.108966	0.444447	0.821415	1.03478	1.22172
1000	0.0081242	0.469391	0.822426	1.00444	1.31153
exact	*0.0930796*	*0.445946*	*0.8*	*1.03846*	*1.37945*

Since the domain is theoretically infinite, we tried larger values for the distances CD and OD, and used the 'natural' approximations $U_{CD} \sim 1$ and $U_{OD} \sim 1$. We achieved slightly better accuracy than that reported in Table 3 with the values $CD = OD = 10$, which this is probably a more realistic scenario than the boundary conditions stated in Figure 3.

5 Boundary element requirements

In general we have tried to use sufficient boundary elements to reflect the geometry of the domain and the boundary conditions, and no more. Using more did not improve accuracy materially in the case of convex domains. In some concave domains, accuracy deteriorated as the number of boundary elements increased. We attribute this effect to a forced high concentration density on some portions of the enclosing circle.

Hence we never needed the large number of nodes required in some other studies. For example, [5], needed a minimum of 72 boundary elements to achieve acceptable accuracy. Liu does confirm our view that too few parameters cannot model D effectively, and that too many increases ill-conditioning unacceptably. Zhang [6] reports similar results. Similarly, Fam [8] solves a rectangular domain problem using dipole sources. The domain boundary is discretised into 12 or 24 boundary points. Our solution achieved the same accuracy with only 8 boundary points.

6 Conclusion

Our distributed source method can produce accurate results for very simple domains and can achieve a significant reduction in the number of parameters (of the concentration function) required, and the necessary number of boundary elements. Of the two classes of concentration function considered (collocation and piecewise linear), the piecewise linear class allows for more concentration function parameters if required. It generally gives less accurate but more reliable results. We have found similar results for other simple convex domains, although achievable accuracy diminishes with increasing complexity of the domain. Within very broad limits, for convex domains, the radius of the circular source and the number of collocation function parameters have little impact on the accuracy of the calculations.

When the domain contains at least one concave element, the configuration of the enclosing circle is more stringent. In addition to the fluid flow example discussed here, we have considered other cases and found, with all of them, that optimal accuracy can only be achieved by using particular circle configurations. The only combination of parameters that works tolerably well in every (concave) case is the combination:

number of concentration function parameters =
number of boundary elements on the convex faces of the discretised domain;

and

circular source radius = n NR(D) where $10 \leq n \leq 20$.

Within this parameter set, $n = 10$ giving marginally better results. Using this combination necessarily involves some trading of accuracy for generality. We stress that this conclusion derives from not only the examples presented here, but also from additional investigation of other domains, with particular attention paid to concave domains. Furthermore, applying the rule suggested above may not give optimal accuracy in any given case. The number of boundary elements should be minimised: there should be sufficient to reflect the geometry of the domain and the boundary conditions, but no more. In particular, the combination "one concentration function parameter per boundary element with infinite radius" would have been appealing.

References

[1] Mitic, P & Rashed, Y.F. Convergence and stability of the Method of Meshless Fundamental Solutions using an array of randomly-distributed sources. Engineering Analysis with Boundary Elements (to appear) 2003

[2] Mitic, P & Rashed, Y.F. The Method of Meshless Fundamental Solutions with sources at Infinity. Proc. 5th International Mathematica Symposium, London, eds. P.Mitic, P.Ramsden & J.Carne. Imperial College Press, , 2003

[3] Alves,C.J.S., Chen, C.S. & Săler,B. The Method of Fundamental Solutions for solving Poisson Problems. Proc Boundary Elements XXIV, Sintra, Portugal, eds. C.A. Brebbia, A.Tadeu & V. Popov. WIT Press, 2002

[4] Fam,G and Rashed,Y. A study on the source points location in the method of fundamental solution. Proc Boundary Elements XXIV, Sintra, Portugal, eds. C.A. Brebbia, A.Tadeu & V. Popov. WIT Press, 2002

[5] Liu, G.R & Gu, Y.T. Boundary meshfree methods based on the Boundary Point Interpolation methods. Proc Boundary Elements XXIV, Sintra, Portugal, eds. C.A. Brebbia, A.Tadeu & V. Popov. WIT Press, 2002

[6] Zhang, J & Yao, Z. Meshless regular hybrid Boundary Node Method. Computer Modelling in Engineering and Sciences, vol. 2, No. 3 pp307-318. Tech Science Press 2001

[7] Poullikkis,A., Karageorghis,A. & Georgiou,G. Methods of fundamental solutions for harmonic and biharmonic boundary value problems. Computational Mechanics 21 pp416-423 Springer 1998

[8] Fam, G. & Rashed, Y. The Method of Fundamental Solutions, a Dipole formulation for potential problems. Proc Boundary Elements XXV, Split, Croatia, eds. C.A. Brebbia, D. Poljak & V. Roge. WIT Press, 2003

[9] Mitic, P. A Generic Circular Source Distribution for solving potential problems using Meshless Methods. Proc. 6th International Mathematica Symposium, Banff, Canada, eds. P.Mitic & J.Carne. Positive Corporation, August 2004

[10] Brebbia, C.A. and Dominguez, J. Boundary Elements: an Introductory Course. McGraw-Hill/CMP 1989

Adaptive error estimation of the Trefftz method for solving the Cauchy problem

C.-T. Chen[1], K.-H. Chen[2], J.-F. Lee[1] & J.-T. Chen[3]
[1]*Department of Hydraulic and Ocean Engineering, National Cheng Kung University, Tainan, Taiwan*
[2]*Department of Information Management, Toko University, Chia-yi, Taiwan*
[3]*Department of Harbor and River Engineering, National Taiwan Ocean University, Keelung, Taiwan*

Abstract

In this paper, the Laplace problem with overspecified boundary conditions is investigated by using the Trefftz method. The main difficulty will appear an obvious divergent result when the boundary condition on an overspecified boundary contaminates artificial errors. The occurring mechanism of the unreasonable result originates from an ill-posed influence matrix. The accompanied ill-posed problem is remedied by using the Tikhonov regularization technique and the linear regularization method respectively, to reconstruct the influence matrix. The optimal parameters of the Tikhonov technique and linear regularization method are determined by adopting the adaptive error estimation technique. The numerical evidence of the Trefftz method is given to verify the accuracy of the solutions after comparison with the results of analytical solution and to demonstrate the validity and instructions of the proposed adaptive error estimation technique. The comparison of the Tikhonov regularization technique and the linear regularization method was also discussed in the example.
Keywords: Trefftz method, adaptive error estimation, Cauchy problem, ill-posed problem, Tikhonov technique, linear regularization method, L-curve concept.

1 Introduction

In 1926, Trefftz [10] presented the Trefftz method for solving boundary value problems by superimposing the basis functions satisfying the governing

WIT Transactions on Modelling and Simulation, Vol 44, © 2007 WIT Press
www.witpress.com, ISSN 1743-355X (on-line)
doi:10.2495/BE070051

equation, although various versions of the Trefftz method, e.g. direct formulations [8] and indirect formulations [7], have been developed. The unknown coefficients are determined by satisfying the boundary condition with the approximate solution. Many applications to the Helmholtz equation [4], the Navier equation [15, 17] and the biharmonic equation [6] had been done. Until recent years, the ill-posed nature in the method was noticed [8, 19] increasingly.

The ill-posed nature (behavior) may be one of the following inverse problems or their combinations: (1) lack the determination of the domain, its boundary, or an unknown inner boundary; (2) lack inference of the governing equation; (3) lack identification of boundary conditions and/or initial conditions (the Cauchy problem); (4) lack determination of the material properties involved; (5) lack determination of the forces or inputs acting in the domain [12]. The Cauchy problem is surveyed in this paper.

Sometimes, unreasonable results occur in the Cauchy problem subjected to the measured and contaminated errors on the overspecified boundary condition because of the ill-posed behavior in linear algebraic system [13, 16]. Mathematically speaking, the Cauchy problem is ill-posed since the solution is very sensitive to the given data. Such a divergent problem could be avoided by using regularization methods [1, 5, 14, 18], e.g., the Tikhonov regularization technique [1] and the linear regularization method [5]. The Tikhonov regularization technique and the linear regularization method had been successfully applied to overcome the ill-posed problem of the Laplace equation [3, 9] and to treat with the divergent problems, since the two methods can obtain the convergent solution more stably and reasonably. In this paper, the comparison of two regularization techniques is made to obtain a better method.

For the Cauchy problem, the influence matrix is often ill-posed such that the regularization technique which regularizes the influence matrix is necessary. Both the Tikhonov technique and the linear regularization method transform into well-posed ones by choosing appropriate parameters for λ and λ^*, respectively [2]. The optimal parameter can be determined according to a local minimal point at error curve (similarly with L-curve shape) by implementing the adaptive error estimation technique. The corner (local minimal point) of the L-curve determines the optimal value of λ which provides the least relative error. The proposed error estimation technique belongs to an adaptive technique and does not need to compare the results with analytical solution. It will be elaborated latterly.

The purpose of this paper is to deal with the Cauchy problem with ill-posed nature of numerical instability by implementing the Trefftz Method in conjunction with the Tikhonov technique, linear regularization method, and adaptive error estimation technique. The technique is employed to obtain the optimal parameter to remedy the ill-posed behavior. Finally, the results of the example contaminated with artificial noise on the overspecified boundary condition are given to illustrate the validity of the proposed technique. Good agreements are observed as comparing analytical solutions. Under no analytical solutions, the numerical examples also demonstrate the validity and instructions of the proposed adaptive error estimation technique.

WIT Transactions on Modelling and Simulation, Vol 44, © 2007 WIT Press
www.witpress.com, ISSN 1743-355X (on-line)

2 Formulation

2.1 Governing equation and over-specified boundary condition

We consider the inverse problem for Laplace equation with overspecified boundary condition as shown in Fig. 1, which satisfies:

$$\nabla^2 u(\underset{\sim}{x}) = 0, \quad \underset{\sim}{x} \in D \tag{1}$$

subjected to the boundary condition on B_1 as

$$u_1(\underset{\sim}{x}) = \bar{u}_1(\underset{\sim}{x}), \quad t_1(\underset{\sim}{x}) = \bar{t}_1(\underset{\sim}{x}), \quad \underset{\sim}{x} \in B_1 \tag{2}$$

where ∇^2 is the Laplacian operator, D is the domain of interesting, $t(\underset{\sim}{x}) = \dfrac{\partial u(\underset{\sim}{x})}{\partial n_{\underset{\sim}{x}}}$ in which n_x is the normal vector at $\underset{\sim}{x}$, B is the whole boundary which consists of the known boundary (B_1) and the unknown boundary (B_2).

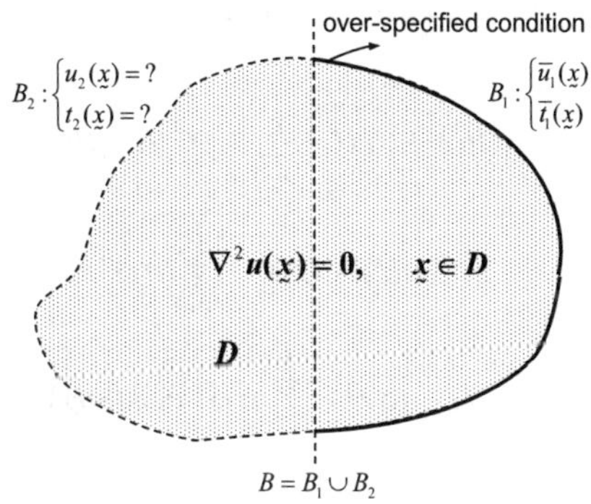

Figure 1: Sketch diagram of inverse problem with over-specified condition.

2.2 Methodology

2.2.1 The Trefftz method

In the Trefftz method, the field solution $u(\underset{\sim}{x})$ is superimposed by the T-complete functions, $A_j(\underset{\sim}{x})$, as follows:

$$u(\underset{\sim}{x}) = \sum_{j=1}^{2N} w_j A_j(\underset{\sim}{x}) \tag{3}$$

where 2N+1 is the number of T-complete functions, w_j is the unknown coefficient, $A_j(\underset{\sim}{x})$ is the T-complementary set which satisfies the Laplace equation. For the interior problem, we choose 1, $\rho^n \sin(n\theta)$ and $\rho^n \cos(n\theta)$ ($n \in N$), to be the bases of the complementary set in two-dimensional problem. Therefore, the eqn (3) can be expressed by

$$u(\rho,\theta) = a_0 + \sum_{n=1}^{N-1} a_n F_n(\rho,\theta) + \sum_{n=1}^{N} b_n G_n(\rho,\theta) \tag{4}$$

where $F_n(\rho,\theta) = \rho^n \cos(n\theta)$ and $G_n(\rho,\theta) = \rho^n \sin(n\theta)$. And the normal differential of the field solution $u(x)=u(\rho,\theta)$, is written as below:

$$t(\underset{\sim}{x}) = \frac{\partial u(\underset{\sim}{x})}{\partial n_{\underset{\sim}{x}}} = \sum_{j=1}^{2N} w_j \frac{\partial A_j(\underset{\sim}{x})}{\partial n_{\underset{\sim}{x}}} = \sum_{j=1}^{2N} w_j B_j(\underset{\sim}{x}) \tag{5}$$

$$t(\rho,\theta) = \frac{\partial u(\rho,\theta)}{\partial n_{\underset{\sim}{x}}} = \sum_{n=1}^{N-1} a_n F_n^*(\rho,\theta) + \sum_{n=1}^{N} b_n G_n^*(\rho,\theta) \tag{6}$$

where

$$F_n^*(\rho,\theta) = \left[\sum_{n=1}^{N-1} n\rho^{n-1}\cos(n\theta)\cdot\cos\theta + \sum_{n=1}^{N-1} n\rho^n \sin(n\theta)\cdot\frac{\sin\theta}{r}\right]\cdot\vec{n}_x,$$

$$+\left[\sum_{n=1}^{N-1} n\rho^{n-1}\cos(n\theta)\cdot\sin\theta - \sum_{n=1}^{N-1} n\rho^n \sin(n\theta)\cdot\frac{\cos\theta}{r}\right]\cdot\vec{n}_y$$

$$G_n^*(\rho,\theta) = \left[\sum_{n=1}^{N} n\rho^{n-1}\sin(n\theta)\cdot\cos\theta - \sum_{n=1}^{N} n\rho^n \cos(n\theta)\cdot\frac{\sin\theta}{r}\right]\cdot\vec{n}_x$$

$$+\left[\sum_{n=1}^{N} n\rho^{n-1}\sin(n\theta)\cdot\sin\theta + \sum_{n=1}^{N} n\rho^n \cos(n\theta)\cdot\frac{\cos\theta}{r}\right]\cdot\vec{n}_y.$$

In order to obtain the unknown coefficients $w_j = a_0$, a_n and b_n, N boundary points on the over-specified boundary $((\rho_1,\theta_1),\ (\rho_2,\theta_2),\cdots,(\rho_N,\theta_N)) \in B_1$ are collocated. Eqn (4) and (6) match the boundary condition on the boundary points to obtain the following linear algebraic system

$$[A]\{x\} = \{b\} \Leftrightarrow \begin{Bmatrix} \{\overline{u}_1\}_{N\times 1} \\ \{\overline{t}_1\}_{N\times 1} \end{Bmatrix} = \begin{bmatrix} [A_1]_{N\times 2N} \\ [B_1]_{N\times 2N} \end{bmatrix} \{w\}_{2N\times 1}, \tag{7}$$

where

$$\{\overline{u}_1\}_{N\times 1} = \begin{Bmatrix} \overline{u}_1 \\ \overline{u}_2 \\ \vdots \\ \overline{u}_N \end{Bmatrix},\quad \{\overline{t}_1\}_{N\times 1} = \begin{Bmatrix} \overline{t}_1 \\ \overline{t}_2 \\ \vdots \\ \overline{t}_N \end{Bmatrix} \tag{8}$$

$$[A_1]_{N\times 2N}=\begin{bmatrix} A_{1,1} & A_{1,2} & \cdots & A_{1,N+1} & A_{1,N+2} & \cdots & A_{1,2N} \\ \vdots & \vdots & \vdots & \ddots & \vdots & \ddots & \vdots \\ A_{N,1} & A_{N+1,2} & \cdots & A_{N,N+1} & A_{N,N+2} & \cdots & A_{N,2N} \end{bmatrix} \tag{9}$$

$$[B_1]_{N\times 2N}=\begin{bmatrix} 0 & B_{1,2} & \cdots & B_{1,N+1} & B_{1,N+2} & \cdots & B_{1,2N} \\ \vdots & \vdots & \vdots & \ddots & \vdots & \ddots & \vdots \\ 0 & B_{N,2} & \cdots & B_{N,N+1} & B_{N,N+2} & \cdots & B_{N,2N} \end{bmatrix} \tag{10}$$

$$\{w\}_{2N\times 1}=\begin{Bmatrix} a_0 \\ a_1 \\ \vdots \\ a_{N-1} \\ b_1 \\ b_2 \\ \vdots \\ b_N \end{Bmatrix} \tag{11}$$

so that the coefficients $w_j = a_0$, a_n and b_n of numerical solutions can be determined by using linear algebraic system solver.

2.2.2 Regularization techniques for the Cauchy problem

2.2.2.1 Tikhonov technique Tikhonov [1] proposed a method to transform this ill-posed problem into a well-posed one. Instead of solving $[A]\{x\}=\{b\}$ directly, the procedures of the Tikhonov technique are written as follows:

$$\text{(I). Minimize } \|x\|^2 \text{, subject to } \|Ax-b\|^2 \le \varepsilon \tag{12}$$

where ε is the prescribed error tolerance.

(II). The proposed problem in eqn (12) is equivalent[9] described as below:

$$\text{minimize } \|x\|^2 \text{, subject to } \|Ax-b\|^2 \le \varepsilon^*, \tag{13}$$

and the Euler-Lagrange equation obtained from reference [9] can be written as

$$(A^T A + \lambda^{(\mathrm{T})} I)x = A^T b \tag{14}$$

where T denotes matrix transposition, $\lambda^{(\mathrm{T})}$ is the regularization parameter (Lagrange parameter), in which the subscript (T) denotes the Tikhonov technique.

2.2.2.2 Linear regularization method The single central idea in the inverse theory is the prescription [5],

$$\text{minimize: } P[x]+\lambda^{(\mathrm{L})}Q[x] \tag{15}$$

where $\lambda^{(\mathrm{L})}$ is the regularization parameter, in which the subscript (L) denotes the Linear regularization method, $P[x]=|A\cdot x-b|^2>0$ and $Q[x]=x\cdot H\cdot x>0$ are two positive functions of x,

$$[\mathrm{H}]_{\mathrm{M}\times\mathrm{M}}=[\mathrm{B}]^T_{\mathrm{M}\times(\mathrm{M}-1)}\cdot[\mathrm{B}]_{(\mathrm{M}-1)\times\mathrm{M}}$$
$$=\begin{pmatrix} 1 & -1 & 0 & 0 & \cdots & 0 & 0 & 0 & 0 \\ -1 & 2 & -1 & 0 & \cdots & 0 & 0 & 0 & 0 \\ 0 & -1 & 2 & -1 & \cdots & 0 & 0 & 0 & 0 \\ \vdots & & & & \ddots & & & & \vdots \\ 0 & 0 & 0 & 0 & \cdots & -1 & 2 & -1 & 0 \\ 0 & 0 & 0 & 0 & \cdots & 0 & -1 & 2 & -1 \\ 0 & 0 & 0 & 0 & \cdots & 0 & 0 & -1 & 1 \end{pmatrix} \tag{16}$$

in which

$$[\mathrm{B}]_{(\mathrm{M}-1)\times\mathrm{M}}=\begin{pmatrix} -1 & 1 & 0 & 0 & \cdots & 0 & 0 & 0 & 0 \\ 0 & -1 & 1 & 0 & \cdots & 0 & 0 & 0 & 0 \\ \vdots & & & & \ddots & & & & \vdots \\ 0 & 0 & 0 & 0 & \cdots & 0 & -1 & 1 & 0 \\ 0 & 0 & 0 & 0 & \cdots & 0 & 0 & -1 & 1 \end{pmatrix} \tag{17}$$

Then, using equation x^2 [5], the minimization principle of eqn (15) is

$$\text{minimize: } P[x]+\lambda^{(\mathrm{L})}Q[x]=|Ax-b|^2+\lambda^{(\mathrm{L})}x\cdot H\cdot x \tag{18}$$

in vector notation,

$$(A^TA+\lambda^{(\mathrm{L})}\mathrm{H})x=A^Tb \tag{19}$$

2.2.3 The adaptive error estimation

To obtain the optimal λ without exact solution, the role of the adaptive tactical procedure is important to handle the inverse problem for Laplace equation with overspecified boundary condition. The method of the proposed adaptive error estimation is described as follows:

According to the ill-posed problem with the artificial contamination subjected to $\bar{u}_1(\underset{\sim}{x}),\ \ \underset{\sim}{x}\in B_1$ and $\bar{t}_1(\underset{\sim}{x}),\ \ \underset{\sim}{x}\in B_1$ as shown in Fig.2, remedied by the regularization method, we can obtain the result $\bar{\bar{u}}_2(\underset{\sim}{x}),\ \ \underset{\sim}{x}\in B_2$. Then the new specified boundary condition is $\bar{\bar{u}}_2(\underset{\sim}{x}),\ \underset{\sim}{x} \in B_2$ and the original boundary condition, $\bar{t}_1(\underset{\sim}{x}),\ \underset{\sim}{x} \in B_1$, in which $\bar{\bar{u}}_2(\underset{\sim}{x})$ is obtained before. The new problem (well-posed) becomes the mixed-type problem as shown in Fig. 3. Furthermore, the result $\bar{\bar{u}}_1(\underset{\sim}{x}),\ \underset{\sim}{x} \in B_1$ is calculated again by using the Trefftz method, and compare it with the original boundary condition $\bar{u}_1(\underset{\sim}{x}),\ \underset{\sim}{x} \in B_1$. Usually, the norm error can be defined as $\left\| \bar{u}_1 - \bar{\bar{u}}_1 \right\| = \left\{ \int \left| \bar{u}_1 - \bar{\bar{u}}_1 \right|^2 \mathrm{d}B_1 \right\}$, which is chosen as the index of sensitivity and λ_1 is chosen as the index of degree of distortion.

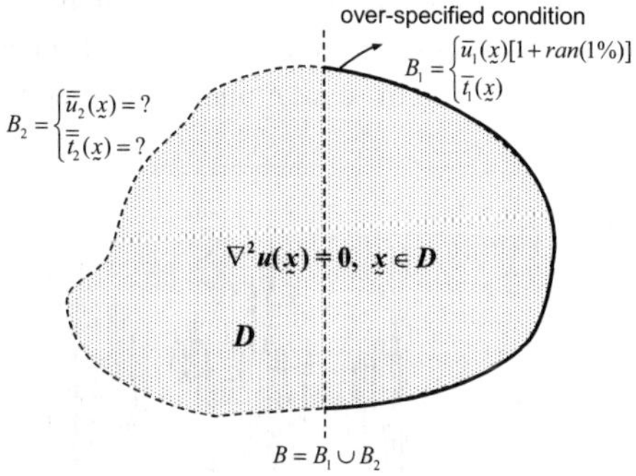

Figure 2: Sketch diagram of the ill-posed problem with the artificial contamination.

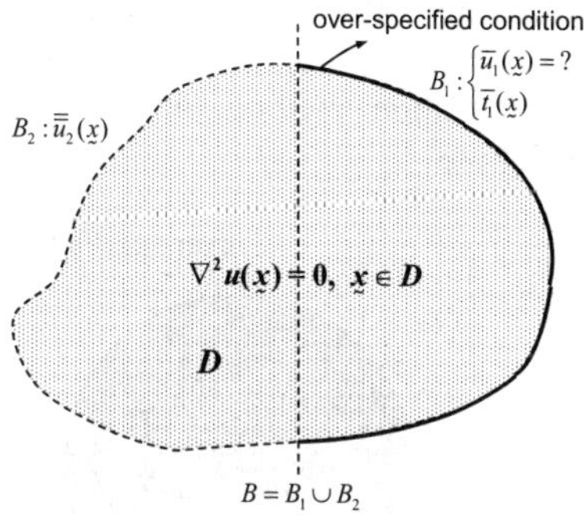

Figure 3: Sketch diagram of mixed-type problem with mixed-type condition.

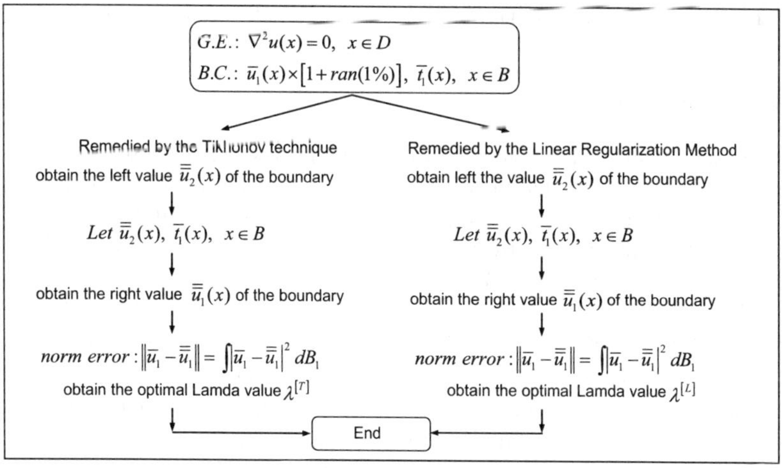

Figure 4: Flowchart of adaptive error estimation.

In the adaptive error estimation, the L-curve shape can be observed and the optimal λ is located on the corner. The corner is a compromise between regularization error (due to data smoothing) and perturbation errors (due to noise disturbance). To clarify the procedure, the flowchart of the adaptive error estimation can be described in Fig. 4.

3 Numerical examples

To illustrate applications of the Trefftz method in conjunction with the Tikhonov technique, the Linear regularization method and the adaptive error estimation for the Laplace equation with overspecified BCs, the case for a circle domain, the radius $R = 1.0$ is chosen as a representation example. Two kinds of treatments in the ill-posed problem are considered: the Tikhonov technique and the Linear regularization method, all for the inverse problem contaminated by noise pollutions.

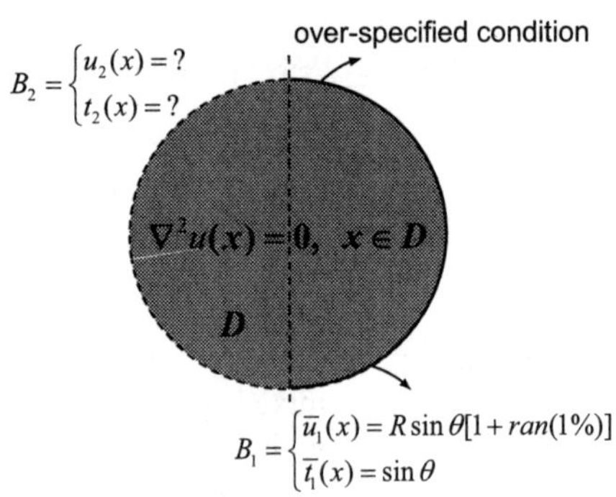

Figure 5: A sketch diagram.

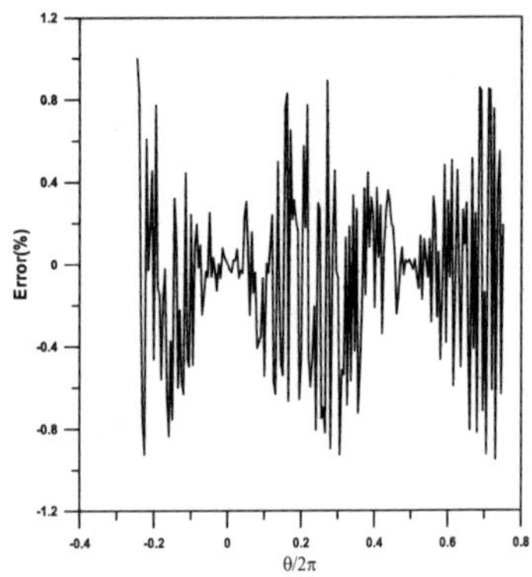

Figure 6: The random error.

3.1 Circular case

The present model of the inverse problem with noises can be shown as Fig. 5. By using random data simulation, we can obtain 1% random errors to contaminate the input boundary data, as described in Fig. 6. If regularization techniques are not employed, the results by using the Trefftz method are unreasonable and divergent as shown in Fig. 7(a), 7(b).

When the Tikhonov technique and the linear regularization method are applied in this case, we can obtain solutions with different values of the regularized parameters of $\lambda^{(\mathrm{T})}$ and $\lambda^{(\mathrm{L})}$ by employing the Tikhonov technique and the linear regularization method respectively, as shown in Fig. 8(a), 8(b), respectively, and the field solutions are shown in Fig. 9(a), 9(b). To obtain the optimal λ, the norm error comparing with exact solution is defined as $\|u - u_e\| = \left\{ \int_a^b |u - u_e|^2 \, \mathrm{d}x \right\}$ and the norm errors versus λ are plotted in Fig. 10,

www.witpress.com, ISSN 1743-355X (on-line)

by using the Tikhonov technique and the linear regularization method respectively. The L-curve shape can be observed and the optimal λ is located on the corner as shown in Fig. 11(a), 11(b).

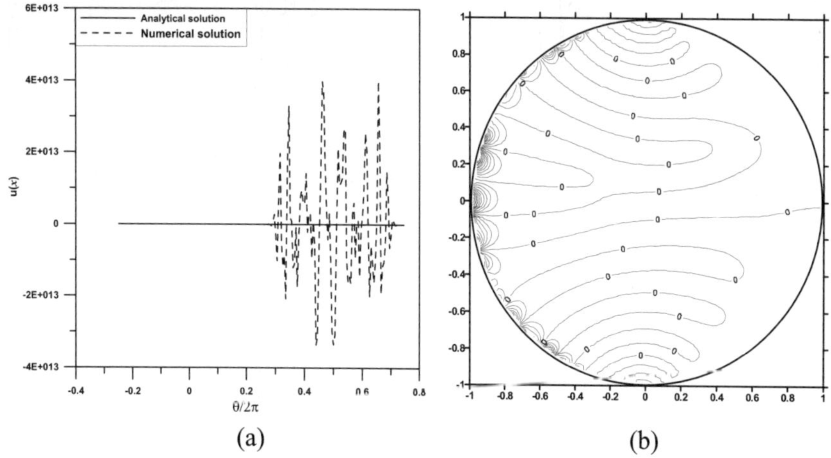

Figure 7: (a) Analytical solution and the boundary potential without regularization techniques, (b) the Field solution without regularization techniques.

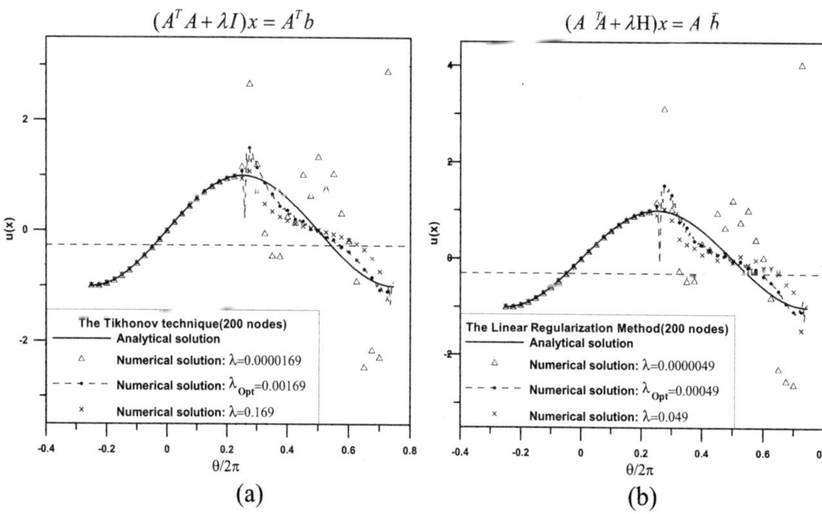

Figure 8: The numerical solution remedied by (a) the Tikhonov technique and (b) the linear regularization method with 3 different lambdas (200 nodes).

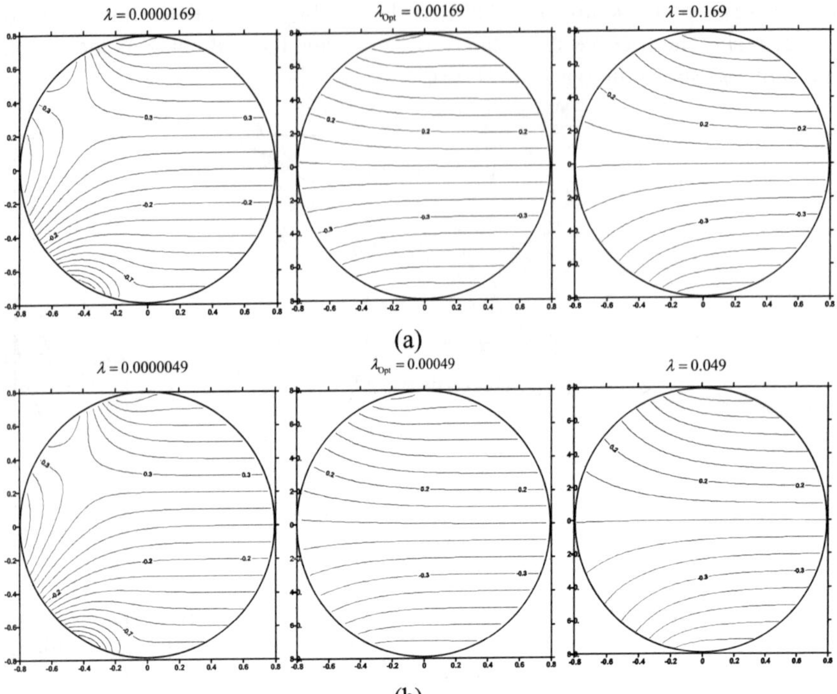

Figure 9: (a) The numerical field solution remedied by the Tikhonov technique and the linear regularization method with 3 different lambdas (200 nodes), (b) the numerical field solution remedied by the linear regularization method with 3 different lambdas (200 nodes).

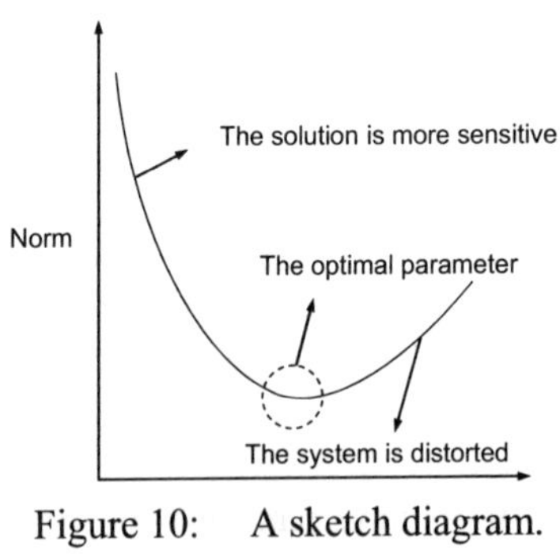

Figure 10: A sketch diagram.

As we expected from the mathematical point of view, a corner is presented in the L-curve shape. If the corner of the L-curve is chosen as an optimal point, the appropriate value is 0.00169 for the Tikhonov technique and 0.00049 for the linear regularization method respectively.

Therefore, we can figure out the norm error of the L-curve shape by the Tikhonov technique is much lower than the linear regularization method shown as Fig. 12. Then the results have been regularized to approximate the analytical solution, as shown in Fig. 13, and the field solutions are shown in Fig. 14(a), 14(b). The regularized result by the Tikhonov technique is more approximate than that by the linear regularization method. However, the exact solution is needed from the definition of norm error.

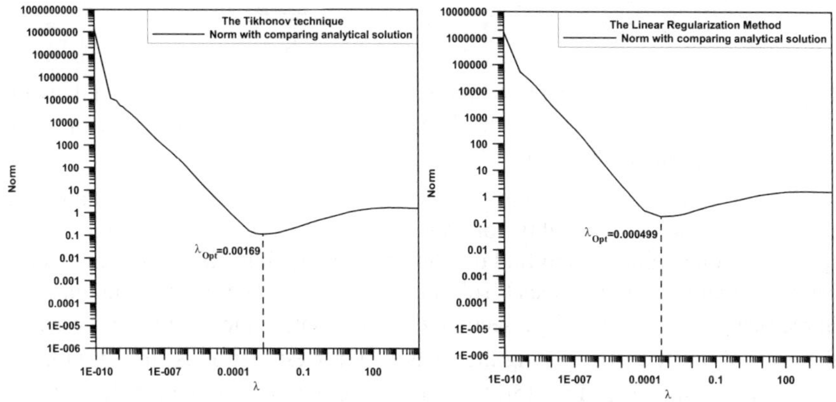

(a) (b)

Figure 11: (a) and (b) The norm deriving from comparing numerical solution with analytical solution by the Tikhonov technique and the linear regularization method.

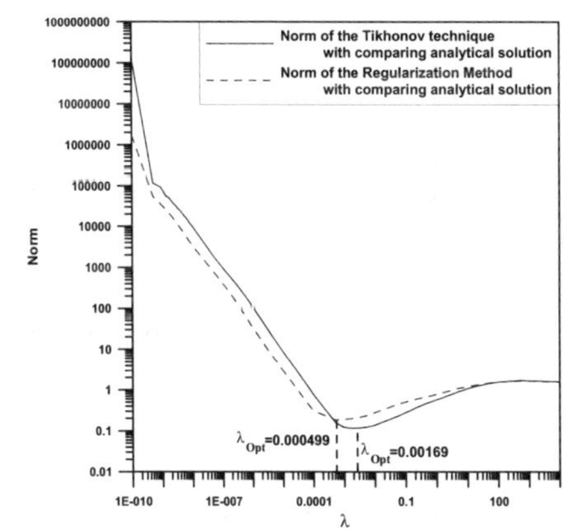

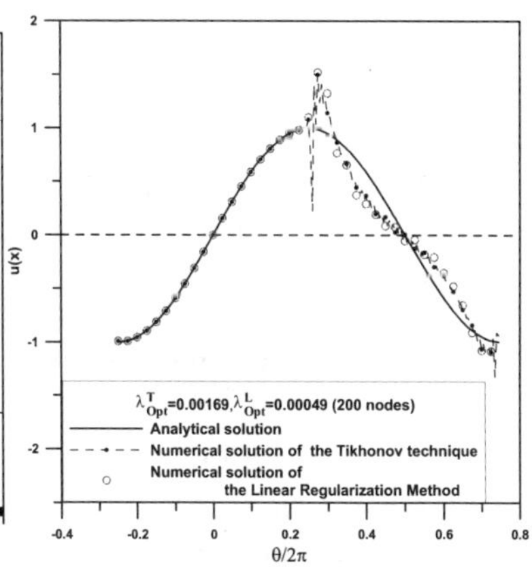

Figure 12: The norm error of the L-curve shape by the Tikhonov technique is much lower than the linear regularization method.

Figure 13: Numerical solution being remedied by the Tikhonov technique and the linear regularization method with the optional lambda (200 nodes).

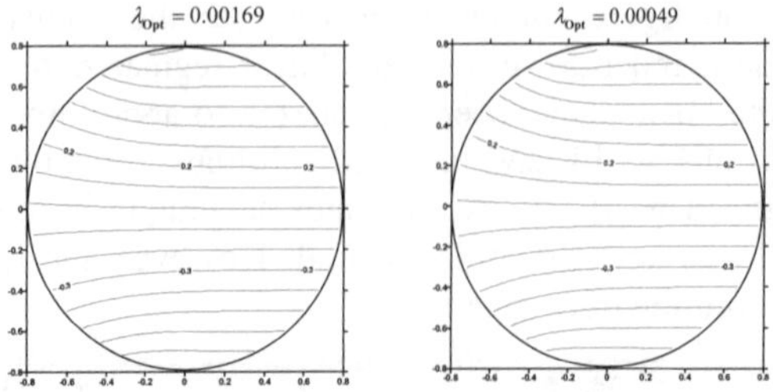

(a) (b)

Figure 14: Numerical field solution being remedied by (a) the Tikhonov technique and (b) the linear regularization method with optional lambda (200 nodes).

We are well aware that many problems usually have no analytic solution. In order to assess the validity of the Tikhonov technique and the linear regularization method and to find out the optimal solution, choosing the adaptive error estimation without exact solution is needed. The new norm error is implemented as defined in the section 2.2.3 and obtain the optimal λ. We find the optimal λ by implementing the adaptive error estimation is similar with the before λ, as shown in Fig. 15(a), 15(b). Therefore, we can derive the optimal result by employing the adaptive error estimation even though no exact solution can be obtained.

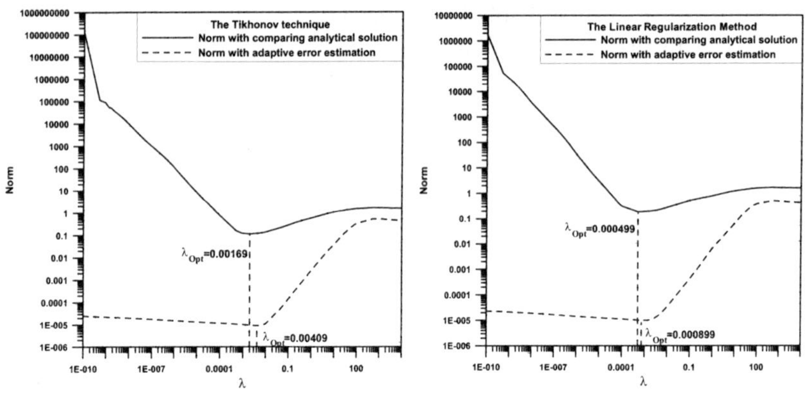

(a) (b)

Figure 15: (a) The optimal result by employing the adaptive error estimation and the Tikhonov technique under no exact solution, (b) the optimal result by employing the adaptive error estimation and the linear regularization method under no exact solution.

4 Conclusion

In this paper, we used the Trefftz method to solve the Laplace equation in a two-dimensional finite domain with overspecified boundary condition.

The numerical instability existing in the solver owing to the regular formulation of this method is encountered. To overcome this difficulty, the regularization techniques using the Tikhonov technique and the linear regularization method, together with the L-curve, plays a role in determining the optimal parameter λ which can maintain the system characteristic and can make the system insensitive to contaminating noise.

Furthermore, the numerical results obtained by using the Tikhonov technique for the case are in very close agreements with the analytical solutions , adaptive error estimation and outperform other regularization techniques.

References

[1] Tikhonov, A.N. & Arsenin, V.Y., *Solutions of Ill-posed Problems*, V.H. Winston and Sons: Washington, D.C., 1977.

[2] Poluektov, A.R., Short communications: a method of choosing the regularization parameter for the numerical solution of ill-posed problems. *Computational Mathematics and Mathematical Physics*, **32(3)**, pp. 397-401, 1992.

[3] Jin, B. A meshless method for the Laplace and Biharmonic equations subjected to noisy boundary data. *Computer Modeling in Engineering and Sciences*, **6(3)**, pp. 253-261, 2004.

[4] Fairweather, G. &Andreas, K., The method of fundamental solutions for elliptic boundary value problems. *Advances in Computational Mathematics*, **9**, pp. 69-95, 1998.

[5] William, H.P., Saul, A.T., William, T.V. & Brian, P.F., *Numerical Recipes in Fortran*, Second edition, Cambridge University Press: New York; 1992.

[6] Jirousek, J. & Wroblewski, A., T-elements: state of the art and future trends. *Archives of Computational Methods in Engineering*, **3-4**, pp. 323-434, 1996.

[7] Chang, J.R., Liu, R.F., Kuo S.R., & Yeih, W., Application of symmetric indirect Trefftz method to free vibration problems in 2D. *International Journal for Numerical Methods in Engineering*, **56(8)**, pp. 1175-1192, 2003.

[8] Chang, J.R., Liu, R.F., Yeih, W.C. & Kuo, S.R., Applications of the direct Trefftz boundary element method to the free-vibration problem of a membrane. *Journal of the Acoustical Society of America*, **112(2)**, pp. 518-527, 2002.

[9] Chang, J.R., Yeih, W. & Shieh, M.H., On the modified Tikhonov's regularization method for the Cauchy problem of the Laplace equation. *Journal of Marine Science and Technology*, **9(2)**, pp. 113-121, 2001.

[10] Chen, J.T., Wu, C.S., Lee, Y.T. & Chen, K.H., On the equivalence of the Trefftz method and method of fundamental solutions for Laplace and biharmonic equations. *Computers and Mathematics with Applications*, (forthcoming).

[11] Chen, J.T. & Chen, K.H., Analytical study and numerical experiments for Laplace equation with overspecified boundary conditions. *Applied Mathematical Modelling*, **22**, pp. 703-725, 1998.

[12] Wu, K.-L., Chen, K.-H., Chen, J.-T. & Kao, J.-H., Regularized meshless method for solving the Cauchy problem. *The 30th National Conference on Theoretical and Applied Mechanics,* Da-Yeh University: Changhwa, Taiwan, R.O.C., December 15-16, 2006

[13] Chen, L.Y., Chen, J.T., Hong, H.K. & Chen, C.H., Application of Cesàro mean and the L-curve for the deconvolution problem. *Soil Dynamics and Earthquake Engineering*, **14**, pp. 361-373, 1995.

[14] Tanaka, M. & Bui, H.D., (Eds.) *Inverse problems in engineering mechanics,* IUTAM symposium, Springer: Berlin, 1992.

[15] Kupradze, V.D., A method for the approximate solution of limiting problems in mathematical physics. *Computational Mathematics and Mathematical Physics,* **4**, pp. 199-205, 1964.

[16] Ivanov, V.K., The Cauchy problem for the Laplace equation in an infinite strip. *Differentsial'nye Uravneniya,* **1(1)**, pp. 131-136, 1965.

[17] Jin, W.G., Cheung, Y.K. & Zienkiewicz, O.C., Trefftz method for Kirchoff plate bending problems. *International Journal for Numerical Methods in Engineering*, **36,** pp. 765-781, 1993.

[18] Yeih, W., *Inverse Problems in Elasticity*, Ph.D. Dissertation, Northwestern University, 1991.

[19] Yeih, W., Liu, R.F., Chang, J.R. & Kuo, S.R., Numerical instability of the direct Trefftz method for Laplace problems for a 2D finite domain. *International Journal of Applied Mathematics and Mechanics*, **2(1)**: 41 - 66, 2006

Section 2
Advanced formulations

New boundary element analysis of acoustic problems with the fictitious eigenvalue issue

M. Tanaka[1], Y. Arai[1] & T. Matsumoto[2]
[1] *Department of Mechanical Systems Engineering, Shinshu University, Japan*
[2] *Division of Mechanical Engineering, Graduate School of Nagoya University, Japan*

Abstract

This paper is concerned with a new approach for avoiding the fictitious eigenfrequency problem to boundary element analysis of three-dimensional acoustic problems governed by Helmholtz equation. It is well known that in solving without any care the external acoustic problem which includes internal sub-domains by means of the boundary integral equation, the solution is disturbed at fictitious eigenfrequencies corresponding to the internal sub-domains. The present paper proposes a new boundary element analysis to circumvent such the fictitious eigenfrequency problem, which is an alternative boundary integral equation approach to the Burton-Miller one. The present approach is implemented, and its validity and effectiveness are demonstrated through numerical computation of typical examples.

1 Introduction

Whenever the acoustic problems which include the sub-domains without vibration are solved by means of the usual boundary integral equation without any care, the so-called fictitious eigenvalue issue is encountered. It is well known that the solution of the external acoustic problem is violated near the eigenfrequencies of the inside sub-domains. In practice, if we locate a few source points in the sub-domains without vibration and solve the system of equations by the method of least squares, we can circumvent the above eigenvalue issue [1–3]. Nevertheless, in finding the optimal shapes of acoustic fields, for example, it is almost impossible to apply the above practical mehtod, as the current shape is changing and the final,

WIT Transactions on Modelling and Simulation, Vol 44,
www.witpress.com, ISSN 1743-355X (on-line)
doi:10.2495/BE070061

converged solution is obtained in an iterative manner. It is true that we cannot find an appropriate number of external points to be added and their appropriate locations for such problems. There is a remedy of this problem, however, which is called the Burton-Miller method. This method employs a linearly combined boundary integral expression of the usual boundary integral equation (OBIE) and the normal derivative boundary integral equation (NDBIE) multiplied with the coupling parameter [4, 5].

The present paper proposes an alternative approach to this method to reduce burden of calculating the coefficients in matrixes of the final system to be solved. If we assume to use higher-order boundary elements, the Burton-Miller expression is used at a smaller number of element nodes, and at the other elemet nodes we employ the NDBIE multipied with the same coupling parameter. Through numerical computations it is demonstrated that the present approarch and the Burton-Miller one provide the almost identical results to circumvent always the fictitious eigenvalue issue.

2 Theory

It is assumed that the acoustic problems to be investigated in this study are in a steady-state vibration and governed by the Helmholtz equation:

$$\nabla^2 p\,(x) + k^2 p\,(x) + f\,(x) = 0 \tag{1}$$

where $p\,(x)$ denotes the sound pressure, $f\,(x)$ the distributed source term, and k the wave number. Denoting C_0 by the sound velocity, the wave number k is expressed by using the angular velocity ω as

$$k = \frac{\omega}{C_0} \tag{2}$$

The boundary conditions are prescribed as

$$p\,(x) = \bar{p}\,(x) \tag{3}$$

$$q\,(x) = \frac{\partial p}{\partial n}\,(x) = \bar{q}\,(x) \tag{4}$$

where $q\,(x)$ is related to the outward normal velocity of a particle $v\,(x)$ and the mass density ρ as follows:

$$q\,(x) = -\mathrm{i}\omega\rho v\,(x) \tag{5}$$

2.1 Regularized boundary integral equation

Under the assumption of a single point sound source with intensity I at the point x^s, the boundary integral equation can be expressed in a regularized form [6] as

$$\int_\Gamma \{q^*\,(x,y) - Q^*\,(x,y)\}\,p\,(x)\,d\Gamma(x) + \int_\Gamma Q^*\,(x,y)\,\{p(x) - p(y)\}\,d\Gamma(x)$$

$$= -\mathrm{i}\omega\rho \int_{\Gamma} p^{*}(x,y)\, v(x) d\Gamma(x) + I p^{*}(x^{s}, y) \tag{6}$$

The fundamental solutions which are denoted by an asterisk are given by

$$p^{*}(x,y) = \frac{1}{4\pi r} \exp(-ikr) \tag{7}$$

$$q^{*}(x,y) = -\frac{1}{4\pi r^{2}} (1 + \mathrm{i}kr) \exp(-ikr) \frac{\partial r}{\partial n}(x) \tag{8}$$

$$Q^{*}(x,y) = -\frac{1}{4\pi r^{2}} \frac{\partial r}{\partial n}(x) \tag{9}$$

where r denotes the distance between the source point y and a field point x.

The boundary integral equation (6), which is called in this paper "OBIE", is usually applied to the standard analysis of acoustic fields and in most cases obtain successful results. Unfortunately, however, in the acoustic fields to be studied in this paper, Eq.(6) suffers the fictitious eigenfrequency problem, and gives "ghost" solutions at an infinite number of eigenfrequencies for internal subdomains without vibration. To improve this situation, we have to prepare another expression of the boundary integral equation and to use it together with Eq.(6).

2.2 Normal derivative boundary integral equation

We now differentiate Eq.(6) with respect to the source point y. Then, we can obtain the following expresssion omitting the source term:

$$\int_{\Gamma} \{q^{*}_{,j}(x,y) - Q^{*}_{,j}(x,y)\}\, p(x)\, d\Gamma(x) + \int_{\Gamma} Q^{*}_{,j}(x,y)\{p(x) - p(y)\}\, d\Gamma(x)$$
$$-\left\{\int_{\Gamma} Q^{*}(x,y)\, d\Gamma(x)\right\} p_{,j}(y) = -\mathrm{i}\omega\rho \int_{\Gamma} p^{*}_{,j}(x,y)\, v(x) d\Gamma(x) \tag{10}$$

We take into account a uniform gradient of the sound pressure p in the above expression, and regularize the boundary integral expression. Then, we can finally derive the following regularied boundary integral eqaution [7, 8]:

$$\int_{\Gamma} \left\{\tilde{q}^{*}(x,y) - \tilde{Q}^{*}(x,y)\right\} p(x)\, d\Gamma(x)$$
$$+ \int_{\Gamma} \tilde{Q}^{*}(x,y)\{p(x) - p(y) - r_{m}(x,y)\, p_{,m}(y)\}\, d\Gamma(x)$$
$$= -\mathrm{i}\omega\rho \int_{\Gamma} \{\tilde{p}^{*}(x,y) - \tilde{u}^{*}(x,y)\}\, v(x) d\Gamma(x)$$
$$-\mathrm{i}\omega\rho \int_{\Gamma} \tilde{u}^{*}(x,y)\{v(x) - n_{m}(x)\, p_{,m}(y)\}\, d\Gamma(x) + I\tilde{p}^{*}(x^{s}, y) \tag{11}$$

where $\tilde{()} = \partial()/\partial n(y)$, and the asterisked functions are related to the fundamental solution. This normal derivative boundary integral equation is called "NDBIE" in this paper.

Together with the two boundary integral equations, we can explain the boundary element methods for solving the acoustic problems under consieration as in the following manner:

1 The standard BEM uses only OBIE for all nodal points,
2 The Burton-Miller Approach adopts OBIE + αNDBIE for all nodal points,
3 The present proposal employs OBIE + αNDBIE at a smaller number of element nodes, and at the other element nodes uses only the expression αNDBIE.

It is intersting to note that according to Ref. [5], we choose the coupling parameter α as $\alpha = \mathrm{i}/k$.

3 Numerical results and discussion

We shall take a breathing sphere with a uniform velocity v on its whole surface, which is located in an infinite acoustic 3-D space. It is well known that the acoustic problem of the external domain with infinity involves the fictitious eigenfrequency issue. The solution by means of the usual boundary integral equation (OBIE) method is disturbed near the eigenfrequencies of the inside sphere itself.

The present study employs quadrilateral boundary elements with second-order polynominal interpolation functions as shown in Fig. 1. When the source point is located at a corner point, the element is divided into two subelements as Type A shown in Fig. 2. On the other hand, when the source point is located at a middle node, the element is divided into three sub-elements as shown as Type B in the figure. The singular integrals are evaluated by such a sub-element method [9].

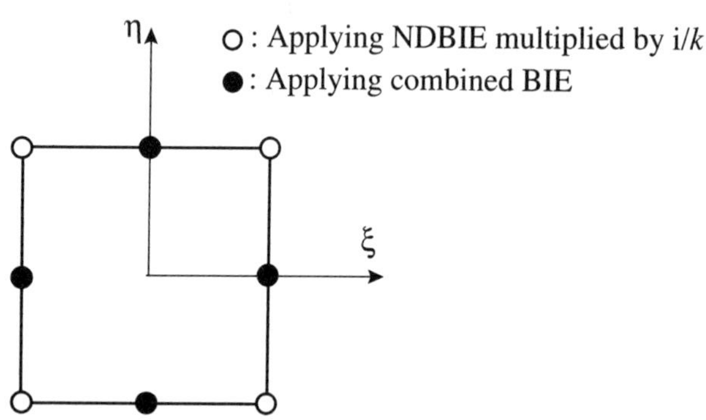

Figure 1: Quadrilateral boundary element with quadratic interpolations.

Boundary element division of the 1/8 part of the spherical surface is shown in Fig. 3. Three evaluation points for the sound pressure p are placed as the measuring point shown in the figure. It is assumed that the mass density $\rho = 1.2\,[\mathrm{kg/m^3}]$ and the sound velocity $C_0 = 340\,[\mathrm{m/s}]$.

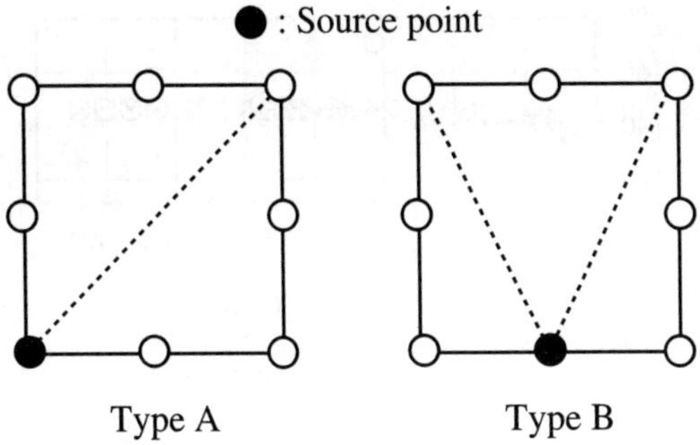

Figure 2: Types of dividing for sub-elements.

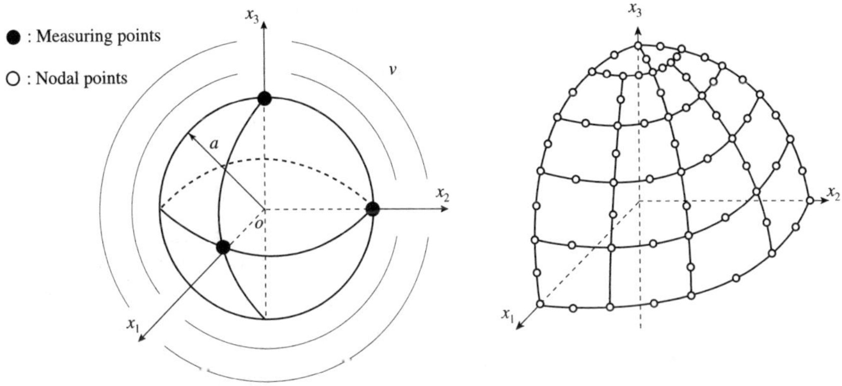

Figure 3: Analysis model 1 and boundary element discretization.

The analytical solution is available in the literature [10] and given by

$$p(r) = v\frac{\mathrm{i}\omega\rho a^2}{(1+\mathrm{i}ka)\,r}\exp\{-\mathrm{i}k(r-a)\} \tag{12}$$

Figure 4 shows the numerical results obtained by the above three methods. There seem to be the fictitious eigenfrequencies near the values of nondimensional wave numbers 3 and 6, as the numerical solutions by the OBIE are disturbed, while the other two methods gives smooth, accurate results. Detailed views near $ka = 3$, and $ka = 6$ are shown in Figs. 5 and 6, respetively.

It can be seen that the present method provides accurate numerical solutions without disturbances due to the fictitious eigenfrequency issue, under a smaller amount of computational burden in comparison with the Burton-Miller method [4]. It is proved from a mathematical point of view that the present method can always provide accurate results without any disturbance due to the fictitious eigenvalue issue.

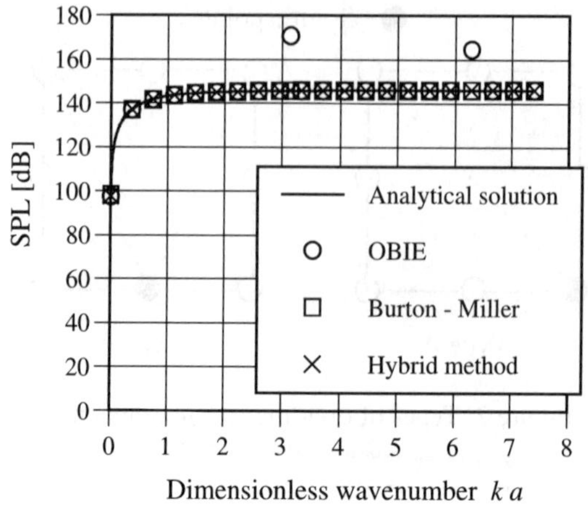

Figure 4: Numerical results obtained by OBIE, Burton-Miller and present Hybrid methods.

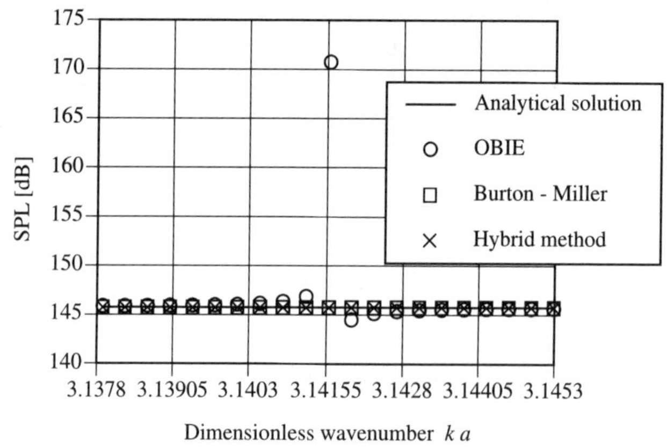

Figure 5: Detailed view of numerical results between $ka = 3.1378$ and 3.1453.

Next, we shall consider the acoustic field between a rigid sphere with radius a_1 and a breathing sphere with radius a_2 concentrically located as shown in Fig. 7. In the numerical computation, it is assumed that $a_1 = 0.25\,\mathrm{m}$ and $a_2 = 0.1\,\mathrm{m}$. The boundary conditions are assumed such that the breathing sphere is subject to the same uniform normal particle velocity as the previous example and the condition on the outside spherical surface is rigid so that the particle velocity vanishes there. The 1/8 part of the two spherical boundary surfaces is divided into the same number of elements as shown in the figure. The numerical results obtained are shown in Fig. 8. Numerical computation is performed by an interval 1 Hz from 1 Hz to 2 kHz. Even in this acoustic problem, the fictitious eigenfrequency problem

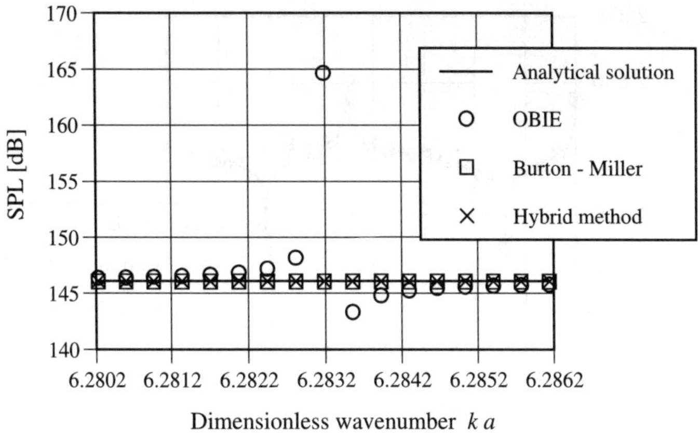

Figure 6: Detailed view of numerical results between $ka = 6.2802$ and 6.2862.

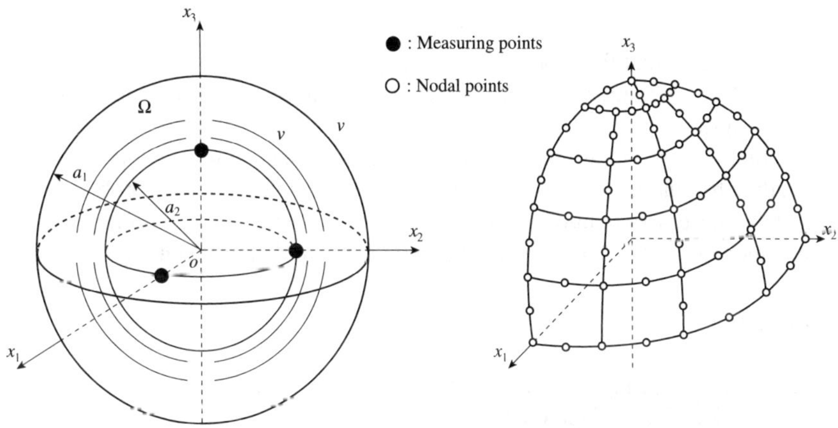

Figure 7: Analysis model 2 and its boundary element discretization.

occurs. That is, near the non-dimensional wavenumber $ka = 3$ in this calculation the OBIE gives disturbed numerical results different to those by the other two methods.

Other 3-D numerical examples and analysis of 2-D acoustic fields with a few examples can be found in authors' separate papers [11, 12].

4 Concluding remarks

The present paper has proposed an alternative approach to the Burton-Miller method to give accurate results without disturbances due to the eigenfrequency issue in the acoustic problems. The usefulness of the proposed method was demonstrated through numerical computations and comparison with other ones.

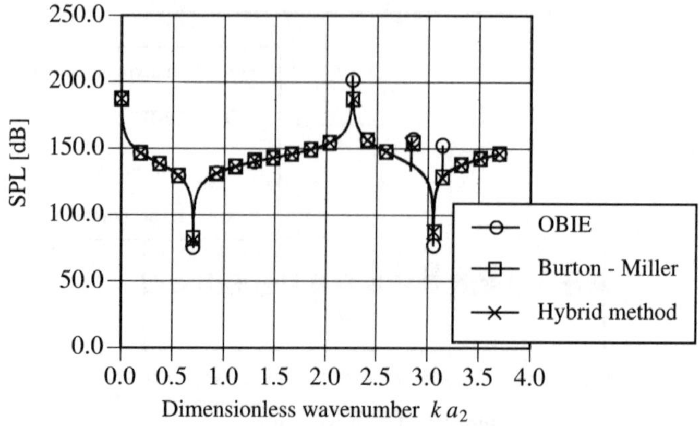

Figure 8: Numerical results obtained by OBIE, Burton-Miller and present Hybrid methods.

It is noted again that the proposed method can give the almost same results as the Burton-Miller one under a smaller burden in calculating the coefficient matrixes in the final system.

As future work along this line, it can be recommended to develop a rigorous procedure for finding the optimal shapes of acoustic fields. For such problems, it is inevitably necessary to have an analysing procedure with a reduced computational task which prevents the fictitious eigenfrequency problem from taking place.

References

[1] Tanaka, M., Matsumoto, T. & Nakamura, M., *Boundary Element Methods*, Baifukan, Tokyo/Japan, 1991.

[2] Kobayashi C.S.(ed), *Wave Analysis and Boundary Element Method*, Kyoto University Press Online, Kyoto/Japan, 2000.

[3] Schenck, H.A., Improved integral formulation for acoustic radiation problems, *J. the Acoustical Society of America*, Vol.44, No.1, pp.41–58, 1968.

[4] Burton, A.J. & Miller, G.F., The application of integral equation methods to the numerical solution of some exterior boundary-value problems, *J. the Royal Society of London, Ser. A*, Vol.323, pp.201–210, 1971.

[5] Cunefare, K.A. & Koopmann, G., A boundary element method for acoustic radiation valid for all wavenumbers, *J. the Acoustical Society of America*, Vol.85, No.1, pp.39-48, 1989.

[6] Matsumoto C.T.D. & Tanaka C.M.C. Alternative discretization technique for regularized boundary integral equation, *Transactions JASCOME*, Vol.1, pp.7–12, 1991.

[7] Arai, M., Adachi, T., & Matsumoto, H. C. Highly accurate analysis by boundary element method based on uniform gradient condition (Application for formulation of classical potential problems), *Transactions of Japan Society of Mechanical Engineers (JSME), Ser. A*, Vol.61, No.581, pp.161–168, 1995.

[8] Matsumoto C.T.D. & Tanaka, M. C. Evaluation of the hypersingular and regularized boundary integral equations for the boundary potential gradients in 2D fieldC *Transactions JSME, Ser. A* , Vol.64, No.619, pp.743–750, 1998.

[9] Yuuki C.R.D. & Kisu C.H., *Boundary Element Methods for Elastic Analysis*, Baifukan, Tokyo/Japan, 1987.

[10] Itou, T., *Basic Acoustical Engineering*, Vol.1, Corona, Tokyo/Japan, pp.268–270, 1990.

[11] Arai, Y., Tanaka, M. & Matsumoto, T., New boundary element analysis of 3-D acoustic fields avoiding the fictitious eigenfreqency problem, *Transactions JSME, Ser. C*, to be printed.

[12] Tanaka C.M., Matsumoto C.T. & Arai, Y., A boundary element analysis for avoiding the fictitious eigenfrequency problem in acoustic field (2nd report: Revised version)C *Transactions JSME, Ser. C*, Vol.72, No.719, pp.2008–2093, 2006.

www.witpress.com, ISSN 1743-355X (on-line)

A BEM formulation of free hexagons based on dynamic equilibrium

P. Procházka
Czech Technical University in Prague, Czech Republic

Abstract

In this paper, a time dependent (dynamical equilibrium) free hexagon DEM is formulated and solved. The main application is found in geomechanics, namely in bumps occurrence in deep mines. The time factor is included in a natural way in the model of discrete elements created by the boundary element method. One of the most important phenomena is the velocity of excavation. In the deep mines the method of depositing packs and its mechanical properties are also decisive. Their mutual coupling can principally influence the safety against bumps. For a correct understanding of the behavior of the rock aggregate (coal seam vs. overburden), the nucleation of cracks finally leading to bumps has to be treated as time dependent, while so far it was observed only from statical equilibrium. According to new experiments and results from accessible literature, a dynamical effect has to be included in thr formulation.

Contact problems leading to bumps occurrence in deep mines have been solved in many of the papers of the present author for the static case. Either Lagrangian multipliers or penalty formulation were used. The new formulation has to be submitted in terms of a penalty, which if high enough (bond effect of adjacent elements) suppresses the influence of time. By including the interface properties with the lumped inertia mass of the elements, complex nucleation can be studied and the information on possible rock bursts is improved. From some examples it was shown in the static case that the behavior at the face of longwall mining is close to that near the crack tip, and the differences in material properties of coal and overburden are also not negligible. These factors are also expected to be important in the case of dynamic problems. Some examples show the application of the procedure proposed.
Keywords: discrete element method, boundary element method, dynamical equilibrium.

WIT Transactions on Modelling and Simulation, Vol 44, © 2007 WIT Press
www.witpress.com, ISSN 1743-355X (on-line)
doi:10.2495/BE070071

1 Introduction

In this paper we discuss a possible solution of the stability of side walls for longwall mining in deep mines. Two phase medium is modeled: rock mass compresses a coal seam, which is positioned at a depth exceeding 700 m.

Numerical methods seem to be the cheapest tool for assessing different types of structures. In the case of the description of underground massif often PFC (particle flow code) [1, 2], is used, which has a long lasting tradition, but for prediction of rock bumps it is a very poor tool. This is caused by the fact that the PFC starts with dynamical equilibrium so that the process described by the PFC is dynamic, which is in contradiction with the real behavior of the rock before bumps occurrence. Moreover, only a one-point touch of adjacent elements cannot describe the true distribution on element boundary displacements and tractions needed for description of continuum. From these considerations it follows that prediction by the PFC has a very poor chance in planning longwall mining.

If the theory of damage should be involved into formulation of the problem to be solved, special treatment is required using continuous methods (FEM, BEM, etc.). The methods, which are extensively used, start with realization of the trial body as a continuum. Named here "Cohesive zone method" [3], "Manifold method" [4–6], for example, which deal with Barenblatt's theory. In the problem of rock bursts such methods are on one side uneasy applicable and on the other side exhibit unreal behavior, according to a couple of test examples. "Smooth Hydrodynamics Method" [7] seems more promising, but problems occur with introducing general boundary conditions

This is why test experiments have been carried out to get knowledge about a reasonable approach for solving the problem. One of a possible experimental treatment was suggested in paper [8], where Araldit and physically similar materials are used. Based on models of such a similarity conception coupled modeling was used in [9–11]. In [9] rock bursts are studied in mines at extreme depths. The free hexagon method is used to determine the bumps state in the rock. In [10] a similar problem is solved for more general cases occurring during mining. In [11] tunnel face stability is assessed. The scale model used in this paper is prepared in collaboration with Muenchen University; Lippman was the coordinator together with J. Vacek.

In [12–14] results from on site measurements are published, recommendations are provided on how to proceed in determining the rock bursts and, what is the most important to us, the way of movement is shown in these publications.

The free hexagon method seems to be one being very promising, as the result from experiments and numerical models are reasonably comparable. The static equilibrium was used in formulation of the free hexagon method. This method has been established in the middle of 90-ties and the fundaments can be found in [15, 16]. In [17] the method is applied to the stability of a tunnel face. The nucleation of cracking is observed at the face, but no larger displacements were calculated; only the failure state was determined. A concise formulation of two methods, static PFC and the free hexagon method, is published in [16] together with couple of applications to various geotechnical problems.

In this paper interface between statical and dynamical states before and at the moment of bumps is characterized. First, the method will be described and basic formulas will be derived, and then some applications to rock bumps will be presented. Time-dependent problem with the D'Alembert forces, which are caused by contact forces of moving particles, simplifies the body of the earth (soil) to a set of hexagons, which are, or are not in mutual contact. The material properties of the hexagons are determined from the state of stresses. The hexagons represent a possible shape of grains the earth consists of. The model proposed in this paper may, contrary to modern numerical methods (FEM, BEM, etc.), enables one to disconnect the medium described by the hexagons, when needed (e.g. providing certain requirement on tensile strength is violated). The most natural contact conditions (Mohr-Coulomb hypotheses) may be simply introduced and, after imposing all such of those contact conditions, the localized damage, or "cracking" can be found out. The stability then depends on the "measure" of the touched zone.

Mechanical behavior inside each element is either linear or non-linear (plastic, viscoelastic, viscoplastic, etc). To describe such behavior, boundary elements are applied. In some papers FEM is used, which causes far more difficulties then the BEM, see, e.g., [17], where tractions are by one degree of polynomials lesser then displacements, although along the boundary abscissas linear relation traction x displacement holds. This is in contradiction with the assertion that along the adjacent boundaries of the elements the tractions and the displacements have the same degree of approximations (splines). On the other hand, it is typically fulfilled when using the BEM that the element boundary displacements and tractions are of the same degree of approximation.

A typical coupled modeling (mathematical and experimental) is published in [18]. In the latter publication the experimental models are based on scale modeling and created from physically equivalent materials. In our case similar modeling is used with such an exception that very particular materials are used, see [7–11].

2 Basic assumptions

Starting with statical equilibrium in the first stage of excavation, after dislocations in the rock continuum and in the coal seam that appears, time dependent dynamical equilibrium has to be considered. Under the assumption that the material properties of both rock and coal are known, hexagon elements are created and linear behavior in them is supposed. Since the elements are considered to be small enough, isotropic case is taken into account, i.e. the elements are homogeneous and isotropic with material characterization given by modulus of elasticity E and Poisson's ratio ν, for example. Classical problem involving generalized Coulomb's friction and exclusion of tensile stress exceeding the tensile strength along the interfaces (possible dislocations) is solved. Typical set up of adjacent elements is illustrated in Fig. 1. In what follows the distribution of mass inside each element is neglected in such a sense that it is concentrated at c.o.g. of the element. Then, first the solution of elastic problem in an element is formulated and the element is put into the neighborhood of adjacent elements. Regular distribution of elements is assumed, i.e. only one matrix relating tractions and boundary displacements will be provided.

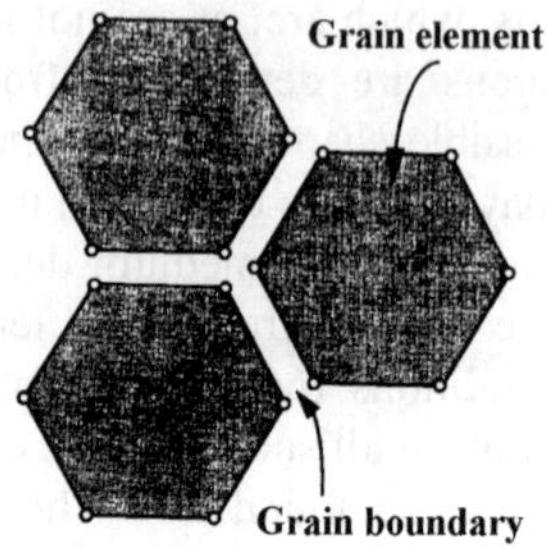

Figure 1: Adjacent grains set up.

3 Boundary element solution in one hexagon

The solution of elasticity on each hexagonal element Ω is approximated by concentration of DOFs to vertices of the hexagon, and distribution of boundary displacements and tractions along edges $\Gamma_s, s = 1,\ldots,6$ of the hexagon Ω is assumed to be linear. Then, generally, integral equations formulate the problem:

$$c_{ik}u_k(\xi) = \int_{\Gamma_s} p_i(\boldsymbol{x})u_{ik}^*(\boldsymbol{x},\xi)\,\mathrm{d}\boldsymbol{x} - \int_{\Gamma_s} u_i(\boldsymbol{x})p_{ik}^*(\boldsymbol{x},\xi)\,\mathrm{d}\boldsymbol{x} + \int_{\Omega} b_i(\boldsymbol{x})u_{ik}^*(\boldsymbol{x},\xi)\,\mathrm{d}\boldsymbol{x} \quad (1)$$

where i and k run 1,2, and s = 1,…,6. In case of regular element distribution is considered, δ_{ik} is Kronecker's delta. In case the regular hexagons are used and linear distribution of both displacements and tractions is used, $c_{ik} = \frac{1}{3}\delta_{ik}$, and then

$$\frac{1}{3}\delta_{ik}u_k(\xi) = \int_{\Gamma_s} p_i(\boldsymbol{x})u_{ik}^*(\boldsymbol{x},\xi)\,\mathrm{d}\boldsymbol{x} - \int_{\Gamma_s} u_i(\boldsymbol{x})p_{ik}^*(\boldsymbol{x},\xi)\,\mathrm{d}\boldsymbol{x} + \int_{\Omega} b_i(\boldsymbol{x})u_{ik}^*(\boldsymbol{x},\xi)\,\mathrm{d}\boldsymbol{x} \quad (1a)$$

while for uniform distribution of both geometrical and statical characterizations along the boundaries we get

$$\frac{1}{2}\delta_{ik}u_k(\xi) = p_i \int_{\Gamma_s} u_{ik}^*(\boldsymbol{x},\xi)\,\mathrm{d}\boldsymbol{x} - u_i \int_{\Gamma_s} p_{ik}^*(\boldsymbol{x},\xi)\,\mathrm{d}\boldsymbol{x} + \int_{\Omega} b_i(\boldsymbol{x})u_{ik}^*(\boldsymbol{x},\xi)\,\mathrm{d}\boldsymbol{x} \quad (1b)$$

$$k = 1, 2, \ s = 1,\dots,6,$$

Formula (1b) is used in this text, i.e. uniform distribution is applied. Knowing the form of kernels denoted by asterisk and substituting approximations for boundary displacements and tractions, matrix equations are obtained:

$$\boldsymbol{Au} = \boldsymbol{Bp} + \boldsymbol{b}, \qquad \boldsymbol{Ku} = \boldsymbol{p} + \boldsymbol{V} \tag{2}$$

where $\boldsymbol{A}$, $\boldsymbol{B}$ and $\boldsymbol{K}$ are square matrices (12 *12), $\boldsymbol{u}$ is the vector of displacement approximations at vertices, $\boldsymbol{p}$ that of tractions and $\boldsymbol{b}$ and $\boldsymbol{V}$ are vectors of volume weight influences. The latter are vectors (1*12).

4 Statical contact conditions

Let us consider two hexagons being in possible contact, see Fig. 2. Introduce a pseudo-cone K, which is defined as:

$$\begin{aligned} &K \equiv \{\boldsymbol{u} \in V, [u]_n \geq 0, p_n \leq p_n^+, \text{if } p_n \geq p_n^+ \Rightarrow p_n = 0, \\ &| p_t | \leq c\,\kappa(p_n^+ - p_n) - p_n \tan\varphi, \\ &\text{if } | p_t | \geq c\,\kappa(p_n^+ - p_n) - p_n \tan\varphi \Rightarrow p_t = p_n \tan\varphi\, \text{sgn}[u]_t\} \end{aligned} \tag{3}$$

where $[u]_n = u_n^2 - u_n^1$, $[u]_t = u_t^2 - u_t^1$, $\boldsymbol{u}$ is split into normal u_n and tangential (shear) u_t components, $\boldsymbol{n}$ is unit outward normal with respect to element 1, V is admissible space of displacements, traction $\boldsymbol{p}$ has now components $\{p_n, p_t\}$, i.e. projections to normal and tangential directions, p_n^+ is the tensile strength, c is the cohesion or shear strength, and φ is the angle of internal friction of the material (rock, coal), κ is the Heaviside function being equal to one for positive arguments and zero otherwise. Here strict sign convention is used: positive sign is tension, while negative one means compression. The pseudo-cone K becomes a cone for $p_n^+ = 0$ and frictionless case.

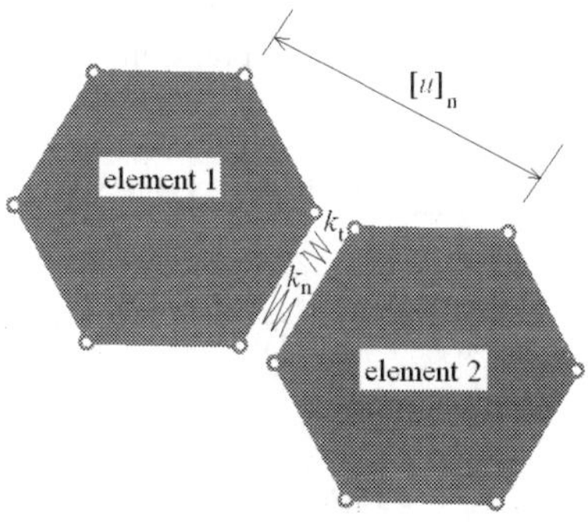

Figure 2: Two hexagons in possible contact.

5 Fischera's conditions

Fischera's conditions have been formerly formulated for K being a cone. In our case the conditions in normal direction can be written as:

$$p_n^+\kappa(p_n^+ - p_n) - p_n \geq 0,\ \ [u]_n \geq 0\,,\ \{p_n^+\kappa(p_n^+ - p_n) - p_n\}[u]_n = 0, \tag{4}$$

Similarly, in the tangential direction it holds:

$$\begin{aligned} &c\,\kappa(p_n^+ - p_n) - p_n\tan\varphi - |\,p_t\,| \geq 0,\quad |\,[u]_t\,| \geq 0\,,\\ &\{c\,\kappa(p_n^+ - p_n) - p_n\tan\varphi - |\,p_t\,|\}\,|\,[u]_t\,| = 0, \end{aligned} \tag{5}$$

The energy of the system can be stored as:

$$\begin{aligned} \Pi = \frac{1}{2}\sum_{\alpha=1}^{N} a_\alpha(\boldsymbol{u},\boldsymbol{u}) - \int_\Gamma \bar{\boldsymbol{p}}^{\mathrm{T}}\boldsymbol{u}\,\mathrm{d}\boldsymbol{x} - \sum_{\beta=1}^{n}\int_{\Gamma_\beta} \{(p_n^+)^\beta \kappa(p_n^+ - p_n^\beta) - p_n^\beta\}[u]_n^\beta\,\mathrm{d}\boldsymbol{x} - \\ - \sum_{\beta=!}^{n}\int_{\Gamma_\beta} \{c^\beta\,\kappa(p_n^+ - p_n^\beta) - p_n^\beta\tan\varphi - |\,p_t^\beta\,|\}\,|\,[u]_t^\beta\,|\,\mathrm{d}\boldsymbol{x} \end{aligned} \tag{6}$$

where α runs over all hexagon elements, $\alpha = 1,\ldots,N$, β runs all contact edges of possible contacts Γ_β, $\beta = 1,\ldots,n$, Γ is the external boundary where $\boldsymbol{p}$ is prescribed, and

$$a_\alpha(\boldsymbol{u},\boldsymbol{u}) = \int_{\Omega_\alpha} (\boldsymbol{\sigma}^\alpha)^{\mathrm{T}}\boldsymbol{\varepsilon}^\alpha\,\mathrm{d}\boldsymbol{x} \tag{7}$$

is the internal energy (bilinear form) inside a hexagon Ω_α, $\boldsymbol{\sigma}^\alpha, \boldsymbol{\varepsilon}^\alpha$ are respectively stresses and strains in Ω_α.

6 Penalty formulation

Setting $p_n = k_n[u]_n$, $p_t = k_t[u]_t$, where k_n, k_t are normal spring and tangential spring stiffnesses, and substituting these expressions in (6) yields

$$\begin{aligned} \Pi = \frac{1}{2}\sum_{\alpha=1}^{N} a_\alpha(\boldsymbol{u},\boldsymbol{u}) - \int_\Gamma \bar{\boldsymbol{p}}^{\mathrm{T}}\boldsymbol{u}\,\mathrm{d}\boldsymbol{x} + \\ + \sum_{\beta=1}^{n}\int_{\Gamma_\beta} \{k_n^\beta([u]_{\mathrm{n}}^\beta)^2 + k_n^\beta[u]_{\mathrm{n}}^\beta\,|\,[u]_t^\beta\,| + k_t^\beta([u]_t^\beta)^2\}\,\mathrm{d}\boldsymbol{x} - \\ - \sum_{\beta=!}^{n}\int_{\Gamma_\beta} \{(p_n^+)^\beta\kappa(p_n^+ - p_n^\beta)[u]_n^\beta + c^\beta\,\kappa(p_n^+ - p_n^\beta)\,|\,[u]_{\mathrm{t}}^\beta\,|\}\,\mathrm{d}\boldsymbol{x} \end{aligned} \tag{8}$$

7 Dynamical response

If each hexagonal element is considered small enough, lump mass dynamical problem can be formulated according to Fig. 2, where for the sake of simplicity the influence of rotation is neglected. Suppose the element 1 possesses a stable position, then element 2 will obey differential equation

$$m\frac{\mathrm{d}^2}{\mathrm{d}t^2}w+k_n w=0\,,\qquad w=[u]_n \tag{9}$$

where m is a mass of the element 1, measured in kg. From the latter equation immediately follows that if k_n is large the inertia forces are suppressed and in each small enough time step no dynamical influence occurs. This assertion will be précised in the next text. The solution of latter equation is known as:

$$w(t)=w_0\frac{\sin\omega\bar{\xi}}{\sin\omega}+w_1\frac{\sin\omega\xi}{\sin\omega},\qquad \xi=\frac{t\text{-}t_0}{h},\qquad \omega=h\sqrt{\frac{k_n}{m}},\qquad \bar{\xi}=1-\xi \tag{10}$$

where $h=t_1-t_0$ is the time step, $w_0=w(t_0)$, $w_1=w(t_1)$, t_0 is the initial time, t_1 is the time in the next time step. At the middle of the time interval, the value of displacement w and the first derivative by time t are derived as:

$$w_{\frac{1}{2}}=w(\xi=\tfrac{1}{2})=\frac{w_0+w_1}{2\cos\frac{\omega}{2}},\qquad \frac{\mathrm{d}}{\mathrm{d}t}w_{\frac{1}{2}}=\frac{\omega(w_1-w_0)}{2h\sin\frac{\omega}{2}} \tag{11}$$

From equations (10) and (11) it follows an important bound estimate on the time step h: $h\le\frac{\pi}{2}\sqrt{\frac{m}{k_n}}$. The only troublesome point remains for $k_n\to 0$. Then linear relation follows from the governing equation and, consequently, the velocity is constant. This is in compliance with the D'Alembert law. The last inequality leads us also to the fact that in case of large penalty k_n no differences in displacements can be expected due to inertia forces.

Using well known approximation formula for second derivative and the above approximate formulas we get:

$$\frac{\mathrm{d}^2}{\mathrm{d}t^2}w(\xi=\tfrac{1}{2})=\frac{1}{4h^2}[w(\xi=1)-2w(\xi=\tfrac{1}{2})+w(\xi=0)] \tag{12}$$

which is an explicit formula for calculating $w(\xi=1)$. Using vector projection to the coordinates system, resulting movement is received. At the moment the c.o.g. of the element is then moved assuming the deformed body as rigid.

8 Examples

A study of a longwall mine has been carried out. Material coefficients of the rock massif Ω have the following values: E = 52 500 MPa, ν = 0.29, the peak values E_p = 38 000 MPa, and ν_p = 0.38, the residual values E_r = 5 000 MPa, the angle of internal friction is 42 degrees, its residual value is 32 degrees, the shear strength c = 0.9 MPa and its residual value is considered as 0.4 MPa. The coal seam is brittle, with E = 5 500 MPa, ν = 0.39, the angle of internal friction and the shear strength vary. In Fig. 3 setting of hexagonal elements is seen, Figs. 4 and 5 display the movements at t = 0.1 sec, h = 0.001 sec, i.e. the starting spring stiffness k_n is derived as 10^{10}.

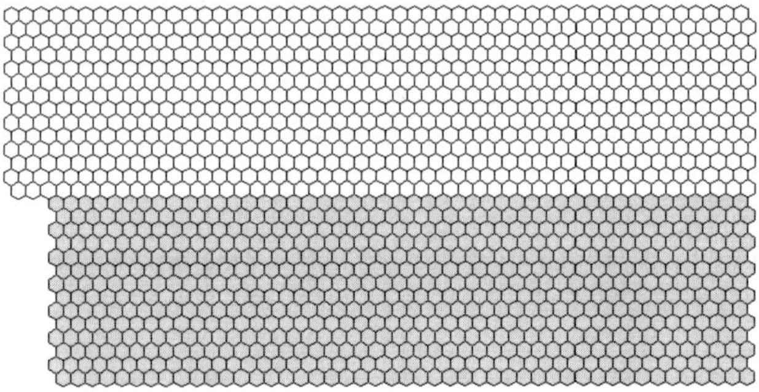

Figure 3: Setting of the hexagonal elements.

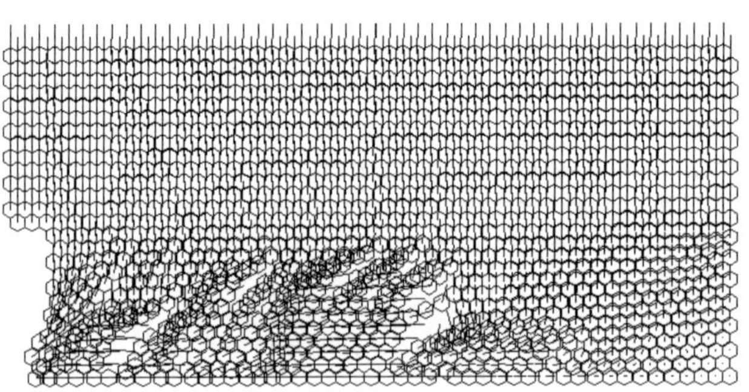

Figure 4: Movements for c = 100 kPa and p_n^+ = 10 kPa.

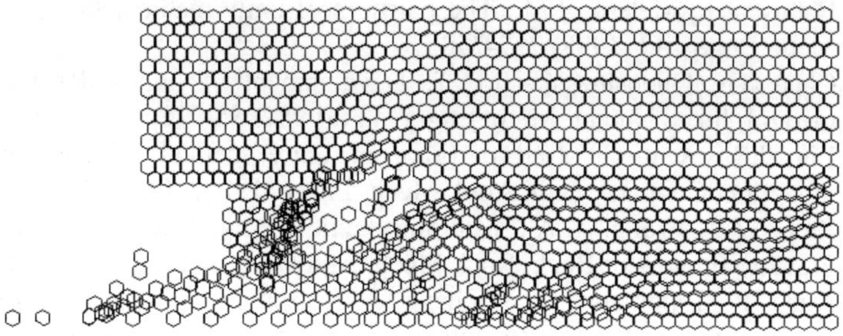

Figure 5: Movements of the particles for c = 10 kPa and p_n^+ = 25 kPa

9 Conclusions

Dynamical behavior of the face of a coal seam at the moment of rock burst and closely after it is studied in this paper. In comparison with the PFC we start with a different shape of particles (to have the possibility to also get stresses in the particles) and with static equilibrium in the state when no bumps occur. After nucleation of cracks, or in other words if small movements are observed, the kinetics of the moved particles is considered. The inertia part of the governing equation starts to prevail and be active. Generally, in contradiction with the PFC dynamical equilibrium is taken into consideration after enough movement of the hexagonal elements. The forces induced along the boundaries of adjacent particles or after mutual touching of extruded particles cause an acceleration of the particles, which defines the way of movement due to D'Alembert forces. Influence of spring stiffness stabilizes the iterative process. If removed, the process can degenerate in unstable convergence of diverge at all.

Acknowledgement

Financial support by GACR, project No. 103/05/0334 is appreciated.

References

[1] Cundall, P.A. A computer model for simulation progressive large scale movements of blocky rock systems. Symposium of the international society of rock mechanics, 132-150, 1971.

[2] Moreau, J.J. Some numerical methods in multibody dynamics: Application to granular materials. *Eur. J. Mech. Solids*, 13, 4, 1994, 93-114.

[3] Elices, M., Guinea, G.V., Gomez, J. & Planas, J. The cohesive zone model: advantages, limitations and challenges. *Engineering Fracture Mechanics* 69, 2002, 137-163.

[4] Babuska, I., Melenk, J.M.: The partition of unity method, Int. J. Numer. Meth. Engrg. 40 (1997) 727-758

[5] Chen, G., Ohnoshi, Y., Ito, T. Development of high-order manifold method, Int. J. Numer. Meth. Engrg. 43 (1998) 685-712

[6] Lin, J.S.: A mesh-based partition of unity method for discontinuity modeling, Comput. Meth. Appl. Mech. Engrg. 192 (2003) 1515-1532

[7] Zhu, W.C., Tang, C.A.: Micromechanical Model for Simulating the Fracture Process of Rock, Rock Mech. And Rock Engrg 37 (1), 25-56.

[8] Kuch, R., Lippmann, H. & Zhang, J. *Simulating coal mine bumps with model material.* Rockbursts and seismicity in mines, Gibowitz & Lasocki (eds.), Balkema, Rotterdam, 1997, 23-25.

[9] Vacek, J. & Procházka, P. Behaviour of Brittle Rock in Extreme Depth. *Our World in Concrete & Structures*. Singapore: CI-Premier, 19, 2000, 653-660.

[10] Vacek, J. & Procházka, P. *Rock Bumps Occurrence during Mining.* Computational Methods and Experimental Measurements X. Southampton: WIT Press, 2001, 437-446.

[11] Procházka, P. & Vacek, J. *Comparative Study of Tunnel Face Stability.* Damage & Fracture Mechanics VII. Southampton: WIT Press, 2002, 163-172.

[12] Haramy, K.Y. and Morgan, T.A. & DeWaele, R.E. *A method for estimating western coal strengths from point load tests on irregular bumps*. 2nd Conf. on Ground Control in Mining, West Virginia University, July 19-21, 1982, 123-136.

[13] Haramy, K.Y., Magers, J.A. & McDonnell, J.P. *Mining under strong roof.* 7th Int. Conf. on Ground Control in Mining, Bureau of Mines, Denver, USA, 1992, 179-194.

[14] Harami, K.Y. & Brady, B.T. *A methodology to determine in situ rock mass failure.* Internal report of Bureau of Mines, Denver, CO, USA, 1995.

[15] Procházka, P. & Válek, M. *The BEM Formulation of Distinct Element Method.* BETECH XXII. Cambridge: WIT Press, 2000, 395-404.

[16] Procházka, P. Application of discrete element methods to fracture mechanics of rock bursts. *Engng. Fract. Mech.* 2003.

[17] Onck, P. & van der Giessen, E. Growth of an initially sharp crack by grain boundary cavitation. *JMPS* 1999, 523-542.

[18] Procházka, P. & Trčková, J. *Coupled modeling of Concrete Tunnel Lining.* Our World in Concrete and Structures, Singapore, 2000, 125-132.

[19] Procházka, P. & Válek, M. *Stability of Tunnel Face Using Coupled DSC & TFA Models*. Damage and Fracture Mechanics VI. Montreal, WIT Press, 2000, 471-480.

[20] Brebbia, C.A., Teles, J.C.F. and Wrobel, L.C. Boundary element techniques. Springer Verlag, Berlin, Heidelberg, NYC, 1984.

Introduction of STEM for stress analysis in statically determined bodies

A. N. Galybin
Wessex Institute of Technology, Southampton, UK

Abstract

This article presents a novel approach to the identification of stress states in statically determined bodies. The approach is based on the stress trajectory concept and therefore it is referred to as the stress trajectories element method, (STEM). Three different variants of STEM are presented and some problems associated with these are posed.
Keywords: boundary value problems, stress trajectories, numerical methods.

1 Introduction

This article is an introduction to a novel numerical method that is currently under development in Wessex Institute of Technology. The method is aimed at the identification of stresses in statically determined bodies by employing stress trajectories.

The concept of stress trajectories comes from photoelasticity, therefore one can adopt the following definition due to Frocht [1]: Stress trajectories are curves the tangents to which represent the directions of one of the principal stresses at the points of tangency. A single stress trajectory is also called an isostatic or a line of principal stresses.

Stresses at each point inside a continuous body represent a second-rank tensor which components satisfy differential equations of equilibrium, DEE, and certain constitutive equations. The latter is of theoretical and/or experimental nature, it is often called "rheology", however only statically determined bodies are considered in this paper. These constitute a broad class and some examples are found in engineering:

– in elasticity, the laplacian applied to the first invariant of the stress tensor should vanish (if body forces are neglected);

WIT Transactions on Modelling and Simulation, Vol 44, © 2007 WIT Press
www.witpress.com, ISSN 1743-355X (on-line)
doi:10.2495/BE070081

- in ideal plasticity, the deviator of the stress tensor is a constant;
- in granular medium, certain linear relationships between the mean stresses and the stress deviator should be fulfilled;
- in rock mechanics non-linear relationships are frequently used.

Classical formulations of boundary value problems, BVPs, for a statically determined body include boundary conditions, BCs, posed in terms of stresses (or other quantities that can be related to stresses one-to-one) on the whole boundary of the body while the number of boundary conditions coincides with the dimension of the body. Under these restrictions, BVPs are usually well-posed, i.e. have unique and stable (with respect to small perturbations in BCs) solutions. The well-posed nature is vital in modern numerical techniques that have been well developed and implemented into computer programs such as finite element methods, boundary element methods, different particle codes and hybrid programs. However, there are a variety of ill-posed problems that are generally defined as problems with non-unique or unstable solutions, [2]. This paper deals with the problems that can be either well- or ill-posed depending on what information about stress trajectories is supplied. Based on theoretical investigations, a universal numerical method is proposed addressing the ill-posed formulations. Stress trajectories of different families are used to form elements or to introduce assumptions within the elements of chosen shapes, therefore the method is further referred to as the stress trajectories element method, STEM.

2 Lame-Maxwell equations of equilibrium

The problems considered are classified with respect to information known in regard to the stress trajectories that can be given: (*a*) everywhere in a domain; (*b*) at discrete points; or (*c*) on the boundary of a domain.

In all cases it is assumed that the domain is in equilibrium, therefore the DEE are valid at each point of the domain including its boundary. These can be presented in different forms, for instance, in the Lame-Maxwell form [3]:

$$\frac{\partial\sigma_1}{\partial s_1}+\frac{\sigma_1-\sigma_2}{\rho_{23}}+\frac{\sigma_1-\sigma_3}{\rho_{32}}+S_1=0,\quad \frac{\partial\sigma_2}{\partial s_2}+\frac{\sigma_2-\sigma_3}{\rho_{31}}+\frac{\sigma_2-\sigma_1}{\rho_{13}}+S_2=0$$
$$\frac{\partial\sigma_3}{\partial s_3}+\frac{\sigma_3-\sigma_1}{\rho_{12}}+\frac{\sigma_3-\sigma_2}{\rho_{21}}+S_3=0 \qquad (1)$$

Here σ_k is a principal stress along the k-th stress trajectory (isostatic), (all σ_k coincide with eigenvalues of the stress tensor), s_k is a coordinate (arc length) along the k-th isostatic, ρ_{ji} is radius of curvature of the i-th isostatic in the plane perpendicular to the j-th isostatic; S_k is a projection of body forces onto the k-th isostatic. In order to distinguish different families of the stress trajectories it is assumed that $\sigma_3\leq\sigma_2\leq\sigma_1$.

For plane problems there are two independent DEE. We further focus on this case that demonstrate all features of the proposed techniques. If body forces are absent than the Lame-Maxwell equations assume the following form

$$\frac{\partial\sigma_1}{\partial s_1}+\frac{\sigma_1-\sigma_2}{\rho_2}+=0,\quad \frac{\partial\sigma_2}{\partial s_2}+\frac{\sigma_1-\sigma_2}{\rho_1}=0 \qquad (2)$$

where the curvature is found as $\rho_k^{-1}=\partial\theta_k/\partial s_k$ (θ_k is inclination of σ_k in a reference coordinate frame).

Problems associated with the essentially different situations (*a*)-(*c*) are discussed in the next sections as well as the different variants of STEM associated with different data. These are summarised in Fig. 1 below.

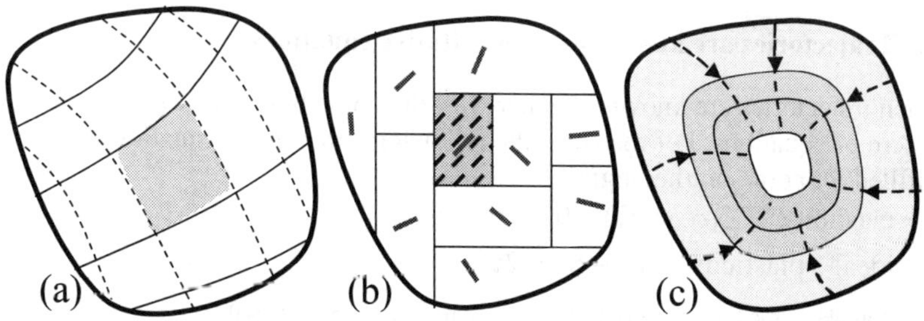

Figure 1: Variants of STEM for different types of data: (a) – trajectories are known, mesh consists of elements formed by trajectories of different families; (b) – data at discrete points, prescribed or adaptive mesh with homogeneous trajectories within elements; (c) – data known on the boundary, adaptive mesh with elements developing from boundary.

3 Stress trajectories are known everywhere in the domain

In this case the curvatures of stress trajectories are also known everywhere in the domain considered. Therefore, equations (2) represent a closed system of partial differential equations for the determination of principal stresses. It is important that the knowledge of constitutive equations is unnecessary in order to obtain unique solution of this system. There is a routine operation very well known in photoelasticity, so called, separation of principal stresses, which provides reconstruction of the stress field on the basis of trajectory patterns. However, the knowledge of rheology yields an overspecified formulation.

3.1 No constitutive equations are given

It is evident that (2) is of hyperbolic type with characteristics coinciding with stress trajectories, e.g. [4]. Thus, 3 types of classical BVPs can be considered:

- Cauchy BVP assumes BCs given on a line that is not a isostatic;
- Goursat BVP assumes BCs given on isostatics of two different families;
- mixed type of two above.

All three BVPs are well known in partial differential equations. However, one more type has to be mentioned: BCs are posed on the boundary that coincides with the stress trajectory of one family (trajectories of the other family are perpendicular to the boundary). Mechanical example for this case is a body which surface is subjected to normal load only (no shear stress). It is clear that different distributions of normal loads on the boundary produce different trajectory patterns within the domain and therefore solutions may exist. For any linear cases (e.g. elasticity) it is also evident that the pattern remains the same if one multiplies the stress magnitudes on the boundary by a non-zero number, which indicates non-uniqueness. Thus, investigation of solvability and uniqueness is one of the tasks of the BVP of this type.

3.2 Trajectories are known with constitutive equations

When trajectories are known together with rheology, this results in overspecified system of equations. For instance, the following constitutive equations should be fulfilled for common rheologies

$$\text{elasticity}: \quad \Delta(\sigma_1+\sigma_2)=0 \tag{3}$$

$$\text{ideal plasticity}: \quad \sigma_1-\sigma_2=2\tau_y \tag{4}$$

$$\text{granular media}: \quad |\sigma_1-\sigma_2|+(\sigma_1+\sigma_2)\tan\phi=\tau_c, \quad \sigma_1+\sigma_2\le 0 \tag{5}$$

Here Δ is laplacian, τ_y is yielding limit, ϕ is frictional angle and τ_c is cohesion. For instance, in the case of elasticity equations and (2) and (3) provide uniqueness of the stress field if boundary tractions are specified. Therefore, stress trajectories found from the solution are uniquely determined but they may differ from the given ones. Although in general the problem has no solution, this example poses certain questions regarding consistency of trajectories and rheology, for instance, as follows:

- determine stresses in the body including its boundary provided that rheology and trajectories are consistent;
- given a trajectory pattern, determine whether it can be realised in a body of given rheology with certain restrictions (e.g., boundedness of stresses);
- determine types of trajectory patterns that are consistent with different rheologies.

Consistency in the case of elasticity has been investigated by Mukhamediev and Galybin [5, 6] who proved that with the exception of some special cases the complete stress tensor is reconstructed from stress orientations non-uniquely. However, the degree of non-uniqueness is determined. Thus, the solution for the 2D stress deviator can have a multiplier that is either a constant (if no restrictions on stress trajectories are imposed) or a real valued bi-holomorphic function (if the angle of inclination of the stress orientations is a harmonic function of coordinates, in this case 5 real constants are arbitrary).

3.3 Homogeneous trajectories

Let us consider the case of homogeneous trajectories that seems to be consistent with all rheologies at least with those mentioned above. In fact, due to $\rho_k=\infty$ the general solution of (2) takes the form $\sigma_1=\sigma_1(s_2)$, $\sigma_2=\sigma_2(s_1)$, and hence either of the conditions (3)-(5) can be satisfied by proper choice of arbitrary functions

$\sigma_1(s_2)$ and $\sigma_2(s_1)$. However, classes of possible stress fields related to homogeneous trajectories are narrow. In elasticity the only allowable stress fields have the form

$$\sigma_1(s_2) = as_2^2 + bs_2 + c, \quad \sigma_2(s_1) = -as_1^2 + ds_1 + e \tag{6}$$

where *a, b, c, d, e* are arbitrary real constants. For the cases of ideal plasticity and granular media the difference σ_1-σ_2 is a specific constant, hence, one arbitrary constant enters into the general solution that constitutes homogeneity of stresses.

3.4 First variant of STEM

When trajectories and rheology are consistent, the problem of stress determination inside the body is the primary task. It is not a BVP, and, as evident from the example for homogeneous trajectories, solution of this problem is non-unique in the general case but depends linearly on a number of arbitrary constants. This circumstance leads to the idea of introduction of a numerical method that is somewhat similar to FEM but with the mesh composed of elements bounded by trajectories of different families, Fig.1a. In this case, two unknown principal stresses (functions σ_1 and σ_2) should be determined within each element. Approximation of these functions within elements should be consistent with rheology, which, similarly to the example for homogeneous trajectories, imposes certain forms of approximations that depend on sought parameters (e.g. coefficients of approximation polynomials). Furthermore, it is evident that other restrictions, e.g. continuity of principal stresses across adjacent elements, decrease the number of independent parameters that have to be determined. On one hand, a certain number of parameters will remain undetermined until additional data are attracted, e.g. stresses magnitudes at discrete points. On the other hand, this approach addresses the ill-posed nature of the problem and allows one to determine the number of conditions that has to be additionally imposed in order to find a unique stress field. It has been emphasised that the problem in this formulation is not a BVP, which means that BCs cannot be specified independently on each element of the boundary. In particular, this also means that none of existing numerical methods (neither FEM nor BEM) are capable of solving the problems with given stress trajectories.

4 Stress orientations are known at discrete points

4.1 Global reconstruction of stresses from discrete stress orientations

A typical approach in this case assumes interpolation using different methods. Examples are found in geophysical applications, e.g. [7–10]. The major defect of this approach is that the pattern of stress trajectories obtained by interpolation is not necessarily consistent with rheology introduced when one attempts to recover the complete stress tensor. Moreover, different interpolation techniques apparently lead to different rheologies; therefore, the choice of interpolation methods may be in conflict with physical meaning.

Consistent approaches have to take rheology into account at the stage of interpolation by employing the Trefftz method. This means that a global solution for the whole domain is sough as a linear combination of independent functions satisfying the governing equations. The problem appears to be a minimisation problem in which calculated stress orientations provide the best fit to data. Its solution is non-unique: although stress trajectories can be found uniquely, stress magnitudes are determined with some degree of arbitrariness. This approach has been suggested in [11–13] and tested for plane elasticity.

4.2 Problems to be investigated

4.2.1 Singular (isotropic) points in stress trajectory fields

If orientations of principal stresses (principal directions) are known at a dense net of nodes inside the domain, a smooth field of stress trajectories could be obtained by standard interpolation methods (with possible inconsistence with rheology as explained in the previous subsection). However, the direct conversion can be significantly complicated by the presence of singular points where the stress deviator vanishes and stress orientations are unidentified ($\sigma_1=\sigma_2$), which suggests the lack of smoothness of the trajectory field. It is a serious obstacle for STEM development. The existence of singular points is well known in photoelasticity (where they are referred to as isotropic points), also numerical calculations of stress trajectories shows the presence of these points in different configurations, see recent results of experimental and numerical investigations by Joussineaua et al [14]. Two types of isotropic points are usually distinguished in photoelasticity although the existence of isotropic points of higher order is theoretically possible [15]. These points can be classified in accordance with the asymptotic behaviour of the stress deviator function [16]. Both these types may appear simultaneously.

The identification of singular points is a separate task that requires special attention. As far as elastic domains are concerned the problem of stress tensor identification from discrete principal orientations can be solved simultaneously with the problem of trajectory field identification. Moreover, the singular points of different types can also be found simultaneously, see examples presented in [11], which demonstrate that for elastic medium all types of singular points (interlocking type, non-interlocking type [15] and points where stresses have infinite gradient) can be recovered with sufficient accuracy.

4.2.2 Other problems

Apart from singular points the following main problems should be addressed:

- investigation of correspondence between interpolation methods and rheology;
- data analysis for the determination of optimal number and types of basis functions in particular cases;
- stability of solutions for different basis functions;
- arbitrariness in the stress tensor for different rheologies;
- development of a variant of the STEM method.

4.3 Second variant of STEM

A possible approach to the latter problem is demonstrated in Fig 1(b). Given that stress orientations are more or less uniformly distributed within the body, it is proposed to introduce such a mesh in which each element contains at least a datum. After that the assumption of homogeneity of stress trajectories within the element can be accepted, which leads to specific expressions for stresses within the element, e.g. (3) for the case of elasticity. Therefore, the total number of parameters to be determined is equal to the number of elements times the number of parameters specific for each rheology. However, the number of independent parameters is much less (it is found from the analysis of arbitrariness in the stress tensor for different rheologies), which means that additional conditions have to be imposed that connects parameters in adjacent elements (e.g., continuity of stress vector). This procedure leaves several parameters to be identified from additional data and all comments made for the STEM in the case of known trajectories remain.

It is, however, impossible to provide continuity of all stress components. At least one of the stress characteristics may be discontinuous across boundaries of adjacent elements. This is a consequence of the introduction of sharp corners in trajectories. For homogeneous domains, continuity of stresses can be prescribed in average or at the vertices of the elements; what assumption is better has to be investigated. In piecewise homogeneous domains the jump of tangential stresses on the interfaces is not zero, therefore if trajectories kink an interface the principal stresses also suffer jumps across the interface satisfying the following conditions (continuity of the stress vector)

$$
\begin{aligned}
(\sigma_1^+ - \sigma_2^+)\sin 2\theta^+ &= (\sigma_1^- - \sigma_2^-)\sin 2\theta^-, \quad \sigma_2^\pm \le \sigma_1^\pm \\
\sigma_1^+ + \sigma_2^+ + (\sigma_1^+ - \sigma_2^+)\cos 2\theta^+ &= \sigma_1^- + \sigma_2^- + (\sigma_1^- - \sigma_2^-)\cos 2\theta^-
\end{aligned}
\tag{7}
$$

where "±" referrer to stress tensor characteristics in two adjacent elements. It follows from the first expression in (7) that the angles θ^+ and θ^- have the same sign ($-\pi/2<\theta\le\pi/2$). Fig 2 illustrates admissible and non-admissible trajectories in two adjacent elements with a common rectilinear interface.

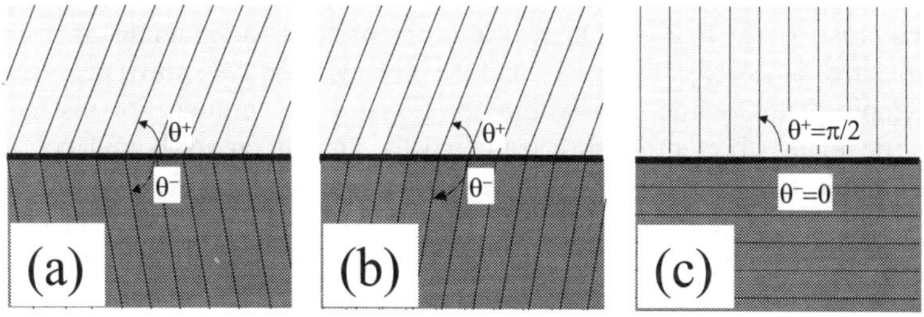

Figure 2: Examples of non-admissible (a) and admissible (b)-(c) stress trajectories of one family.

5 Stress orientations given on the boundary

5.1 Non-uniqueness of solutions

It is evident that orientations of principal stresses remain if any constant mean stress is superimposed on a particular solution of DEE or if the stress deviator is multiplied by any positive constant (equilibrium also remains). This indicates non-uniqueness in the BVP formulated in terms of given principal directions (of the stress tensor) on the boundary. However the number of possible solutions is finite if this BC is complemented by an additional BC, say by continuity of the stress vector. One type of BC for plane elastic body has been investigated in [17], where the curvatures of stress trajectories have been used as the second condition. It has been shown that this BVP can have finite number of linearly independent solutions (stress states) or have no solutions, which depends on, so-called, index of the problem that is determined from the analysis of the principal directions on the boundary (curvatures of stress trajectories do not affect the index). The number of solutions is uniquely identified by the index, which suggests that the use of other types of the second BC will not make the problem well-posed. This hypothesis has to be thoroughly investigated by analysing different BCs for different rheologies.

5.2 Third variant of STEM

Another major task assumes the formulation of a variant of the STEM method for different BVPs. An adaptive mesh with elements developing from the boundary (Fig.1(c)) seems to be the best option because at each step directed inwards the stress orientations are known, therefore trajectories are approximated by tangents, which eventually lead to a piecewise linear system of stress trajectories within the body. As has been mentioned, non-uniqueness and possible instability of solutions considerably restricts the direct application of conventional numerical methods (as FEM or BEM). In the STEM approach the investigation of stability is amongst the major tasks.

5.3 Applications

Geophysical applications include determination of stresses in stable blocks of the earth's crust. Here BVPs formulated in terms of stress orientations are of great significance. Stress orientations around the globe are always known near margins of tectonic plates, while data within some plates are limited. For instance, in Antarctic plate stress orientations are mostly known on its boundary, which makes application of conventional back analysis for the identification of tectonic stresses in this plate impossible. However, it has been reported [12] that the approach based on the Trefftz method for complex potentials in elasticity followed by minimisation technique provides stable results in reconstruction of the stress trajectories within Antarctica.

A numerical example for the identification of possible elastic fields within the Antarctic plate is shown in Fig. 3. Data include 170 stress orientation measurements shown by segments in Fig. 1(a) (supplied by the WSM project). Continuity of tractions across the boundary have been assumed.

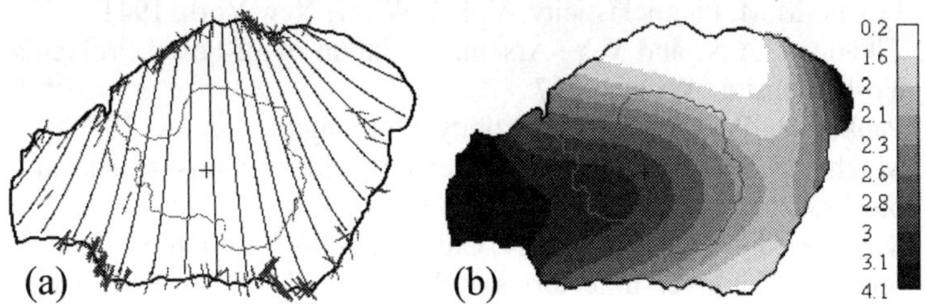

Figure 3: Elastic stresses in the Antarctic plate: stress trajectories (a) and normalised maximum shear stress (b).

There are cases when direct data on stress measurements are unavailable at all, for instance, in crusts of other planets, although investigation of stresses is significant for modelling tectonics, e.g. [18–19]. The only reliable information in these cases is obtained from observations of fracture patterns on the surface. This can be used as an input to identify the orientations of principal stresses, which eventually will lead to mathematical formulations in terms of principal orientations.

At small scales there is a strong influence of measuring devises, e.g. indenters. Therefore the development of non-direct methods for small-scale measurements is also vital as for remote stress field determination mentioned above.

6 Conclusive remarks and summary

This article presents main tasks in the programme for the development of the STEM for stress identification in statically determined bodies. Different variants of STEM are proposed addressing three main types of data. The summary of the problems is presented below.

Case 1. Stress trajectories known on boundary with

Unknown Rheology: BVPs for hyperbolic DEE (Cauchy, Goursat, mixed and non-classical)

Given Rheology: overspecified problems, consistency of rheology and trajectories, investigation of solvability.

Case 2. Stress trajectories known at discrete points with given rheology: ill-posed and minimisation problems (not a BVP), no uniqueness.

Case 3. Stress trajectories known everywhere in domain with given rheology: extra BC required, ill-posed BVP with non-unique solutions (solvability depends upon the index that can be determined from BCs).

Acknowledgement

The author acknowledges the support of EPSRC through Research Grant EP/E032494/1.

References

[1] Frocht, M.M. Photoelasticity, Vol. 1. Wiley, New York, 1941
[2] Tikhonov, A.N. and V.Y. Arsenin. Solution of Ill-Posed Problems, New York: Winston, Wiley, 1977.
[3] Papkovich, P.F. Theory of elasticity. Oborongiz, Moscow, 1939.
[4] Mukhamediev, ShA. 1991. Retrieving field of stress tensor in crustal blocks. Izvestiya Earth Physics. 27: 370–7.
[5] Mukhamediev, Sh.A. and A.N. Galybin, 2004. Solution of a plane elastic problem with given trajectories of the principal stresses. Doklady Physics. 49 (5), 311-314
[6] Mukhamediev, Sh.A. and A.N. Galybin, 2007. Determination of Stresses from the Stress Trajectory Pattern in a Plane Elastic Domain. Math. Mech. of Solids, 12, 75-106
[7] Hansen K.M. and V.S. Mount, 1990. Smoothing and extrapolation of crustal stress orientation measurements J.Geophys.Res. 95(B), 1155-1165
[8] Lee, J.-C. and J. Angelier 1994. Paleostress trajectory maps based on the results of local determinations: the "Lissage" program. Computers and Geosciences. 20, 161-191
[9] Bergerat, F. and J. Angelier, 1998. Fault systems and paleostresses in the Vestfirdir Peninsula. Geodinamica Acta (Paris) 11 (2-3), 105-118.
[10] Badawy, A. and F. Horvath, 1999. Recent stress field of the Sinai subplate region. Tectonophysics. 304, 385–403
[11] Galybin, A.N. and Sh.A. Mukhamediev, 2004. Determination of elastic stresses from discrete data on stress orientations. Int. Journal of Solids and Structures. 41 (18-19), 5125-5142
[12] Galybin, A.N. and Sh.A. Mukhamediev, 2004. On the problem of stress reconstruction from discrete orientations of principal stresses. Bollettino di Geofisica Teorica ed Applicata. 45 (1) supplement, 338-342
[13] Mukhamediev, Sh.A., A.N. Galybin and B.H.G. Brady, 2006. Determination of stress fields in elastic lithosphere by methods based on stress orientations. Int Journal of Rock Mechanics and Mining Sciences. 43 (1), 66-88
[14] Joussineaua, G de, J-P Petit and B. D.M. Gauthier, 2003. Photoelastic and numerical investigation of stress distributions around fault models under biaxial compressive loading conditions. Tectonophysics 363, 19– 43
[15] Kuske, A. & G. Robertson Photoelastic stress analysis. Wiley, London 1974
[16] Karakin A.V. and Sh.A. Mukhamediev, 1994. Singular points in nonuniform field trajectories of the principle tectonic stress. Physics Solid Earth. 29 (11), 956-965.
[17] Galybin, A.N. & Sh.A. Mukhamediev, 1999. Plane elastic boundary value problem posed on orientation of principal stresses, JMPS. 47, 2381-2409
[18] Hoppa, G. V., Tufts, B.R., Greenberg, R. and P.E. Geissler, 1999. Formation of Cycloidal Features on Europa. Science. 285, 1899-1902
[19] Sandwell, D.T., Johnson, C.L.. Bilotti, F and J. Suppe, 1997. Driving forces for limited tectonics on Venus. Icarus. 129, 232-244.

Null-field integral equations and their applications

J.-T. Chen & P.-Y. Chen
Department of Harbor and River Engineering,
National Taiwan Ocean University, Keelung, Taiwan

Abstract

In this paper, a null-field equation approach is proposed to deal with boundary value problems containing circular boundaries. The mathematical tools, degenerate kernels and Fourier series, are utilized in the null-field integral formulation. Although we employ the null field equations, we can exactly collocate the point on the real boundary. Thus, the singularity is novelly avoided since the kernel is expressed in a degenerate form. Five gains of well-posed model, singularity free, boundary layer effect free, exponential convergence and mesh-free approach are achieved. To demonstrate the validity of the present formulation, some applications are considered: (1) torsion problem, (2) bending problem, (3) SH wave impinging successive canyons. It is found that previous results by other investigators are not consistent with ours. After comparing with other independent solutions, the accuracy and efficiency of our approach is acceptable.
Keywords: null-field integral equation, degenerate kernel, Fourier series.

1 Introduction

Engineering problems can be formulated as mathematical models. In order to solve the boundary value problems, researchers and engineers have paid more attention on the development of boundary integral equation method (BIEM), boundary element method (BEM) and meshless method than domain type methods, finite element method (FEM) and finite difference method (FDM). Among various numerical methods, BEM is one of the most popular numerical approaches for solving boundary value problems. Although BEM has been

WIT Transactions on Modelling and Simulation, Vol 44, © 2007 WIT Press
www.witpress.com, ISSN 1743-355X (on-line)
doi:10.2495/BE070091

involved as an alternative numerical method for solving engineering problems, five critical issues are of concern.

(1) Treatment of singularity and hypersingularity. It is well known that BEM are based on the use of fundamental solutions to solve partial differential equations. These solutions are two-point functions which are singular as the source and field points coincide. Several regularizations for hypersingularity were offered to handle it in direct and indirect ways. In the present approach, we employed the degenerate kernel to represent the fundamental solution for problems with circular boundaries. The singularity and hypersingularity disappeared in the boundary integral equation after describing the potential into two parts.

(2) Boundary-layer effect. Boundary-layer effect in BEM has received attention in the recent years. In real applications, data near boundary can be smoothened since maximum principle exists for potential problems. Nevertheless, it also deserves study to know how to manipulate the nearly singular integrals in applied mathematics. How to eliminate the boundary-layer effect in BEM is vital for researchers.

(3) Convergence rate. Regarding to constant, linear and quadratic elements, the discretization scheme does not take the special geometry into consideration. It leads to the slow convergence rate. For example, Fourier series is suitable for boundary densities on circular boundaries. Although previous researchers have employed the Fourier series expansion, no one has ever introduced the degenerate kernel in boundary integral equations to tackle their problems. Mathematicians have proved that the exponential convergence instead of the algebraic convergence in the BEM can be achieved by using the degenerate kernel and Fourier expansion.

(4) Ill-posed model. Null-field approach or fictitious BEM free of calculating the singular and hypersingular integrals yields an ill-conditioned system. To approach the fictitious boundary to the real boundary or to move the null-field point to the real boundary can make the system well-posed. However, singularity appears in the meantime. We may wonder is it possible to push the null-field point on the real boundary but free of facing the singular and hypersingular integrals. The answer is yes and can be found in this paper.

(5) Mesh on boundary is still necessary.

The five issues, singularity free, the suppression of boundary-layer effect, exponential convergence, well-posed model and mesh-free will be examined in this paper.

Engineering problems with circular boundaries are often encountered, *e.g.* missiles, aircraft, naval architecture, etc., either to reduce the weight of the whole structure or to increase the range of inspection as well as piping purposes. Analytical approach using bi-polar coordinate [1] was developed for two-hole problems. Complex variable techniques were also employed for the annular case. For a problem with several holes, many numerical methods, *e.g.* finite element method (FEM) and boundary element method (BEM) were resorted to solve. To develop a systematic approach for engineering problems with circular boundaries is not trivial.

Null-field integral equation approach is used widely for obtaining the numerical solutions of engineering problems. Various names, *e.g.* T-matrix method [2] and extended boundary condition method (EBCM) [3], have been coined. A crucial advantage of this method consists in the fact that the influence matrix can be computed easily. Although many works for acoustic and water wave problems have been done, we focus on the solid mechanics here.

In this paper, we review the recent development of the null-field integral equation approach [4-9] for boundary value problems (BVPs) with circular boundaries. The key idea is the expansion of kernel functions and boundary densities in the null-field integral equations. Vector decomposition technique using the adaptive observer system is required for nonfocal cases. Applications to the Laplace and Helmholtz problems are addressed. Several examples were demonstrated to see the validity of the new formulation.

2 Null-field integral equation approach for boundary value problems

Suppose there are N randomly distributed circular boundaries bounded to the domain D and enclosed with the boundary, B_k ($k = 0, 1, 2, \cdots, N$) as shown in Figure 1. We define

$$B = \bigcup_{k=0}^{N} B_k \,. \tag{1}$$

In mathematical physics, boundary value problems can be modelled by the governing equation,

$$L\ u(x) = 0\,,\ x \in D\,, \tag{2}$$

where L may be the Laplace, Helmholtz, biharmonic or biHelmholtz operator, $u(\mathrm{x})$ is the potential function and D is the domain of interest. For the 2-D Laplace and Helmholtz problems, the integral equation for the domain point can be derived from the third Green's identity, we have

$$2\pi u(\mathrm{s}) = \int_B T(\mathrm{s},\mathrm{x}) u(\mathrm{s}) dB(\mathrm{s}) - \int_B U(\mathrm{s},\mathrm{x}) t(\mathrm{s}) dB(\mathrm{s}), \mathrm{x} \in D\,, \tag{3}$$

$$2\pi \frac{\partial u(\mathrm{x})}{\partial \mathrm{n}_\mathrm{x}} = \int_B M(\mathrm{s},\mathrm{x}) u(\mathrm{s}) dB(\mathrm{s}) - \int_B L(\mathrm{s},\mathrm{x}) t(\mathrm{s}) dB(\mathrm{s}), \mathrm{x} \in D\,, \tag{4}$$

where s and x are the source and field points, respectively, $t = \partial u/\partial \mathrm{n}$, B is the boundary, n_x denotes the outward normal vector at the field point x and the kernel function U(s,x), is the fundamental solution, and the other kernel functions, T(s,x), L(s,x) and M(s,x), are defined in the dual boundary integral method (BIEM) [9]. It is noted that more potentials are needed in eqns (3) and (4) for biharmonic and biHelmholtz cases. By moving the field point to the boundary, the eqns (3) and (4) reduce to

$$\pi u(\mathrm{x}) = C.P.V.\int_B T(\mathrm{s},\mathrm{x})u(\mathrm{s})dB(\mathrm{s}) - R.P.V.\int_B U(\mathrm{s},\mathrm{x})t(\mathrm{s})dB(\mathrm{s}), \mathrm{x} \in B, \quad (5)$$

$$\pi \frac{\partial u(\mathrm{x})}{\partial \mathrm{n}_{\mathrm{x}}} = H.P.V.\int_B M(\mathrm{s},\mathrm{x})u(\mathrm{s})dB(\mathrm{s}) - C.P.V.\int_B L(\mathrm{s},\mathrm{x})t(\mathrm{s})dB(\mathrm{s}), \mathrm{x} \in B, \quad (6)$$

where *C.P.V.*, *R.P.V.* and *H.P.V.* denote the Cauchy principal value, Riemann principal value and Hadamard principal value, respectively. By collocating the field point x outside the domain (including the boundary), the null-field integral equations yield

$$0 = \int_B T(\mathrm{s},\mathrm{x})u(\mathrm{s})dB(\mathrm{s}) - \int_B U(\mathrm{s},\mathrm{x})t(\mathrm{s})dB(\mathrm{s}), \mathrm{x} \in D^c \cup B, \quad (7)$$

$$0 = \int_B M(\mathrm{s},\mathrm{x})u(\mathrm{s})dB(\mathrm{s}) - \int_B L(\mathrm{s},\mathrm{x})t(\mathrm{s})dB(\mathrm{s}), \mathrm{x} \in D^c \cup B, \quad (8)$$

by choosing appropriate forms of degenerate kernels, where D^c is the complementary domain.

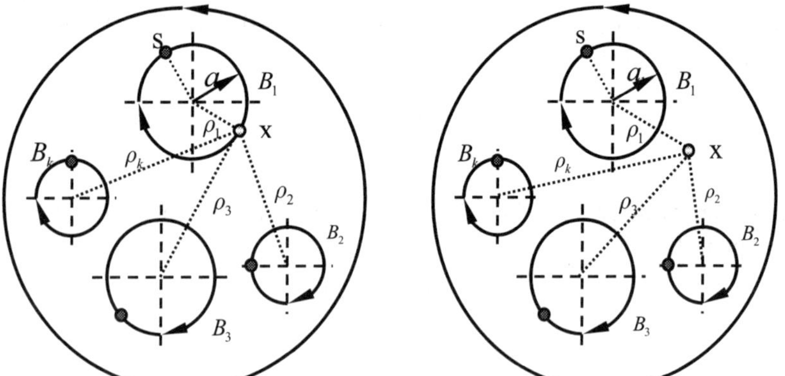

Figure 1: Sketch of null-field and domain points in conjunction with the adaptive observer system (left: collocation on the boundary point, right: collocation on the interior point).

3 Expansions of the fundamental solution and boundary density

Instead of directly calculating the *C.P.V.*, *R.P.V.* and *H.P.V.* in eqns (5) and (6), we obtain the linear algebraic system from the null-field integral equations of eqns (7) and (8) through the kernel expansion.

Based on the separable property, the kernel function *U*(s,x) can be expanded into the separable form by dividing the source and field points:

$$U(\mathrm{s,x}) = \begin{cases} U^i(\mathrm{s,x}) = \sum_{j=1}^{\infty} A_j(\mathrm{s})B_j(\mathrm{x}), \ |\mathrm{s}| \geq |\mathrm{x}|, \\ U^e(\mathrm{s,x}) = \sum_{j=1}^{\infty} A_j(\mathrm{x})B_j(\mathrm{s}), \ |\mathrm{x}| \geq |\mathrm{s}|, \end{cases} \quad (9)$$

where the $A(\mathrm{x})$ and $B(\mathrm{x})$ can be found for the Laplace [4–8], Helmholtz [9], biharmonic [5] and biHelmholtz operators and the superscripts "i" and "e" denote the interior ($|\mathrm{s}| \geq |\mathrm{x}|$) and exterior ($|\mathrm{s}| < |\mathrm{x}|$) cases, respectively. To classify the interior and exterior regions, Figure 2 shows for one-, two- and three-dimensional cases. For the degenerate forms of T, L and M kernels, they can be derived according to their definitions. We apply the Fourier series expansions to approximate the potential u and its normal derivative t on the B_k circular boundary

$$u(\mathrm{s}_k) = a_0^k + \sum_{n=1}^{\infty} (a_n^k \cos n\theta_k + b_n^k \sin n\theta_k) \quad (10)$$

$$t(\mathrm{s}_k) = p_0^k + \sum_{n=1}^{\infty} (p_n^k \cos n\theta_k + q_n^k \sin n\theta_k) \quad (11)$$

where a_n^k, b_n^k, p_n^k and q_n^k ($k = 0, 1, 2, \cdots, N$) are the Fourier coefficients and θ_k is the polar angle measured with respect to the x-direction.

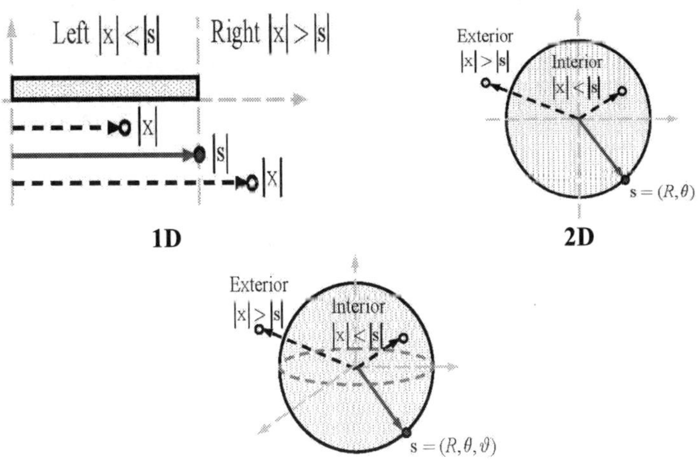

Figure 2: The degenerate kernel for the one-, two- and three-dimensional problems.

After collocating the null-field points in the null-field integral equation of eqn (7), the boundary integrals through all the circular contours are required. It is worth noting that the origin of the observer system is located on the center of the corresponding circle under integration to entirely utilize the geometry of circular boundary for the expansion of degenerate kernels and boundary densities. Figure 1 shows the boundary integration for the circular boundaries in the adaptive observer system. By collocating the null-field point x_k exactly on the kth circular boundary for eqn (7) in Figure 1, we have

$$[U]\{t\}=[T]\{u\} \tag{13}$$

where $[U]$ and $[T]$ are the influence matrices with a dimension of $(N+1)(2m+1)$ by $(N+1)(2m+1)$, $\{u\}$ and $\{t\}$ denote the column vectors of Fourier coefficients with a dimension of $(N+1)(2m+1)$ by 1 in which m indicates the truncated terms of Fourier series. For the circular-inclusion problem, multi-domain approach by taking the free body of each interface between the matrix and inclusions should be introduced. Therefore, an exterior problem for the matrix and several interior problems for each inclusion are needed to be solved by employing the null-field approach. The continuity of displacement and equilibrium of traction should be considered on the interface between the matrix and inclusions [8, 9]. Then, the resulted linear algebraic system is obtained. After the boundary unknowns are obtained, the field potential can be easily obtained according to eqn (3).

4 Illustrative examples

Case 1: A circular bar with multiple circular holes under torsion (Laplace equation)

A circular bar with multiple equal circular holes removed is under torque at the end [10, 11]. Table 1 shows the comparison of the torsional rigidities G of three cases with different geometries of circular holes. The present solutions show improvement over Ling's results [10] in every case. The discrepancy in the second example in Table 1 may ascribe to the Ling's lengthy calculation in error as pointed out by Caulk [11].

Table 1: Torsional rigidity in Ling's examples [11].

Case	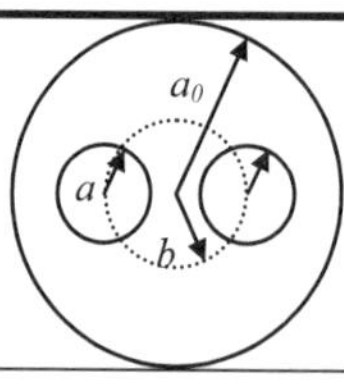 $a/a_0=2/7, b/a_0=3/7$	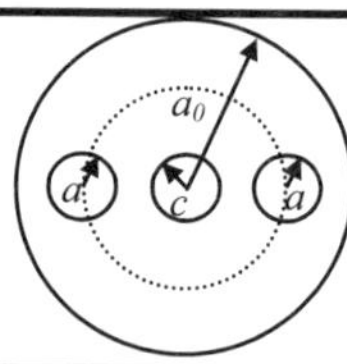 $c/a_0=1/5, a/a_0=1/5,$ $b/a_0=3/5$	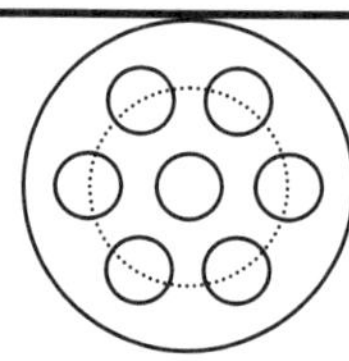 $c/a_0=1/5, a/a_0=1/5,$ $b/a_0=3/5$
Caulk [11]	0.8713	0.8732	0.7261
Ling's results [10]	0.8809	0.8093	0.7305
Present method ($m=10$)	0.8712	0.8732	0.7244

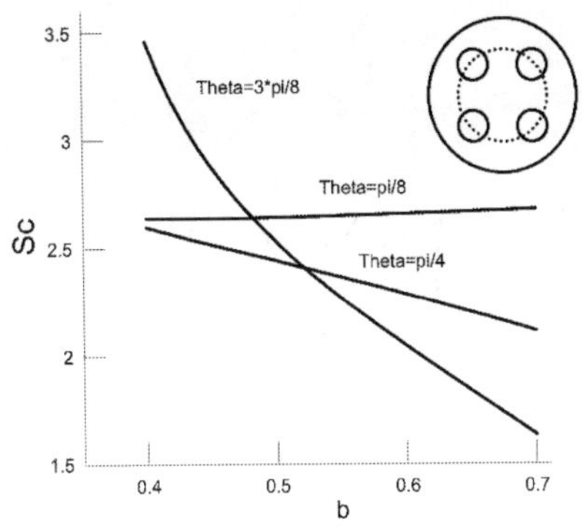

(a)

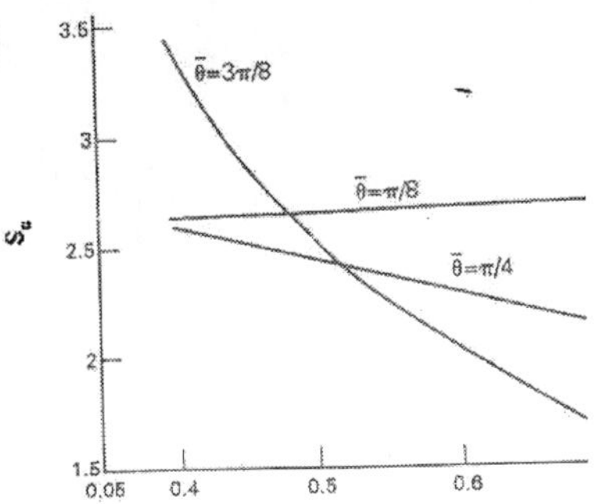

(b)

Figure 3: Stress concentration versus b for $a=0.12$, $R=1.0$ and three different values of $\bar{\theta}=\pi/8$, $\bar{\theta}=\pi/4$ and $\bar{\theta}=3\pi/8$. (a) Sc at the point B (present method); (b) Sc at the point B (Naghdi's result [12]).

Case 2: A circular beam with four circular holes under bending (Laplace equation)

Naghdi [12] and Bird and Steele [13] both calculated the stress concentration for the four equal-sized circular holes problem under bending. Bird and Steele [13] stated that the deviation by the Naghdi's data is 11%. The grounds for this discrepancy were not identified in their paper. Our numerical results agree well with the Naghdi's data as shown in Figure 3.

Case 3: Two canyons subject to the incident SH-wave (Helmholtz equation-a half plane problem)

Tsaur *et al.* [14] and Fang [15] both calculated the response of two canyons subject to the incident SH-wave. Tsaur *et al.* [14] pointed out that the error of Fang [15] is due to wrong use of orthogonal properties. Good agreement is made after comparing with the results of Tsaur *et al.* [14] as shown in Figure 4.

5 Conclusions

A semi-analytical approach was proposed for solving BVPs with circular boundaries. Some recent results were reviewed. The key idea is that we can collocate on the real boundary although we employ the concept of null field equations. Not only the singularity is transformed to the series sum but also the boundary-layer effect is eliminated. In order to verify the formulation, applications to the Laplace and Helmholtz problems were done. Five gains of well-posed model, singularity free, boundary layer effect free, exponential convergence and mesh-free approach were achieved.

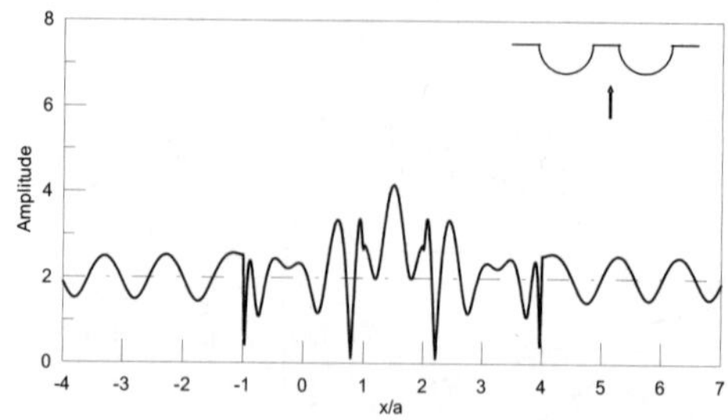

(a) $\gamma = 0^{\circ}$

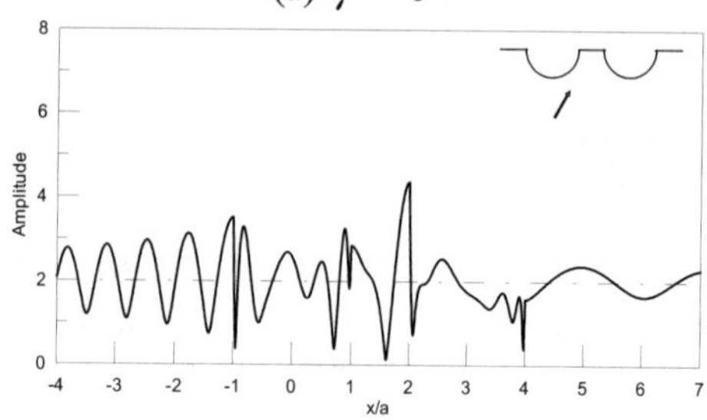

(b) $\gamma = 30^{\circ}$

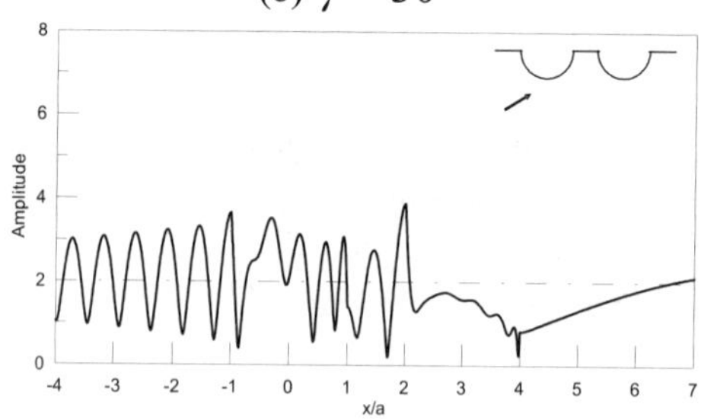

(c) $\gamma = 60^{\circ}$

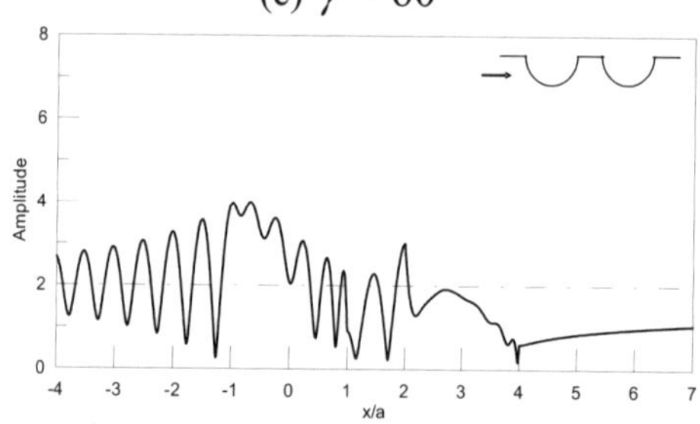

(d) $\gamma = 90^{\circ}$

Figure 4: Surface displacements of two canyons problem ($\mu^{I} / \mu^{M} = 10^{-8}$ and $\eta = 2$).

References

[1] Lebedev, N.N., Skalskaya, I.P. & Uflyand, Y.S., *Worked problem in applied mathematics*. Dover Publications: New York, 1979.

[2] Waterman, P.C., Matrix formulation of electromagnetic scattering. *Proc. IEEE*, **53**, pp. 805-812, 1965.

[3] Doicu, A. & Wriedt, T., Extended boundary condition method with multiple sources located in the complex plane. *Optics Communications*, **139**, pp. 85-91, 1997.

[4] Chen, J.T., Shen, W.C. & Wu, A.C., Null-field integral equations for stress field around circular holes under anti-plane shear. *Engineering Analysis with Boundary Elements*, **30**, pp. 205-217, 2005.

[5] Chen, J.T., Hsiao, C.C. & Leu, S.Y., Null-field integral equation approach for plate problems with circular holes. *ASME Journal of Applied Mechanics*, **73**, pp. 679-693, 2006.

[6] Chen, J.T., Shen, W.C. and Chen, P.Y., Analysis of circular torsion bar with circular holes using null-field approach. *Computer Modelling in Engineering Science*, **12**, pp. 109-119, 2006.

[7] Chen, J.T. & Wu, A.C., Null-field approach for multi-inclusion problem under anti-plane shears. *ASME Journal of Applied Mechanics*, Accepted, 2007.

[8] Chen, J.T. & Wu, A.C., Null-field approach for piezoelectricity problems with arbitrary circular inclusions. *Engineering Analysis with Boundary Elements*, Accepted, 2006.

[9] Chen, J.T., Null filed integral equation approach for boundary vale problems with circular boundaries, *Keynote lecture of ICCES 2005*, India, 2005.

[10] Ling, C.B., Torsion of a circular tube with longitudinal circular holes. *Quarterly of Applied Mathematics*, **5**, pp. 168-181, 1947.

[11] Caulk, D.A., Analysis of elastic torsion in a bar with circular holes by a special boundary integral method. *ASME Journal of Applied Mechanics*, **50**, pp. 101-108, 1983.

[12] Naghdi, A.K., Bending of a perforated circular cylindrical cantilever. *International Journal of Solids and Structures*, **28**, pp. 739-749, 1991.

[13] Bird, M.D. & Steele, C.R., A solution procedure for Laplace's equation on multiply connected circular domains. *ASME Journal of Applied Mechanics*, **59**, pp. 398-404, 1992.

[14] Tsaur, D.H., Chang, K.H. & Lin, J.G., Response of multiple semi-circular cylindrical vanyons subject to plane SH-waves, *Asia Pacific Review of Engineering Science and Technology*, **2(2)**, pp. 251-266, 2004. (in Chinese)

[15] Fang, Y.G., Scattering of plane SH-waves by multiple circular-arc valleys at the two-dimensional surface of the earth. *Earthquake Engineering and Engineering Vibration*, **15(1)**, pp. 85-91, 1995. (in Chinese)

References

Section 3
Dual reciprocity method

Hybrid BEM for the early stage of a 3D unsteady heat diffusion process

A. Peratta & V. Popov
Wessex Institute of Technology, Southampton, UK

Abstract

This article presents a hybrid three dimensional dual reciprocity boundary element method (BEM) for solving the early stage of a 3D time dependent convective heat transfer equation in non-homogeneous media. The method addresses the problem in which the initial distribution of temperature presents a discontinuous jump at the interface between two regions of very dissimilar diffusion coefficients. The goal of this hybrid formulation is to sort out the numerical inconvenience of different time scales and apparent large flux during early time by combining the 3D BEM with a series of 1D semi-analytical profiles of the time dependent heat diffusion equation, employing a two-level finite difference time integration scheme. This article presents the theoretical background, and one 3D test example. The formulation provided offers convenient advantages for relatively large scale models and complex 3D geometries.

1 Introduction

The problem of transient heat transport through 3D non-homogeneous media represents a real challenge for any standard numerical approach. A particular difficult situation is when the initial distribution of temperature presents a discontinuous jump, and this discontinuity is located at the interface between two regions A and B, as shown in Fig. 1(a), of very dissimilar diffusion coefficients, ie. differing in few to many orders of magnitude. Suppose that A is a region of very low conductivity with initially high temperature and characteristic size L_A, embedded in region B, highly conductive, which is initially at lower temperature. As time passes by, there is a faint energy release from the high temperature region and thermal energy is quickly conveyed by convection and diffusion throughout region B. In view of the different diffusivities a boundary layer-type profile of

WIT Transactions on Modelling and Simulation, Vol 44,
www.witpress.com, ISSN 1743-355X (on-line)
doi:10.2495/BE070101

characteristic thickness $d_s(t)$ will develop close to the interface Γ in region A. This scenario presents two different problems. First, the high contrast between the transport properties at both sides of the interface, which introduces time and length scales of very different orders of magnitude; and second, the discontinuous jump of temperature, which introduces a "nearly infinite" thermal flux at early stage of the process. Most time marching integration schemes combined with the conventional finite element method, boundary elements methods (BEMs), finite volume methods, or mesh-less methods are likely to fail when dealing with these kind of problems, being a common symptom large first-time step errors, numerical instabilities, excessive numerical dispersion or diffusion results.

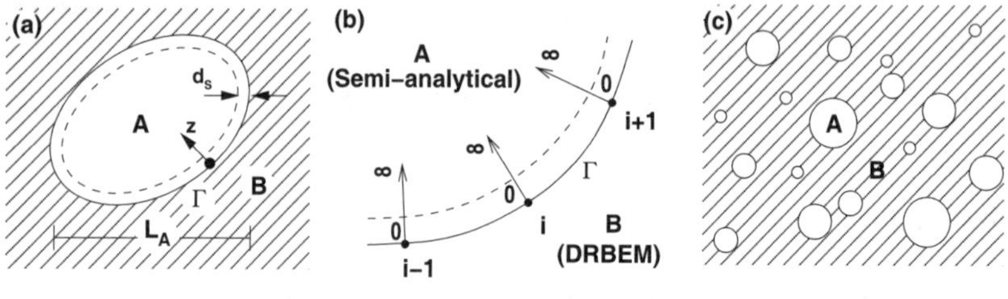

Figure 1: (a) Region of low diffusivity A embedded in a region of high diffusivity B. (b) Each degree of freedom $i = 1, \dots, N$ at the interface Γ is associated with a semi-analytical profile given by (3).

Numerical modelling with BEM [1, 2] is very attractive in the sense that it avoids volume discretisation and at the same time it uses the fundamental solution of the leading differential operator in the equation to solve. BEM applied to the C-DHT equation has been widely developed in the last decades [3, 4], and a large variety of efficient formulations were established. An interesting approach for improving the accuracy at early time stage has been proposed by Grigoriev and Dargush in [5], where a high-order BEM has been established in terms of a singular flux formulation. Other approaches involve the use of time or space scaling in order to zoom into the boundary layer developed close to the interface and perform asymptotic matching with the far field solution. But the computational implementation of these sub-scaling techniques for arbitrary 3D problems might become too complex. The aim of this work is to propose another approach to solve the time dependent convective-diffusive heat transfer (C-DHT) equation in relatively complex 3D situations with a hybrid Dual Reciprocity BEM (DRBEM) employing a simple finite difference two-time-levels time marching scheme. In the proposed hybrid approach, a standard DRBEM strategy for region B, is coupled at the interface Γ with several independent semi-analytical profiles defined in region A. These profiles should be capable of capturing the solution close to Γ thus avoiding the use of high order time integration schemes, or boundary layer scaling techniques. In addition, the method should employ a minimum amount of adjusting parameters, so that to reduce the computational burden as much as possible. The integration domain is discretised with a mixed unstructured mesh, where some regions may be decomposed into many sub-domains and some others may be discretised only in their boundary. The former is known as multi-domain region while the latter is identified with single domain region. The assembly of multi and single-domain regions into the same problem provides a suitable pre-processing flexibility which allows the treatment of complicated 3D geometries. Usually, the DRBEM yields a system of equations whose condition number grows with a certain power law of the number of degrees of freedom [6], thus imposing an upper practical limit to the size of the model to solve, this problem for DRBEM has not been completely sorted out so far. We define the *local transient time scale* (τ_L) of a region of the domain of characteristic size L by taking the maximum between the convective (τ_A), the diffusive (τ_D), and the reaction time scales ($1/k_r$), according to $\tau_L = \max(\tau_A, \tau_D, 1/k_r)$, where $\tau_A := L/v$ and $\tau_D = 4L^2/\alpha$, where, $v = |\mathbf{v}|$ is the average absolute value of the velocity in the region, and α is the average local diffusivity. A "slow-transport region" (STR) is defined as the part of the FPM with very low diffusivity such that τ_L is much higher than the local transient time scales defined for any other region in the integration domain or to the time scale of practical interest in the problem (t_h). We define a "fast-transport region" (FTR) as that part of the FPM characterised by high values of conductivity such that $\tau_L(FPM) \ll \tau_L(STR)$. STR's can also be regarded as inclusions embedded in a larger FTR region, as illustrated in Figure 1(c). When STRs are in close contact to FTRs the solution in the former can be expressed in terms of an assembly of one dimensional asymptotic semi-analytical test profiles of the diffusion equation, as sketched in Fig. 1(b). In this approach, the smaller the ratio t_h/τ_L the more accurate the approximation.

2 Governing equations

The heat transfer equation considered for FTR (B-type) regions is given by:

$$\frac{\partial T_f}{\partial t} + \nabla \cdot \mathbf{q} = -k_r T_f + \nabla \cdot \mathbf{q}_m, \tag{1}$$

where k_r is a reaction constant, t is time, T is temperature, the subscript f stands for the FTR, and the heat flux is given by: $\mathbf{q} = \mathbf{v}T_f - \alpha \nabla T_f$. The source term $\mathbf{q}_m$ on the right hand side of (1) represents the flux exchanged with the STR in close contact to the present FTR, and α is the thermal diffusivity of the region. The temperature in the STR is also described by the 3D C-DHT equation. However, in view of the large difference between diffusivities [7], it is enough to solve a 1D local profile of (1) given by:

$$\frac{\partial T_s}{\partial t} - \alpha_s \frac{\partial^2 T_s}{\partial z^2} = -k_r T_s, \tag{2}$$

where subscript s stands for variables in STR (Region A in Fig. 1) and z is the normal distance from the interface to the STR (see Fig. 1a). This approximation is valid as long as $t_h \ll \tau_L$ or $L_A \gg d_s(t_h)$ in the STR, meaning that the characteristic length of the STR (L_A), see Fig. 1, is several orders of magnitude larger than the penetration depth $d_s(t) = \sqrt{\alpha_s t}/2$ at time t_h into the STR. The STR equation is solved by means of a Semi-Analytical Method (SAM), originally presented by Vinsome and Westerveld [8], later adapted by Birkholzer et al [9] to solute transport problems and recently applied by Peratta and Popov for arbitrary fractured porous media [7]. The idea is to represent the temperature profile along z by means of a reasonably flexible function containing few adjusting parameters to be determined by imposing suitable conservation equations at the interface. The main advantage of this approach is that it can be evaluated in a simple and fast way, thus lowering computational burden. The trial function is defined as follows:

$$T_s(z,t) - T_i = \left(T_f - T_i + p_1 z + p_2 z^2\right) \exp\left[-\frac{z}{d_s}\right], \tag{3}$$

where T_i is temperature in STR at initial time (assumed to be uniform), T_f is the time varying temperature at the interface STR-FTR, and p_1 and p_2 are two time varying best fit parameters to be adjusted by imposing the conservation of energy and continuity of T throughout the interface (Γ). This surface is discretised into N_e Boundary Elements (BEs), each one containing a certain number of collocation points (i), as represented in Fig 1, depending on the interpolation degree of the BEM. Thus, parameters p_1 and p_2 will be different for each collocation point and they must be recalculated at each time step for every grid point according to local conditions at Γ. Hence, p_1 and p_2 can be regarded as new degrees of freedom and their values define a unique local temperature profile for the adjacent STR associated to each freedom node at Γ.

Next, we derive the two local equations used to determine p_1 and p_2. The idea is to insert (3) into the C-DHT for the STR (2) and to evaluate it in the interface ($z = 0$). The time derivative of T in either STR or FTR is approximated by the two-time level finite difference scheme defined by:

$$\frac{\partial T}{\partial t} \approx \frac{T^{m+1} - T^m}{\delta t}; \quad \text{and} \quad \begin{pmatrix} T \\ \partial_n T \end{pmatrix} = \theta \begin{pmatrix} T \\ \partial_n T \end{pmatrix}^{m+1} - \bar{\theta} \begin{pmatrix} T \\ \partial_n T \end{pmatrix}^{m} \tag{4}$$

where $\theta \in [0,1]$; $\bar{\theta} := (\theta - 1)$; $\partial_n T := \hat{n} \cdot \nabla T$; and $\hat{n}$ is the normal vector at the interface. The first and second derivatives of T_s at the interface ($z = 0$) are, respectively:

$$\left.\frac{\partial T_m}{\partial z}\right|_{z=0} = p_1 - \frac{T_f - T_i}{d_s}, \quad \text{and} \quad \left.\frac{\partial^2 T_m}{\partial z^2}\right|_{z=0} = 2p_2 - 2\frac{p_1}{d_s} + \frac{T_f - T_i}{d_s^2}. \tag{5}$$

By inserting (4) and (5) for $z = 0$ into (2), we obtain the first equation:

$$\frac{T_f^{m+1} - T_f^m}{\alpha_s \delta t} = \frac{T_f^{m+1} - T_i}{d_s^2} - 2\frac{p_1}{d_s} + 2p_2 - \frac{k}{\alpha_s} T_f^{m+1} \tag{6}$$

The second equation is provided by the energetic balance in a local 3D control volume attached to a freedom node at the interface. It remains valid as long as the height δz of the control volume is considerable larger than the penetration depth and considerably smaller than the size of the STR, ie. $L_{STR} \gg \delta z \gg d_s$. The time derivative of the energy contained in a differential volume of unitary cross section and height dz in the STR can be written as:

$$\frac{\partial}{\partial t}\int_V (T_s - T_i)\,dV = -\alpha_s \left.\frac{\partial T_s}{\partial z}\right|_{z=0} + k_r \int_V (T_s - T_i)\,dV, \tag{7}$$

where it was considered that $\partial T_s/\partial z = 0$ when $z \gg d_s$. Next, it is convenient to define the following energy integral:

$$I(t) := \int_V (T_f - T_i)\,dV = \int_z (T_s(z,t) - T_i)\,dz, \tag{8}$$

which is proportional to the internal energy in dV, and can be expressed in terms of p_1 and p_2 by means of eq (3) as: $I(t) = (T_f - T_i)\,d_s + d_s^2 p_1 + 2d_s^3 p_2$. Then, the energy balance equation can be written in the following way:

$$I^{m+1} = I^m - \alpha_s \delta t \left(p_1 - \frac{T_f - T_i}{d_s}\right), \tag{9}$$

where $I^m = I(t_m)$ and $I^{m+1} = I(t_m + \delta t)$, are the energy integrals at times t_m and $t_m + \delta t$, respectively. Finally, (6) and (9) form a system of two equations with two unknowns (p_1, p_2) for each degree of freedom at Γ of the form:

$$\begin{bmatrix} 2\frac{\alpha_s}{d_s} & -2\alpha_s \\ \alpha_s + \xi d_s^2 & 2\xi d_s^3 \end{bmatrix} \begin{pmatrix} p_1 \\ p_2 \end{pmatrix} = \mathbf{B}\left(T_f^{m+1}, T_f^m, T_i, 1\right)^t \tag{10}$$

and matrix $\mathbf{B} \in R^{2\times 4}$ on the right hand side term is given by:

$$B = \begin{bmatrix} \theta\mu - \frac{1}{\delta t} & -\bar{\theta}\mu + \frac{1}{\delta t} & -\frac{\alpha_s}{d_s^2} & 0 \\ -d_s\xi + \frac{\alpha_s \theta}{d_s} & \frac{\bar{\theta}\alpha_s}{d_s} & d_s\xi - \frac{\alpha_s}{d_s} & I^m(k_r - \xi) \end{bmatrix} \tag{11}$$

being $\xi := 1/\delta t - k_r\theta$ and $\mu := k_r + \alpha_s/d_s^2$. The temperature profile in the STR, can be obtained by solving the system (10) for the parameters p_1 and p_2. Note that T_s^{m+1} becomes function of the temperature in the interface (T_f) at Γ, at present and previous time levels, the initial temperature in the block T_i and the energy content in the volume I.

3 DRBEM

The time dependent C-DHT in the FTR is solved with the DRBEM [4, 10]. First, the C-DHT is cast into a Poisson-like equation $\alpha_f \nabla^2 T_f(\mathbf{x}) = \rho(\mathbf{x}, t)$; $\mathbf{x} \in \Omega$; with an arbitrary source term $\rho(\mathbf{x}, t)$ and suitable Dirichlet, Neumann or Robin boundary conditions at the boundary $\partial(\Omega)$.

Then, in the context of the BEM, the matrix form of the discretised integral formulation can be expressed in the following way:

$$\alpha_f\left(\mathbf{H}\,\mathbf{T_f}-\mathbf{G}\,\partial_\mathbf{n}\mathbf{T_f}\right)=\mathbf{S}\boldsymbol{\rho},\tag{12}$$

where $\mathbf{H}\in R^{M\times M}$ and $\mathbf{G}\in R^{M\times N}$ are the standard BEM matrices based on the Green's function of Laplace equation, $\mathbf{S}\in R^{M\times M}$ is the DRM matrix, N is the number of freedom nodes used to discretise Γ, $M=N+L$, and L is the number of DRM nodes in Ω. The column arrays $\mathbf{T_f}\in R^{(M\times 1)}$ and $\partial_\mathbf{n}\mathbf{T_f}\in R^{(N\times 1)}$ specify T_f and $\partial T_f/\partial\hat{n}$ in each freedom node and the DRM matrices are defined as follows [4, 11]:

$$\mathbf{S}:=\left(\mathbf{H}\hat{\mathbf{U}}-\mathbf{G}\hat{\mathbf{Q}}\right)\mathbf{F}^{-1},\qquad \mathbf{S}\in R^{M\times M}\tag{13}$$

where $\mathbf{F}=f_{ij},\in R^{(M+4)\times(M+4)}$ is the 3D augmented thin plate splines radial basis functions matrix defined by the set $\{r_{ij},1,x_j,y_j,z_j\}$. Here, r_{ij} is the distance between i and j freedom nodes, whereas $1,x_j,y_j,z_j$ is the augmentation polynomial at point j. Then, $\hat{\mathbf{U}}=\hat{u}_{ij},\in R^{M\times(M+4)}$ and $\hat{\mathbf{Q}}=\partial\hat{u}_{ij}/\partial\hat{n}_i\in R^{N\times(M+4)}$ are the usual DRM matrices, whose elements obey the following relationship $\nabla^2\hat{u}_{ij}:=f_{ij}$. Next, the generalised right hand side term for the C-DHT ρ is given by:

$$\rho(\mathbf{x},t)=\frac{\partial T_f}{\partial t}+\mathbf{v}\cdot\nabla T_f+k_rT_f.\tag{14}$$

Finally, the matrix form of the discretised integral formulation can be expressed in the following way:

$$\mathbf{HT_f}-\mathbf{G}\partial_\mathbf{n}\mathbf{T_f}=\frac{\mathbf{S}}{\alpha_f}\left[\frac{\mathbf{T_f^{m+1}}-\mathbf{T_f^{m}}}{\delta t}+\sum_{p=1}^{3}\left(\mathbf{V}_p\cdot\mathbf{T}_p\right)T_f+k_r\mathbf{T_f}\right],\tag{15}$$

where the following matrices were employed [6, 11]: $\mathbf{V}_p=\operatorname{diag}\{v_p\}$ and $\mathbf{T}_p=\nabla_p\mathbf{F}\cdot\mathbf{F}^{-1}$; where $\mathbf{V}\in R^{M\times M\times 3}$ and $\mathbf{T}\in R^{M\times M\times 3}$, see ref. [11] for details.

4 Numerical implementation

The linear system established by eq. (10) provides the parameters p_1 and p_2 of each freedom node in Γ. These parameters define a unique 1D profile of temperature in the STR that best matches the required conservation equations. However, in a large scale calculation it might be more useful to deal with temperatures (T_f^{m+1}) and normal fluxes (q_m^{m+1}) rather than with p_1 and p_2. The change of variables: $(p_1,p_2)\rightarrow(T_f,q_m)^{m+1}$ can be done by expressing the normal flux exchanged between STR and FTR in terms of p_1 as: $q_m = v_n T_f - \alpha_s \frac{\partial T_s}{\partial n} = \alpha_s \left(p_1 - \frac{T_f - T_i}{d_s}\right)$, so that to obtain:

$$p_1 = \frac{q_m}{\alpha_s} + \frac{T_f - T_i}{d_s}. \tag{16}$$

Then, combination of eqs. (2), (5) and (16), yields:

$$p_2 = \frac{q_m}{\alpha_s d_s} + \frac{T_f - T_i}{2d_s^2} + \frac{1}{2\alpha_s}\frac{\partial T_s}{\partial t} - \frac{k_r}{2\alpha_s} T_f, \tag{17}$$

or expressed in a more suitable way:

$$\begin{pmatrix} p_1 \\ p_2 \end{pmatrix} = \mathbf{E} \begin{pmatrix} T_f^{m+1} \\ q_m^{m+1} \end{pmatrix} + \mathbf{F}, \tag{18}$$

where

$$\mathbf{E} = \frac{1}{\alpha_s} \begin{pmatrix} \frac{\alpha_s}{d_s} & 1 \\ \frac{\alpha_s}{2d_s^2} + \frac{1}{2\delta t} - \frac{k_r}{2} & \frac{1}{d_s} \end{pmatrix} \quad \text{and} \quad \mathbf{F} = -\frac{T_i}{d_s} \begin{pmatrix} 1 \\ \frac{1}{2d_s} + \frac{1 d_s}{2\alpha_s \delta t} \frac{T_f}{T_i} \end{pmatrix}. \tag{19}$$

The use of (18) in (10) provides a more suitable expression that relates temperature with normal flux at the interface:

$$A_1^* T_f^{m+1} + A_2^* q_m^{m+1} = B^* \tag{20}$$

where

$$A_1^* = (1-\theta)\frac{\alpha_s}{d_s} + \left(3 + \frac{d_s^2}{\alpha_s \delta t} - \frac{d_s^2 k_r}{\alpha_s}\right) d_s \xi, \qquad A_2^* = 1 + 3\frac{\xi d_s^2}{\alpha_s},$$

$$B^* = (1-\theta)\frac{\alpha_s}{d_s} T_f^m + 3 d_s \xi T_i + I^m (k_r + \xi) + \frac{d_s^3 \xi}{\alpha_s \delta t}.$$

Thus the semi-analytical part of the method introduces one equation (20) per each freedom node located at Γ. Note that the coefficients $A_{1,2}^*$ and the right hand side term RHS^* involved in (20) depend only on the energy integral I and the field values at the previous time step (m), as well as the material properties and the initial temperature T_i, but they no longer depend on the unknown T_f^{m+1}. Next, assembly of eq. (15) yields the following system of linear equations:

$$\mathbf{M_1 T_f^{m+1}} - \theta \mathbf{G q_f^{m+1}} = \bar{\mathbf{M}}_1 \mathbf{T_f^m} - \bar{\theta} \mathbf{G q_f^m} \tag{21}$$

where

$$\begin{pmatrix} \mathbf{M_1} \\ \bar{\mathbf{M}}_1 \end{pmatrix} := \begin{pmatrix} \theta \\ \bar{\theta} \end{pmatrix} \alpha_f \mathbf{H} - \begin{pmatrix} \theta \\ \bar{\theta} \end{pmatrix} \mathbf{S} \sum_{p=1}^{3} (\mathbf{V}_p \cdot \mathbf{T}_p) - \begin{pmatrix} \xi \\ \bar{\xi} \end{pmatrix} \mathbf{S} \tag{22}$$

and $\bar{\xi} := 1/\delta t - k_r \bar{\theta}$. After applying proper boundary conditions, eq.(21) yields a system of equations of the form $\mathbf{Ux} = \mathbf{b}$, where array $\mathbf{x}$ contains the unknowns

q and T, and b is composed of the right hand side term of eq (21) and the corresponding boundary conditions. Finally, eqs (20) and (21) are solved together, either by assembling all the equations in the same system of equations or by an iterative technique between STR and FTR systems, in order to solve the coupled FTR-STR problem.

5 Results

Some benchmark examples with this approach have already been conducted in our recent paper [7], particularly for problems of flow and solute transport through fractured and non-fractured porous media. This section will present one of those results in the context of a heat transfer problem.

5.1 Test problem for a slab

The test example consists of a slab of length $L = 21m$ and square cross section of $1m$ by $1m$, constant along L, as shown in Fig. 2. The block is composed of two regions: STR and FTR, corresponding to $\alpha_{STR} = 10^{-10}m^2/s$ and $\alpha_{FTR} = 10^{-6}m^2/s$, respectively. The former extends from $y = 0$ to $y = 1m$ and the latter from $y = 1m$ to $y = L = 21m$. The interface Γ is located at $y = L_s = 1m$. A convective term corresponding to $v = 9.98^{-11}m/s$ in y direction has been switched on. The surface at $y = 0$ is kept at constant temperature $T = 1$, while all the other surfaces are considered as adiabatic. The initial condition is given by: $T(\mathbf{x}, 0) = 1$ if $\mathbf{x} \in$ [STR] and $T(\mathbf{x}, 0) = 0$ anywhere else. Hence an the initial distribution yields an apparent infinite diffusive flux at $y = 1$. The coupled problem STR-FTR is assembled into a unique linear system of equations. The STR has one sub-domain with its boundary discretised into 182 linear discontinuous triangles. The FTR is decomposed into 539 sub-domains, each sub-domain is a linear discontinuous tetrahedron like the ones employed by the FEM. Figure 2(a) shows the meshed slab of the two regions.

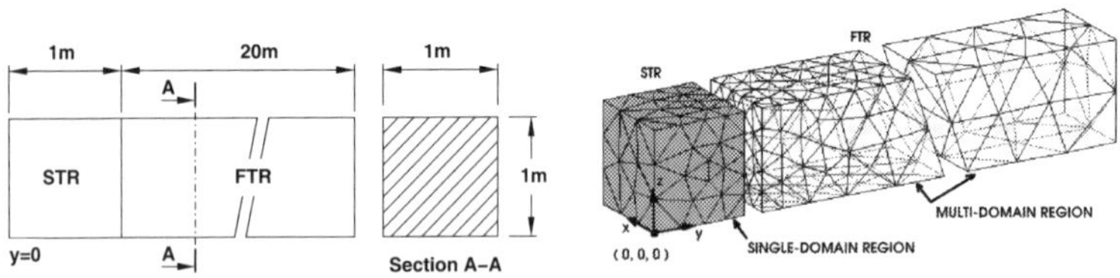

Figure 2: Dimensions of the test example, and 3D mesh of the model.

The volume mesh (involving 282 geometrical nodes was created with a FEM-like mesh generator [12, 13] based on the advance frontal method [14]. The derivatives of the DRM matrix F were used in order to compute the gradient of the temperature for the FTR region: $\partial T(\mathbf{x})/\partial x_p|_{x_i} = \sum_{j=1}^{N+4} \partial f_j(\mathbf{x})/\partial x_p|_{x_i} \alpha_j$

where the summation extends over the total number of RBFs associated to a given sub-domain (see [11] for more details). In the case of linear tetrahedrons there are three discontinuous freedom nodes per element, therefore $N = 12$, and the $\mathbf{F}$ matrix in the DRM has dimensions of 16×16. The solution for the temperature profile has been compared to a one dimensional Eulerian FVM reference code, based on cell-center formulation with MacCormak time integration scheme and constant time step. The convective term is calculated with an upwind discretisation scheme and the diffusive term with centered differences. The temperature profiles obtained with the SAM compared to the reference FVM along y coordinate at time levels $t_A = 0s$, $t_B = 5 \times 10^3 s$ and $t_C = 5 \times 10^4 s$ are shown in Fig. 3. The time step for the SAM $\delta t = 5 \times 10^3 s$ was kept constant along the time iterations. Good agreement is being observed between both results.

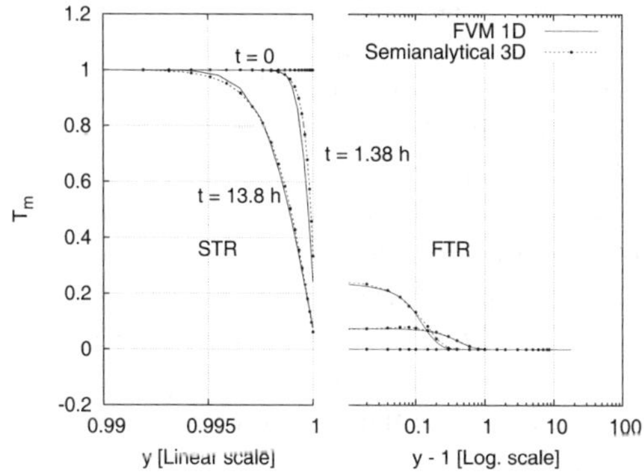

Figure 3: Temperature profile along the test slab in the STR(left) and FTR(right) regions at different time levels.

6 Conclusions

A hybrid semi-analytical approach coupled to the DRBEM for solving the early stage of 3D time dependent heat transport problems in non-homogeneous media has been proposed and tested with a simple 3D example involving a large aspect ratio composite body. The approach is suitable for 3D heat transfer problems dominated by diffusion, and of particular use when the initial distribution of temperature presents a discontinuous jump at the interface between two regions whose diffusivity constants differ in many orders of magnitude. By means of a semi-analytical profile, a linear equation that relates the temperature with its normal flux has been derived. In a block regarded as a sub-domain of low diffusivity in comparison with the surrounding media, this relation leads to two matrices which play the same role as the matrices resulting from the single and double layer integrals in the BEM, but with only one non-zero coefficient per row, respectively. Hence, they can be easily coupled to conventional BEM systems of equations in order to solve more complex geometrical situations.

References

[1] C.A. Brebbia and J. Dominguez. *Boundary Elements, An Introductory Course. Second Edition.* Computational Mechanics Publications, McGraw-Hill, New York Colorado San Francisco Springs Mexico Montreal Oklahoma City San Juan Toronto, 1992.

[2] C.A. Brebbia, J. C. Telles, and L. C. Wrobel. *Boundary Elements Techniques.* Springer-Verlag, Berlin, Heidelberg New York and Tokio, 1984.

[3] W. Florez, H. Power, and F. Chejne. Numerical solution of thermal convection problems using the multidomain boundary element method. *J. Num. Meth. for Partial Differential Equations*, 18(4):469–489, 2002.

[4] L.C. Wrobel and D. B.De Figueiredo. A dual reciprocity boundary element formulation for convection-diffusion problems with variable velocity fields. *Engng Analysis with Boundary Elements*, 8(6):312–319, 1991.

[5] M.M. Grigoriev and G. G. Dargush. Higher-order boundary element methods for transient diffusion problems. part ii: Singular flux formulation. *Int. J. Num. Meth. in Engng*, 55:41–54, 2002.

[6] A. Peratta. *BEM applied to Flow and Transport in Fractured Porous Media.* PhD thesis, University of Wales - Wessex Institute of Technology, Southampton, UK, December 2004.

[7] A. Peratta and V. Popov. Hybrid BEM for the early stage of unsteady transport process. *Int. J. for Num. Methods in Engng*, 2006. in press.

[8] P.K.W. Vinsome and J. Westerveld. A simple method for predicting cap and base rock heat losses in thermal reservoir simulations. *J. of Canadian Pet. Tech.*, 21:1861–1874, 1985.

[9] J. Birkholzer, G. Rouvé, K. Pruess, and J. Noorishad. An efficient semi-analytical method for numerical modeling of flow and solute transport in fractured media. In G. Gambolati, A Rinaldo, Brebbia, and W.G. Gray, editors, *Comp. Meth. in Subsurface Hydrology, Conference on Computational Methods in Water Resources-Held I*, pages 235–243. Springer-Verlag, 1990.

[10] V. Popov and H. Power. DRM-MD approach for the numerical solution of gas flow in porous media, with application to landfill. *Engng Analysis with Boundary Elements*, 23(2):175–188, 1999.

[11] A. Peratta and V. Popov. A new scheme for numerical modelling of flow and transport processes in 3d fractured porous media. *Advances in Water Resources*, 29:42–61, 2006.

[12] CIMNE, Int. Center for Num. Meth. in Engng., Barcelona, Spain. *GID, The personal pre/postprocessor Manual.*

[13] GID resources. Website. http://gid.cimne.upc.es.

[14] R. Lohner and P. Parikh. Generation of three-dimensional unstructured grids by the advancing front method. *Int. J. Num. Meth. in Fluids*, 8:1135–1149, 1988.

Evaluation of strong shear thinning non-Newtonian fluid flow using single domain DR-BEM

M. Giraldo[1], H. Power[2] & W. Flórez[1]
[1] *Instituto de Energía y Termodinámica, Universidad Pontificia Bolivariana, Medellín, Colombia*
[2] *School of Mechanical, Materials and Manufacturing Engineering, University of Nottingham, Nottingham, UK*

Abstract

The dual reciprocity method (DRM) has been successfully employed along with the boundary element method (BEM) to simulate non linear flow phenomena such as convective momentum transport and shear thinning fluids. In the latter however, domain partitioning has been necessary to achieve convergence when the power law index is below 0.8. This paper shows how a single domain DR-BEM formulation for non Newtonian low Reynolds number flows can be implemented in order to obtain accurate results for lower values of the power law index. Some of the characteristics of this implementation are the use of quadratic elements and an iterative solution of the non linear system of equations using a modified Newton–Raphson method. Along with the implementation, two radial basis functions (RBFs) were used and compared on two classical problems of inelastic non Newtonian flow: couette mixing and slit flow. Solutions obtained are also compared to results from a multi-domain dual reciprocity method (MD-DRM) for equal meshes. Results showed that using the above mentioned strategies, single domain DR-BEM can accurately predict the flow field in inelastic non Newtonian flow for values of the power law index as low as 0.5. It is also worth noting that the accuracy of the single domain strategy was shown to be higher than MD-DRM, although the latter clearly reduced computational resource consumption.
Keywords: single domain DRM, power law fluids, higher order elements, iterative methods for non linear systems.

WIT Transactions on Modelling and Simulation, Vol 44, © 2007 WIT Press
www.witpress.com, ISSN 1743-355X (on-line)
doi:10.2495/BE070111

1 Introduction

Study of non linear phenomena such as non Newtonian fluid flow using domain methods is accurate but requires large amounts of computational resources and are often accompanied with complex remeshing algorithms. The BEM has the advantage of greatly reducing the computational cost of numerical simulations by only discretizing the boundary of the problem thus reducing the problem dimensionality by one [1], but are often seen in disadvantage to domain methods when dealing with non linear problems. Non linear terms are usually treated in BEM as a domain integral of a pseudo body force [2], that was originally solved using cell integration, a type of domain meshing that makes BEM lose its boundary only nature and increase computational cost beyond FEM [3]. A more recent development is the DRM which uses RBF interpolation to expand the non linear term and then apply the usual divergence theorem to convert domain integrals into equivalent boundary integrals [4].

The main difficulty associated with shear thinning and shear thickening fluids is the dependence of the viscosity on domain variables, namely velocity gradients, leading to highly non linear Partial Differential Equation (PDF). To solve these type of flows, [2] uses a DRM approximation along with domain sub divisions. The resulting approach called multi-domain dual reciprocity (MD-DRM) showed to have good accuracy while significantly reducing resource consumption in comparison to other subdomain approaches such as the Green element method (GEM) [5]. Both works coincide with in the necessity of subdomains for the simulation of highly non linear problems.

The objective of this paper is to show how an adequate implementation of the DR-BEM along with a specific iterative scheme for the non linear system of equations can give accurate results for the flow of an inelastic non Newtonian fluid obeying the Power Law rheological model [6] without the need of domain partitioning, situation that can be an advantage if moving boundaries are considered. Two numerical examples are used to test the performance of the proposed formulation: couette and slit flow, both having analytical solutions to which the results obtained are compared [6]. The use of these examples also allows for different considerations as the first has only Dirichlet boundary conditions while the second is a mixed boundary condition problem (pressure driven flow). This implementation is also tested using two different radial basis functions (RBF), the traditional thin plate spline (TPS) and a more up-to-date compact support RBF with different values of the compactness parameter.

Work is divided as follows. Governing equations and integral formulation are initially presented, followed by the details of the numerical implementation of the resulting equations. The next section shows the results obtained for the different problems and finally conclusions are drawn.

2 Direct boundary integral formulation for non Newtonian fluids

The fluid inside the a closed Lyapunov surfaces S must obey the non Newtonian Stokes system of equations (1). The use of Stokes equation for this case is supported in the fact that common non Newtonian fluids have viscosities well over $10^3 Pa \cdot s$, assuring that $Re << 1$.

$$\frac{\partial u_i}{\partial x_i} = 0 \qquad \frac{\partial \sigma_{ij}}{\partial x_j} = 0 \tag{1}$$

where

$$\sigma_{ij} = -p\delta_{ij} + \eta\left(\dot{\gamma}\right)\left(\frac{\partial u_i}{\partial x_j} + \frac{\partial u_j}{\partial x_i}\right) \tag{2}$$

In equations (1) and (2), $\vec{u}$ is the velocity, p the pressure, δ_{ij} the Kronecker delta, $\eta\left(\dot{\gamma}\right)$ is the viscosity of the fluid.

An adequate rheological model for $\eta\left(\dot{\gamma}\right)$ must be selected in order to accurately simulate a given physical system. In the case of inelastic non Newtonian fluids, the power law model has shown to be a powerful tool bringing together accuracy and simplicity [6].

$$\eta\left(\dot{\gamma}\right) = k\dot{\gamma}^{n-1} \tag{3}$$

Boundary conditions vary according to the problem that is being solved. In the case of couette flow, the external cylinder is stationary while the internal one rotates at a constant angular velocity of value 1. For slit flow, the superior and inferior surfaces are stationary, while at the entrance and exit the perpendicular velocities are made nil while the tractions are given only by a pressure difference between them. For the cases shown, Δp is also 1.

Integral representations for the velocity in BEM are only valid for a constant viscosity, therefore a small modification must be made to the internal tractions (2). The idea is to subtract and add the stress tensor multiplied by an arbitrary constant μ^N allowing to redefine the momentum equation as:

$$\frac{\partial \sigma_{ij}^N}{\partial x_j} = -\frac{\partial \tau_{ij}^e}{\partial x_j} \tag{4}$$

where

$$\tau_{ij}^e = \left(k\dot{\gamma}^{n-1} - \mu^N\right)\left(\frac{\partial u_i}{\partial x_j} + \frac{\partial u_j}{\partial x_i}\right); \ \sigma_{ij}^N = -p\delta_{ij} + \mu^N\left(\frac{\partial u_i}{\partial x_j} + \frac{\partial u_j}{\partial x_i}\right) \tag{5}$$

Using Green's second identity for stokes flow [7] to expand the Newtonian traction (σ_{ij}^N) and using the properties of the fundamental solutions, the integral representation of the velocity field for an arbitrary point in a closed domain filled with a non Newtonian fluid, which for the present case is a point $x \in \Omega_i$, is given by:

$$Cu_i(x) - \int_S K_{ij}(x,y)\, u_j(y) dS_y = \frac{1}{\mu^N} \int_\Omega u_i^j(x,y) \left(-\frac{\partial \tau_{jk}^e}{\partial x_k} \right) d\Omega - \frac{1}{\mu^N} \int_S u_i^j(x,y)\, \sigma_{jk}^N(\vec{u}(y)) n_k(y)\, dS_y \quad (6)$$

where C is a constant dependant on the position of the point. For internal points $C = 1$ and for point a smooth boundary $C = 1/2$.

The Stokeslet for two dimensions is given by:

$$u_i^j(x,y) = -\frac{1}{4\pi} \left[\ln\left(\frac{1}{r}\right) \delta_{ij} + \frac{(x_i - y_i)(x_j - y_j)}{r^2} \right] \quad (7)$$

r being the Euclidean distance between point x and y, $r = |x - y|$. The corresponding normal derivative or Stresslet is given by (in two dimension):

$$K_{ij}(x,y) = -\frac{1}{\pi} \frac{(x_i - y_i)(x_j - y_j)(x_k - y_k)\, n_k(y)}{r^4} \quad (8)$$

3 Dual reciprocity approximation

In order to avoid domain meshing, the dual reciprocity method [4] is used to expand the domain integral. The basis of this method is to approximate the non homogeneous term using interpolation functions [2]:

$$-\frac{\partial \tau_{jk}^e}{\partial x_k} = \sum_{m=1}^{p} f(x, y^m)\, \alpha_l^m \delta_{il} \quad (9)$$

$f(x, y^m)$ is a known set of functions dependent only on geometry and α_l^m is an unknown vector of coefficients to be determined by collocation on y^m $(m = 1, 2, 3, ..., p)$ points in the domain of interest. Using this approximation and Green's identities, the domain integral becomes:

$$\int_\Omega u_i^j(x,y) \left(-\frac{\partial \tau_{jk}^e}{\partial x_k} \right) d\Omega_i = \sum_{m=1}^{p} \alpha_l^m \left[\hat{u}_k^l(x, z^m) - \int_S K_{kj}(x,y)\, \hat{u}_j^l(y, z^m)\, dS_y + \int_S u_i^k(x,y)\, \hat{t}_j^l(y, z^m)\, dS_y \right] \quad (10)$$

3.1 Choice of interpolating function

One of the most popular interpolating functions used is the thin plate spline:

$$f^m(x) = r^2(x, y^m)\log(r(x, y^m)) + P_1(x) \tag{11}$$

This function has shown to be an accurate and simple alternative for use with boundary integral methods [2, 3]. For this case, the augmented polynomial P_1 is composed of the functions x_1, x_2 and 1. The particular solutions of the first term of the second degree GTPS are ($\bar{x}_i = x_i - y_i^m$):

$$\hat{u}_i^j(x, y^m) = \frac{1}{96}\left[\left(5r^4\log r - \frac{7}{3}r^4\right)\delta_{ij} - \bar{x}_i\bar{x}_j\left(4r^2\log r - \frac{5}{3}r^2\right)\right] \tag{12}$$

$$\hat{t}_i^j(x, y^m) = \frac{1}{96}\left[\begin{array}{l} 8r^2(\bar{x}_j n_i + \bar{x}_i n_j + \bar{x}_l n_l \delta_{ij})\left(2\log r - \frac{1}{3}\right) \\ -4\bar{x}_i\bar{x}_j\bar{x}_l n_l\left(4\log r + \frac{1}{3}\right) \end{array}\right] \tag{13}$$

Since the polynomials are an integral part of the GTPS, the particular solutions of the complete GTPS is equal to functions given in (12) and (13) plus the summation of the particular solutions to each of the augmentation polynomials. It is worth noting that these polynomials are only dependant on point x, not the other collocation nodes y^m; this fact also holds true for the auxiliary flow fields.

The values of the auxiliary fields and derivatives of the augmented polynomials used with the second degree GTPS are the following:

For $f(x) = x_1$

$$\hat{u}_i^j(x, y^m) = \frac{1}{24}\left[\begin{array}{l} x_1^3(3\delta_{ij} - 2\delta_{1i}\delta_{1j} - \delta_{2i}\delta_{2j}) + \\ 3x_1x_2^2(\delta_{ij} - \delta_{1i}\delta_{1j}) - 3x_1^2x_2(\delta_{1i}\delta_{2j} + \delta_{2i}\delta_{1j}) \end{array}\right] \tag{14}$$

$$\hat{t}_i^j(x) = \frac{1}{8}\left[\begin{array}{l} 3x_1^2(n_1\delta_{ij} + n_j\delta_{1i} + n_i\delta_{1j}) - \\ 2x_1^2(2n_1\delta_{1i}\delta_{1j} + n_1\delta_{2i}\delta_{2j} + n_2\delta_{1i}\delta_{2j} + n_2\delta_{1j}\delta_{2i}) + \\ x_2^2(n_1\delta_{ij} + n_j\delta_{1i} + n_i\delta_{1j} - 2n_1\delta_{1i}\delta_{1j}) + \\ 2x_1x_2(n_2\delta_{ij} + n_j\delta_{2i} + n_i\delta_{2j}) - \\ 4x_1x_2(n_2\delta_{1i}\delta_{1j} + n_1\delta_{1i}\delta_{2j} + n_1\delta_{1j}\delta_{2i}) \end{array}\right] \tag{15}$$

Particular solutions for $f(x) = x_2$ are of the same form, but wherever a sub-index 1 appears, it must be replaced by 2 and vice versa. Finally for $f = 1$

$$\hat{u}_i^j(x) = \frac{1}{16}\left[3|x|^2\delta_{ij} - 2|x_i||x_j|\right] \tag{16}$$

$$\hat{t}_i^j(x) = \frac{1}{4}\left[x_j n_i + x_i n_j + x_l n_l \delta_{ij}\right] \tag{17}$$

The second choice of interpolating function is a compact support radial basis function (CS-RBF) [8] which has the characteristic of assigning a value of 0 to points outside a circle of radius C. In [9] CS-RBFs are used along with DRM to solve the Navier–Stokes system of equations with good results, although as is the case in this paper, the compactness of the function is not used and parameter C is chosen to be bigger than the total domain. The specific function used in this work was (non-zero part):

$$f^m(x, y^m) = \left(1 - \frac{r(x, y^m)}{C}\right)^4 \left(4\frac{r(x, y^m)}{C} + 1\right) \tag{18}$$

The auxiliary flow fields for this CS-RBF are (Terms corresponding to $r > C$ are omitted):

$$\hat{u}_i^j(x, y^m) = \begin{bmatrix} \left(\frac{3}{16}r^2 - \frac{25}{48}\frac{r^4}{C^2} + \frac{24}{35}\frac{r^5}{C^3} - \frac{35}{96}\frac{r^6}{C^4} + \frac{32}{441}\frac{r^7}{C^5}\right)\delta_{ij} - \\ \bar{x}_i\bar{x}_j\left(\frac{1}{8} - \frac{5}{12}\frac{r^2}{C^2} + \frac{4}{7}\frac{r^3}{C^3} - \frac{5}{16}\frac{r^4}{C^4} + \frac{4}{63}\frac{r^5}{C^5}\right) \end{bmatrix} \tag{19}$$

$$\hat{t}_i^j(x, y^m) = \begin{bmatrix} \left(\frac{1}{4} - \frac{5}{3}\frac{r^2}{C^2} + \frac{20}{7}\frac{r^3}{C^3} - \frac{15}{8}\frac{r^4}{C^4} + \frac{4}{9}\frac{r^5}{C^5}\right)\bar{x}_j n_i + \\ (\bar{x}_i n_j + \bar{x}_l n_l \delta_{ij})\left(\frac{1}{4} - \frac{5}{3}\frac{r^2}{C^2} + \frac{20}{7}\frac{r^3}{C^3} - \frac{15}{8}\frac{r^4}{C^4} + \frac{4}{9}\frac{r^5}{C^5}\right) \\ -2\bar{x}_i\bar{x}_j\bar{x}_l n_l\left(-\frac{5}{6C^2} + \frac{12}{7}\frac{r}{C^3} - \frac{5}{4}\frac{r^2}{C^4} + \frac{20}{63}\frac{r^3}{C^5}\right) \end{bmatrix} \tag{20}$$

4 Integral equation discretization

Traditionally, the approaches used to approximate both the geometry and the densities have been constant and linear elements. Elements of higher order, such as quadratic, have found only limited application in a reduced number of problems. In this work, both geometrical and functional discretization had been made using quadratic elements. The interpolation scheme for a function $X(\xi)$ is given by:

$$X(\varepsilon) = \psi_1(\varepsilon)X^{(1)} + \psi_2(\varepsilon)X^{(2)} + \psi_3(\varepsilon)X^{(3)} \tag{21}$$

where $X^{(1)}, X^{(2)}, X^{(3)}$ are the values of $X(\xi)$ on the three nodes of the element. Interpolation functions $\vec{\psi}$ are show in (22).

$$\psi_1 = \tfrac{1}{2}\varepsilon(\varepsilon - 1); \quad \psi_2 = (1 - \varepsilon)(1 + \varepsilon); \quad \psi_3 = \tfrac{1}{2}\varepsilon(\varepsilon + 1) \tag{22}$$

The resulting set of integrals are regular and can be evaluated using standard Gaussian quadratures. In order to avoid the weak singularity present when calculating the integral kernels when integrating over the same element where the source point is located, Telles' transformation is initially used, followed by rigid body motion in the case of the double layer kernel [1].

5 Iterative solution of the non linear system of equations

After the integral equations are assembled, a system of equations arises for the variables u_i on points $x \in \Omega_i$, and either u_i or $\sigma_{ij}n_j$ on $\xi \in S$ depending on boundary conditions. Given the non linear character of this system caused by the dependence of τ_{ij}^{e} on the nodal velocities, an iterative procedure must be set in place to determine the actual solution. For simplicity a new system of equations is defined as:

$$[A]\left[\vec{X}\right] = \left[\vec{C}\right] + \left[\vec{B}\left(\vec{X}\right)\right] \tag{23}$$

where matrix A is constituted with elements of matrices H and G depending on boundary conditions. Vector X are the unknowns and C the boundary conditions multiplied by the corresponding elements of H and G, and B the non linear term.

The solution of this complete system of equations can be sought in different ways. Simulations showed that traditional schemes such as Piccard iterations only achieved convergence when the power law index is above 0.8. In order to achieve convergence in values well below 0.8 as those reported in this paper, a Newton–Raphson method with Backtracking was employed directly on (23) without distinction if a given unknown was a velocity or a traction. Following the usual procedure, the following equation is found, where $W\left(\vec{X}\right)$ is the function to minimize.

$$W\left(\vec{X}\right) \approx W\left(\vec{X}^{(t)}\right) + \frac{1}{\phi} J_X\left(\vec{X}^{(t)}\right) \cdot \left(\vec{X}^{(t+1)} - \vec{X}^{(t)}\right) \tag{24}$$

The Jacobian matrix (J_X) was calculated numerically using centered finite differences with the data available from iteration t. ϕ is a smoothing parameter for the iterative method. The program works initially with $\phi = 1$, if the problem diverges, then equation (25) is used.

$$\phi_{optimal} \approx \left(\frac{\left\| W\left(\vec{X}^{(t+1)}\right) \right\|^2}{\left\| W\left(\vec{X}^{(t)}\right) \right\|^2} + 1 \right)^{-1} \tag{25}$$

6 Accuracy assessment

This section presents two problems, couette and slit flow for non Newtonian power law fluids in order to test the performance of the proposed implementation, along with the RBF's and different values of the support parameter C in the case of the CS-RBF (Eq. 18).

6.1 Couette flow

This problem has an analytical solution for the tangential velocity. The analytical solution is given by [6]:

$$u_t = \frac{\omega}{\left(\frac{r_e}{r_i}\right)^{2/n} - 1} \times \frac{r_e^{2/n} - r^{2/n}}{r^{(2/n)-1}} \tag{26}$$

where u_t is the tangential velocity at a radius r, ω is the angular velocity of the internal cylinder of radius r_i, and r_e is the radius of the external cylinder. n is the Power Law Index of the rheological model. A simple mesh consisting of 24 quadratic elements on the outer surface, 24 quadratic elements on the inner surface and 240 internal collocation points will be employed.

Table 1 shows good results of the numerical method when compared to the analytical solutions. It is worth noting that all simulations generated an error below 1%. Best performance by far was obtained with the CS-RBF using a value of $C = 15.0$, 50% larger than the maximum distance between two points in the domain. At smaller values of n GTPS showed a greater error compared to CS-RBF, but total errors are still within acceptable parameters. In all cases, MD-DRM was the least accurate of the tested approximations, but the time used to find such solutions was significantly smaller than the one for the single domain approaches. It is important to note however that the MD-DRM has been tested for values of n as low as 0.2 with good results, but these require denser meshes.

Table 1: Implementation performance on non Newtonian couette problem.

	$n = 0.8$		$n = 0.6$		$n = 0.5$	
	L2 error (%)	Iter.	L2 error (%)	Iter.	L2 error (%)	Iter.
GTPS	0.73868	18	2.0531	10	3.048	16
CS 10.01	0.54706	6	1.5845	13	2.4178	18
CS 15.0	0.5464	8	1.5797	18	2.4115	20
CS 20.0	0.54755	7	1.5797	16	2.4106	17
MD-DRM	1.9076	6	4.7367	11	6.3728	14

By graphically comparing the results from the single domain approximation CS 20.0 with the MD-DRM approach (Figure 1(a)) it is clear how the single domain solution is quite more accurate in the zones of greater non linearity (left of the plot). Single domain approaches are in this case the best choice in terms of accuracy, although it is important to consider that given the matrix sparsity of MD-DRM solutions found using this approach are significantly faster allowing for the use of denser meshes with which the solution can improve its accuracy.

6.2 Slit flow

An analytical solution is available for the velocity profile [6]:

$$u_1 = \frac{n}{2n+2}\left(\frac{1}{2k}\frac{\Delta p}{L}\right)^{1/n} h^{(1/n)+1}\left[1 - \left(\frac{2\,|x_2|}{h}\right)^{(1/n)+1}\right] \tag{27}$$

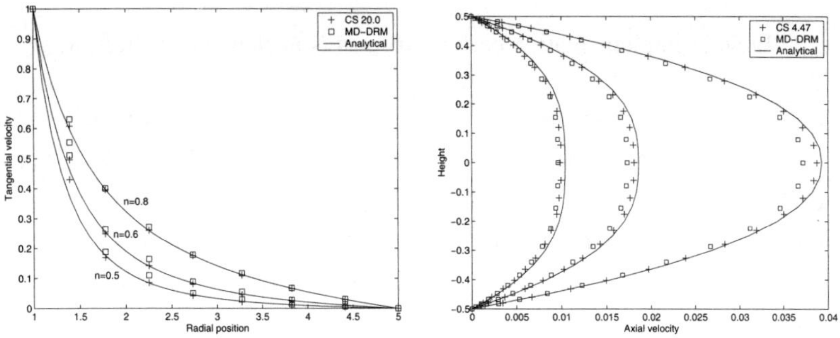

(a) (b)

Figure 1: Velocity profiles for Non Newtonian (a) couette and (b) slit flow.

where L is the channel length, h its height and Δp the imposed pressure difference. The tested mesh consisted of 48 quadratic elements on the outer surface, 385 internal collocation points.

Table 2 shows that all the single domain solutions exceed the accuracy of MD-DRM for $n = 0.8$. The case of CS 3.35 (50% larger than the maximum distance between two points in the domain) is worth signaling out, as the total L2 error is an order of magnitude below the remaining approximations. As the power law index is reduced, as expected all solutions increase their L2 errors, making some of the single domain approaches show higher errors than MD-DRM for $n = 0.5$ but overall showing that the correct selection of the interpolation function can give very accurate results without domain partitioning. However, it is evident that the DR-BEM works better of Dirichlet boundary conditions (i.e. couette flow) than for mixed boundary conditions.

In Figure 1(b) it can be seen that in this case the biggest errors for the single domain approaches are located in the center of the domain, not near the walls where the higher gradients exists. This situation shows that a single domain DR-BEM tends to overshot the viscosity increase and therefore decrease the total flow compared to the analytical solutions. As in the former section, the single domain approach has a greater accuracy than MD-DRM.

7 Conclusions

The use of single domain DRM for inelastic non Newtonian flows has been traditionally considered not accurate for values of the power law index below 0.8 making the use of domain partitioning a requirement. This paper shows how the use of higher order elements combined with an adequate iteration scheme for the resulting non linear system of equations can improve the results obtained for this problem without the need of domain partitioning. Results for couette and slit flow showed that the use of this strategies improves the accuracy of single domain DRM above that of MD-DRM for values of the power law index as low as 0.5. Even

Table 2: Implementation performance on Non Newtonian slit flow.

	$n = 0.8$		$n = 0.6$		$n = 0.5$	
	L2 error (%)	Iter.	L2 error (%)	Iter.	L2 error (%)	Iter.
GTPS	6.0912	6	20.457	9	35.693	34
CS 2.24	1.3936	30	10.557	13	54.355	19
CS 3.35	0.10092	37	173.73[1]	16	2.2979	21
CS 4.47	3.8419	20	8.9166	22	16.016	16
MD-DRM	11.238	6	15.52	7	18.144	8

though MD-DRM is less time consuming and can simulate cases of lower values of n, the requirement of remeshing if moving boundaries are considered makes the use of single domain DRM a justified and accurate choice.

References

[1] Brebbia, C., Telles, J. & Wrobel, L., *Boundary Element Techniques*. Springer Verlang: New York, 1984.

[2] Flórez, W. & Power, H., Multi-domain mass conservative dual reciprocity method for the solution of the non-newtonian stokes equations. *Applied Mathematical Modelling*, **26**, pp. 397–419, 2002.

[3] Power, H. & Botte, V., An indirect boundary element method for solving low reynolds number navier-stokes equations in a three-dimensional cavity. *International Journal for Numerical Methods in Engineering*, **41**, pp. 1485–1505, 1998.

[4] Partridge, P., Brebbia, C. & Wrobel, L., *The dual reciprocity boundary element method*. Computational Mechanics Publications: Southampton, 1992.

[5] Tran-Canh, D. & Tran-Cong, T., Bem-nn computation of generalised newtonian flows. *Engineering Analysis with Boundary Elements*, **26**, pp. 15–28, 2006.

[6] Bird, R.B., Armstrong, R.C. & Hassager, O., *Dynamics of polymeric liquids*, volume 1: Fluid mechanics. John Wiley & sons: New York, 2nd edition, 1987.

[7] Ladyzhensaya, O., *The mathematical theory of viscous incompressible flow*. Gordon and Breach: New York, 1963.

[8] Chen, C., Brebbia, C. & Power, H., Dual reciprocity method using compactly supported radia basis functions. *Communications in numerical methods in engineering*, **15**, pp. 137–150, 1999.

[9] Shuaib, N.H., *Numerical simulation of thin film flow including shear and gravitational effects*. Phd, University of Nottingham, 2005.

DRM-MD approach for bound electron states in semiconductor nano-wires

R. Gospavic[1], V. Popov[1] & G. Todorovic[2]
[1]*Wessex Institute of Technology, Ashurst Lodge, Southampton, UK*
[2]*Faculty of Civil Engineering, Belgrade, Serbia*

Abstract

Using the boundary element dual reciprocity method-multi domain (DRM-MD), bound electron states and the corresponding wave functions in semiconductor quantum wires embedded in a matrix were considered. The single circular and rectangular as well as the two near circular quantum wires were analysed. In the case of two coupled quantum wires, the dependence of the resulting wave function and eigenenergies as a function of the distance between wires was calculated. The DRM-MD gave a linear electron state model and the developed numerical approach accurately captured the symmetry breaking and splitting of the degenerated energy states due to the presence of additional wire. According to the symmetry of the structures a suitable mesh reduction was employed and different modes were considered separately. For a case of hetero structures, domain decomposition was used.
Keywords: dual reciprocity-method, bound states, wave function, quantum wire.

1 Introduction

The semiconductor nanostructures due to their unique physical properties have wide range of potential applications in electronic and optoelectronic devices [1–5]. Latest advances in fabrication technology of semiconductors make it possible to obtain nanostructures in a controllable way and with a wide range of geometries and shapes [6–8], which resulted in a tremendous increase of research interests in nanotechnology and nanostructures in recent years.

As the dimensions of these devices approach nano-scales, quantum effects become significant and quantum mechanical treatment of the problem is

WIT Transactions on Modelling and Simulation, Vol 44,
www.witpress.com, ISSN 1743-355X (on-line)
doi:10.2495/BE070121

necessary. Accordingly, there is an increased need for development of efficient numerical models that can be used during the design process.

In the case of quantum wires the free carriers are confined in two directions by potential barriers, forming so called 1D electron gas [9, 10]. The lateral dimensions of such structures are much smaller then their length, i.e. they could be considered as wires with infinite length. The electrons are confined to a small region inside the cross section of wires while they can move freely along the wires. If the diameter of wires is getting smaller then the electron states become quantized at discrete energy levels. Such a structures have a number of unique physical characteristics related to their electrical and optical properties [9, 10], which make them suitable for various applications.

In this work the DRM-MD approach was employed for the case of constant band effective mass and potential, which produces linear generalized eigenvalue problems and in that sense has advantage relative to the classical BEM which produces a non-linear ones [12, 13]. An additional advantage of the presented numerical method is that it could be used not only for constant potential and electron effective mass in each region, the case of the classical BEM approach [16], but it can also be used for arbitrary spatial potential distribution and coordinate and energy dependent electron effective masses. In this case a nonlinear eigenvalue system of equations must be solved.

2 The physical model

The physical model used to describe the bound states of conductor band electrons inside quantum wires could be expressed by the Schrödinger equation in the one-band effective mass approximation [11, 15]:

$$-\frac{\hbar^2}{2}\nabla\left(\frac{1}{m^*(x,y,z)}\nabla\Psi(x,y,z)\right)+V(x,y)\cdot\Psi(x,y,z)=E\cdot\Psi(x,y,z) \qquad (1)$$

where $\Psi(x,y,z)$ denotes a wave function belonging to energy level E, $m^*(x,y,z)$ is band effective mass which is in general a function of spatial coordinates, $V(x, y, z)$ is barrier potential, $\hbar = h/2\pi$ and h is Planck's constant. [9, 10, 16].

The z-axis was oriented along the quantum wire axis and it was assumed that the barrier potential depends only on x and y coordinates and that a charge carrier travels freely along z-axis with preset value of the wave number k_z. It was also assumed that the electron effective mass is constant inside each region and there is a jump in the effective mass only on the junction between different materials. According to these assumptions, the method of variable separation could be used in order to transform the governing equation (1) into a set of equivalent equations as follows:

$$\begin{aligned}
&\Psi(x,y,z)=\psi_T(x,y)\cdot\psi_z(z)\\
&\Delta_T\psi_T(x,y)-\left(\frac{2\cdot m^*}{\hbar^2}(V(x,y)-E)+k_z^2\right)\cdot\psi_T(x,y)=0\\
&\frac{d\psi_z(z)}{dz}+k_z^2\cdot\psi_z(z)=0\Rightarrow\\
&\psi_z(z)=e^{ik_z\cdot z};\quad \Psi(x,y,z)=\psi_T(x,y)\cdot e^{ik_z\cdot z}
\end{aligned} \tag{2}$$

where $\Delta_T=\frac{\partial^2}{\partial x^2}+\frac{\partial^2}{\partial y^2}, k_x=\sqrt{\frac{2\cdot m^*}{\hbar^2}}, E=E_Z+E_T, E_T=E-\frac{\hbar^2\cdot k_z^2}{2\cdot m^*}$,

and k_z , ψ_z, E_T , E the electron wave number, the wave functions for the free electron in z direction, the electron energy of motion in transversal direction (xy plane) and the total energy respectively.

At the interface between different regions (contour Γ_1 in figure 1) the next matching conditions exist [9, 10, 11, 16]:

$$\psi_1\big|_{\Gamma_1}=\psi_2\big|_{\Gamma_1},\ \frac{1}{m_1^*}\frac{\partial\psi_1}{\partial n}\bigg|_{\Gamma_1}=\frac{1}{m_2^*}\frac{\partial\psi_1}{\partial n}\bigg|_{\Gamma_1}, \tag{3}$$

where m_1^* and m_2^* denote effective mass in corresponding regions, and n is the normal on the boundary between regions.

Also, on the boundary far from the quantum wires, at distance R, the wave function and its gradient vanish i.e. the following boundary conditions on contour Γ_2 could be used [11]:

$$\psi\big|_{\Gamma_2}=0;\qquad \frac{\partial\psi}{\partial n}\bigg|_{\Gamma_2}=0\,. \tag{4}$$

3 The DRM-MD approach

The first equation in (2) could be transformed using Green's second identity into the following equation [16]:

$$\begin{aligned}
&\chi(\vec{r}_1)\cdot\psi(\vec{r}_1)+\oint_{\Gamma_2}\left(q^*(|\vec{r}_1-\vec{r}_2|)\cdot\psi(\vec{r}_2)-u^*(|\vec{r}_1-\vec{r}_2|)\cdot q(\vec{r}_2)\right)\cdot d\Gamma_1\\
&=\int_\Omega b(\vec{r}_2)\cdot u^*(|\vec{r}_1-\vec{r}_2|)\cdot d\Omega
\end{aligned} \tag{5a}$$

$$b(x,y)=\left(\frac{2\cdot m^*}{\hbar^2}(V(x,y)-E)+k_z^2\right)\cdot\psi(x,y);$$

$$q=\frac{\partial \psi}{\partial n};\quad q^{*}=\frac{\partial u^{*}}{\partial n}$$

$$\chi(\vec{r}_1)=\begin{cases}1, & \vec{r}_1\in\Omega/\partial\Omega\\ \theta/2\pi, & \vec{r}_1\in\partial\Omega\\ 0, & \vec{r}_1\notin\Omega\end{cases}\qquad \Gamma=\partial\Omega;\quad \vec{r}_1=(x_1,y_1);\ \vec{r}_2=(x_2,y_2),\tag{5b}$$

where θ is interior angle at boundary point $\vec{r}_1$ in radians, Ω is the problem domain, Γ is the boundary contour and u^* denotes the fundamental solution for 2D Laplace's operator [17].

The domain integral in (5) could be avoided by using interpolation of the term b over the problem domain Ω by approximation functions [18].

For the 2D case the most convenient choice for RBF functions are augmented thin plate spline functions (ATPS) [19]. In order to transform the domain integral in (5) into boundary one, an auxiliary function is introduced [12, 13].

After spatial discretization of the boundary contour Γ, using linear elements for interpolating unknown wave function ψ and its normal derivative q, boundary integrals could be evaluated and using collocation technique this procedure could be repeated for each boundary as well as interior nodes. In this way a linear generalized eigenvalue problem is obtained. After integral evaluations, (5) can be written in following matrix form [12]:

$$\begin{aligned}&H\cdot\psi-G\cdot Q=\left(H\cdot\hat{U}-G\cdot\hat{Q}\right)\times F^{-1}\times b;\\ &\psi=[\psi_1,\cdots,\psi_{N_\psi}];\quad Q=[q_1,\cdots,q_{N_q}]\\ &b=\left(\frac{2\cdot m^*}{\hbar^2}\cdot V+k_z^2-\frac{2\cdot m^*}{\hbar^2}\cdot E\right)\cdot\psi,\end{aligned}\tag{6}$$

where ψ_i and q_i are unknown nodal values of the wave function and its normal derivate, respectively. Matrix $\boldsymbol{F}$ is obtained from (7) and (8). The above relation corresponds to the case of the constant potential and effective mass in each region separately.

Similar sets of equations are obtained in each of the sub-domains and a system matrix can be assembled. In this way the generalized eigenvalue problem is obtained.

4 Numerical results

In all presented numerical result the software ARPACK was used for finding bounded energies and wave functions of the corresponding linear eigenvalue problems. In all examples the field variables and derivatives were represented using linear interpolation over the elements while on the corners and curved contours, linear discontinues elements were used [12]. For the DRM interpolation the ATPS radial basis functions were used. In all numerical examples the considered structures consist of two media and according to the physical model given in (1) the electron effective mass has different values in each region.

4.1 Case study 1

For the case of a cylindrical wire with constant potentials inside and outside the wire, the bounded energies could be expressed in analytical form [9].

If the potential is a symmetrical function of the coordinates, the wave functions are either even or odd functions of the x and y coordinates and could be classified in four modes, where indexes i and j in the wave function ψ_{ij} indicate the symmetry regarding x and y coordinates respectively (i, j=S for a symmetrical function and i, j=A for asymmetrical one). All these wave functions are orthogonal and thus they are base states, i.e. any bound state could be represented as a linear combination of these four base states.

In Table 1 the numerical results for the bounded energies obtained by the DRM-MD were compared with the analytical ones, in the case of cylindrical wires. The numerical parameters have the following values: m_1^*=0.00665m$_e$; m_2^*=0.00858m$_e$; V_0=320meV; k_z=0 nm; r=5nm where m_e is the free electron mass. These parameters correspond to the GaAs/$Ga_{0.63}Al_{0.37}As$ structure [11].

Table 1: Comparison of numerical results obtained using the DRM-MD with six sub-domains and analytical solution for the case of cylindrical wires.

mode	DRM-MD [meV]	Analytical [meV]
SS	75.983	76.06
	317.8111	317.2
AA	317.6100	317.2
SA	188.8149	188.5605
AS	188.8122	188.5605

Due to the Hamiltonian's symmetry the quantum states ψ_{SA} and ψ_{AS} are degenerated, i.e. both states have the same energy even though the wave functions are different and each linear combination of these states are also bounded states with same energy and correspond to the rotation of the coordinate system. Also, the ψ_{AA} state and the second states in SS mode (ψ_{SS2}) are degenerated and the same explanation holds.

4.2 Case study 2

In this case two identical cylindrical wires with constant potential inside and outside the wires were considered. The material parameters, the wave number in the axial direction and the dimensions of the wires were the same as in the pervious case. The calculations were performed for two different distances between wires (d =4 and 10 nm) and for SS mode. In table 2 the obtained numerical results are summarized.

Table 2: Numerical results for bounded energies in SS mode obtained using the DRM-MD for the case of two identical cylindrical wires for different distance d [nm] between them (radius of wires = 5 nm, number of sub-domains for each of the cases is denoted with SD).

mode	energy [meV]; d=4nm; 5-SD	energy [meV] d=10nm; 7-SD
SS	75.259	75.464
	185.668	188.146
	313.713	316.335

The normalized wave functions and contour plots for the first state in *SS* mode for the case of *d*=4nm, are shown in figures 1(a)-(b).

4.3 Case study 3

In this case study one rectangular wire with dimensions 20 x 10 nm was considered. The material parameters were the same as in the previous cases and the potential barrier between different materials was 276 meV and $k_z = 0$ nm [11].

Domain decomposition with ten sub-domains was used with total number of nodes 1739 of which 798 on the boundary and 941 interior nodes. The wave function and contour plot for the fourth quantum state in *SS* mode are presented in Figure 2(a)-(b).

5 Conclusions

The boundary element DRM-MD approach was developed and applied for calculation of bounded electron states of quantum wires. In order to describe the conduction electrons bound states inside the quantum wires the one-band effective mass approximation was used for solving the Schrödinger equation. The present approach employed linear interpolation of field variables over the boundary elements.

This approach has advantage in respect to the classical BEM approach as it leads to a linear eigenvalue problem compared to the classical BEM which results in a non linear eigenvalue problem. Additional advantage compared to the classical BEM is that the DRM-MD approach can be used not only in the case of constant potential and effective mass inside each region but also for the case when potential and effective mass vary in space [14].

The developed numerical scheme was tested on three case studies: (i) a single cylindrical wire, (ii) two identical cylindrical wires, and (iii) a single rectangular quantum wire. In the case of a single cylindrical wire the results were compared with analytical solution and good agreement was observed.

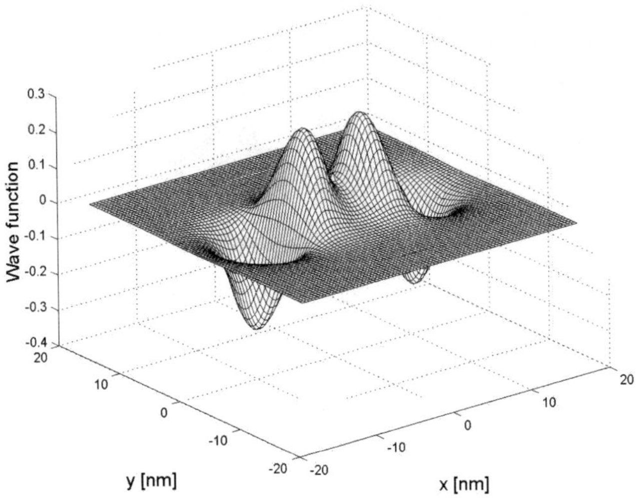

(a)

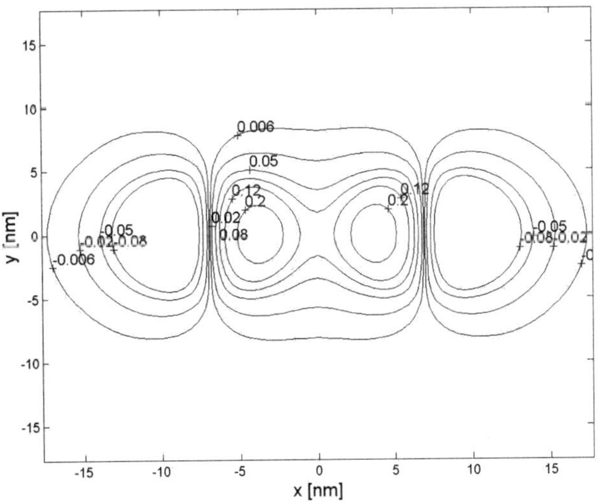

(b)

Figure 1: The normalized wave function and contour plot for the first state in SS mode (table 2) for the case of two identical cylindrical quantum wires (radius of wires = 5 nm, distance between wires = 4 nm, potential barrier 320 meV, bound energy 75.259 meV).

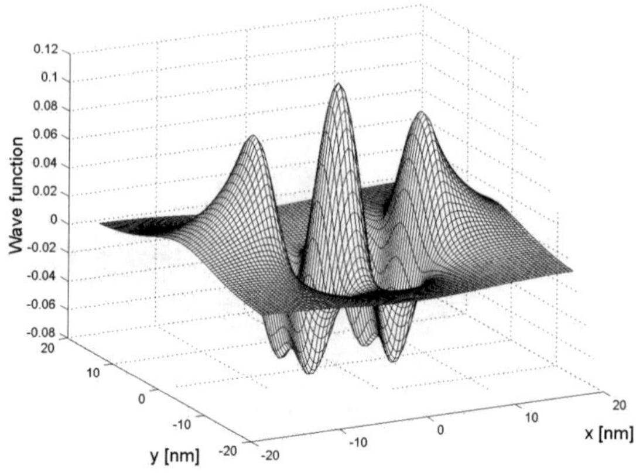

(a)

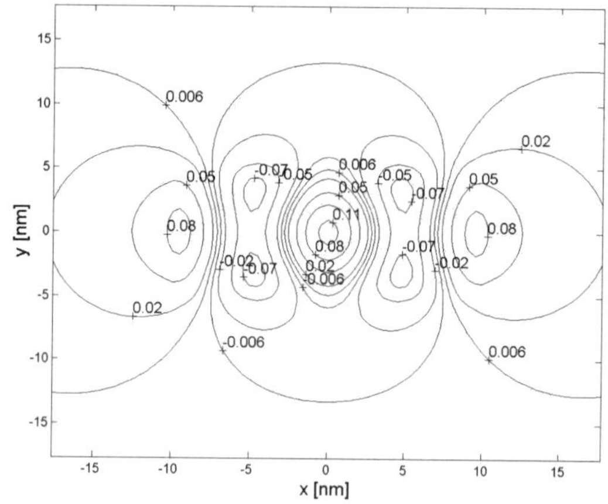

(b)

Figure 2: The normalized wave function and contour plot for the fourth state in SS mode (table 2) for the case of one rectangular quantum wire (dimension 20 x 10 nm, potential barrier 276 meV, bound energy 272.801 meV)

In the case of the two identical wires two different distances between the wires were considered. In all three cases the domain decomposition technique was used, the shapes and sizes of the subdomains were adjusted according to the physical sub-domains of the considered structure and potential and effective mass were considered to be constant in each sub-domain.

In the case of symmetric potential the problem domain was reduced only to the first quadrant of the rectangular coordinate system leading to high reduction in the mesh used.

The developed numerical approach could be used for more complex structures such as an arbitrary number of quantum wires arranged in lattice, a case of potential or effective mass varying in space or energy-dependent effective mass.

References

[1] C. Weisbuch, B. Vinter, Quantum Semiconductor structures: Fundamentals and Applications Academic Press, San Diego, 1991.

[2] P. Havu, M.J. Puska, R.M. Nieminen, Electron transport through quantum wires and point contacts, *Physical Review B* **70**, 233308 (1-4), 2004.

[3] T. Morimoto, Y. Iwase, N. Aoki, T. Sasaki, Y. Ochiai, Nonlocal resonant interaction between coupled quantum wires, *Applied Physics Letters* **82 (22)**, 3952-3954, 2003.

[4] R. Bhat, E. Kapon, S. Simhony, E. Colas, D.M. Hwang, N.G. Stoffel, M.A. Koza, Quantum wire lasers by OMCVD growth on nonplanar substrates, *Journal of Crystal Growth* **107**, 716-723 1991.

[5] J.L. Merz, P.M. Petroff, Making quantum wires and boxes for optoelectronics devices, *Materials Science and Engineering* **B9**, 275-284, 1991.

[6] T. Takebe, T. Watanabe, K. Fujita, Fabrication of quantum wires and dots on GaAs (111)A patterned substrates by molecular beam epitaxy, *Superlattices and Microstructures* **24 (1)**,1-6 1998.

[7] Ch. Heyn, C. Klein, S. Kramp, S. Beyer, S. Grünther, D. Heitmann, W. Hansen, Fabrication of quantum wires by in-situ ion etching and MBE overgrowth, *Journal of Crystal Growth* **227-228,** 980-984, 2000.

[8] E. Giovine, E. Cianci, V. Foglietti, A. Notargiacomo, F. Evangelisti, Nanofabrication of quantum wires on (100) Si and SiGe by shifted-resist pattern and anisotropic wet etching, *Microelectronic Engineering* **53**, 217-219, 2000.

[9] P. Harrison, Quantum Wells, Wires and Dots (Wiley, New York, 2000).

[10] D.K. Ferry, S.M. Goognick, Transport in Nanostructures (Cambridge University Press, Cambridge, 1997.

[11] E.P. Pokatilov, V.A. Fonoberov, S.N. Balaban, V.M. Fomin, Electron states in rectangular quantum well wires (single wires, finite and infinite lattices), *J. Phys.: Condens. Matter* **12 (42)**, 9035-9052, 2000.

[12] V. Popov, H. Power, DRM-MD approach for the numerical solution of gas flow in porous media, with applications to landfill, *Eng Anal Bound Elem*. **23**, 175-188, 1999.

[13] V. Popov, H. Power. The DRM integral equation method: an efficient approach for the numerical solution of domain dominant problems, *International Journal for Numerical Methods in Engineering*, **44**, 327-353, 1999.

[14] F. Gelbard, K.J. Malloy, Modeling Quantum Structures with the Boundary Element Method, *Journal of Computational Physics,* **172**, 19-39, 2001.

www.witpress.com, ISSN 1743-355X (on-line)

[15] D. Indjin, G. Todorovic, V. Milanovic, Z. Ikonic, On numerical solution of the Schrödinger equation: the shooting method revised, *Computer Physics Communication*, **90**, 87-94, 1995

[16] G. Bastarad, Wave mechanics applied to semiconductor heterostructure, Les editions de physique, Les Ulis Cedex, 1988.

[17] C.A. Berbbia, J.C. Tells, L.C. Wrobel, Boundary element techniques, Springer, Berlin, Heidelberg, New York, Tokyo, 1984.

[18] P.W. Patridge, C.A. Brebbia, L. Wrobel, The dual reciprocity boundary element method Computational Mechanics Publication, Southampton and Elsevier Applied Science, London & New York, 1992.

[19] M.D. Buhmann, Radial basis functions: theory and implementations, University Press, Cambridge, 2003.

WIT Transactions on Modelling and Simulation, Vol 44, © 2007 WIT Press
www.witpress.com, ISSN 1743-355X (on-line)

Section 4
Computational issues

Comparison of radial basis functions in evaluating the Asian option

F. Zhai, K. Shen & E. Kita
Graduate School of Information Sciences, Nagoya University, Japan

Abstract

Some researchers have presented the application of radial basis function approximation to the evaluation of option contracts.

In a previous study, the authors described the evaluation of Asian options by using radial basis function approximation. The numerical results indicated that the computational accuracy depended on the radial basis function and the reciprocal multi-quadric function was better than the multi-quadric one.

So, in this study, some radial basis functions are applied to the evaluation of the Asian option of one asset. We compare the multi-quadric, the reciprocal multi-quadric, and Gaussian functions. The results show that the reciprocal multi-quadric function and Gaussian function give better numerical results and the reciprocal multi-quadric function is better than the others.

1 Introduction

Recently, the financial derivatives are dealt widely and the importance is expanded. The importance of the derivative transaction is increasing for the adequate sharing of the financial risk. The option transaction is one of the most important financial derivatives and therefore, several schemes have been presented by many researchers for their pricing [1, 2].

Several financial options have been developed; European option, American option, Look-Back option, Exotic option and so on. In the previous study [3], the authors described the formulation for evaluation of the Asian option. The results showed that the computational accuracy of the present algorithm depended on the radial basis function to be employed for the evaluation. In this study, three kinds of the radial basis functions are applied for evaluation of Asian option and then, the computational accuracy and cost are compared.

WIT Transactions on Modelling and Simulation, Vol 44, © 2007 WIT Press
www.witpress.com, ISSN 1743-355X (on-line)
doi:10.2495/BE070131

The remaining of the paper is organized as follows. The algorithm for the valuation of the Asian option is described in section 2. The numerical examples are shown in section 4. Finally, the obtained results are summarized in section 5.

2 Formulation

2.1 Governing equation and strike condition

First, we will define the time-average value of the asset price S as the function:

$$I = \int_0^t S(\tau) d\tau. \tag{1}$$

In the European-type average strike option, the payoff depends on the difference between the time-average value and the asset price on the expiration date. The governing differential equation of the option is given as:

$$\frac{\partial V}{\partial t} + S\frac{\partial V}{\partial I} + \frac{1}{2}\sigma^2 S^2 \frac{\partial^2 V}{\partial S^2} + rS\frac{\partial V}{\partial S} - rV = 0 \tag{2}$$

If the function R is defined from the asset price S as

$$\begin{aligned} R &= \frac{1}{S}\int_0^t S(\tau) d\tau \\ &= \frac{I}{S}, \end{aligned} \tag{3}$$

the price V is given as

$$V(S, R, t) = SH(R, t). \tag{4}$$

Substituting equations (3) and (4) to equation (2), we have

$$\frac{\partial H}{\partial t} + FH = 0, \tag{5}$$

where the operator F is defined as

$$F = \frac{1}{2}\sigma^2 R^2 \frac{\partial^2}{\partial R^2} + (1 - r)R\frac{\partial}{\partial R}. \tag{6}$$

The payoff condition of the average strike option on the expiration date $t = T$ is defined as follows, in the case of European-call type,

$$\max\left(S - \frac{1}{T}\int_0^t S(\tau)d\tau, 0\right) \tag{7}$$

and, in the case of European-put type,

$$\max\left(\frac{1}{T}\int_0^t S(\tau)d\tau - S, 0\right). \tag{8}$$

where $\max(a_1, a_2)$ means the bigger one among a_1 and a_2.

Now, we consider the pricing of the average strike option in the call-type.

Substituting equations (3) and (4) to (7), we have the payoff condition on the expiration date $t = T$;

$$SH(R,T) = S \max\left(1 - \frac{R}{T}, 0\right),$$

and therefore,

$$H(R,T) = \max\left(1 - \frac{R}{T}, 0\right). \tag{9}$$

Finally, the governing equation and the boundary condition of the average strike option are given by equation (5) and (9), respectively.

2.2 Solution using RBF

Discretizing the equation (5) with the Crank-Nicolson Scheme, we have

$$\frac{H(t+\Delta t) - H(t)}{\Delta t} + (1-\theta)FH(t+\Delta t) + \theta FH(t) = 0 \tag{10}$$

where the parameter θ is taken in the range of $0 \leq \theta \leq 1$.

Defining the parameters $H(t) = H^m$ and $H(t + \Delta t) = H^{m+1}$, we have

$$AH^{m+1} = BH^m \tag{11}$$

where

$$A = 1 + (1-\theta)\Delta t F$$
$$B = 1 - \theta \Delta t F.$$

The price H governed with the equation (5) is approximated with the RBF function as

$$H = \sum_{n=1}^{N} \lambda_n \phi_n \tag{12}$$

where N and λ_j denote the total number of data points and the unknown parameters, respectively.

Substituting equation (12) to equation (11), we have

$$A \sum_{n=1}^{N} \lambda_n^{m+1} \phi_n = B \sum_{n=1}^{N} \lambda_n^{m} \phi_n$$

$$\sum_{n=1}^{N} A\phi_n \lambda_n^{m+1} = \sum_{n=1}^{N} B\phi_n \lambda_n^{m} \tag{13}$$

www.witpress.com, ISSN 1743-355X (on-line)

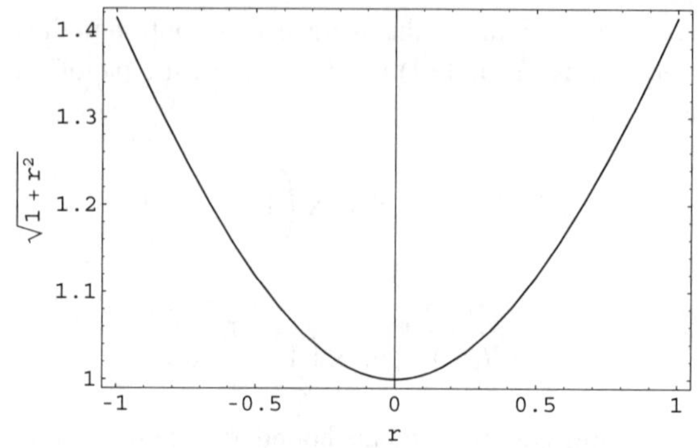

Figure 1: Multi-quadric RBF.

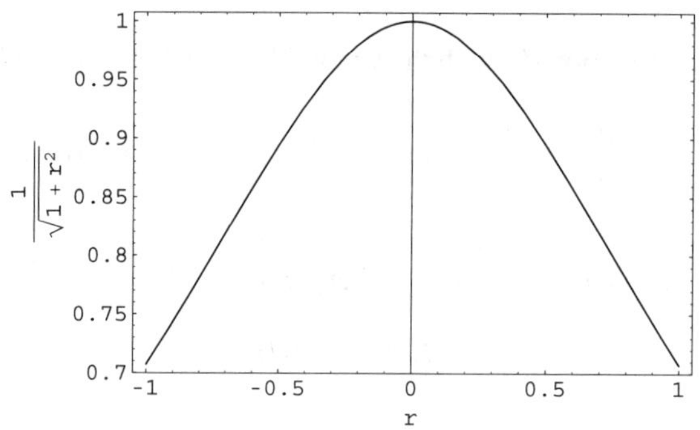

Figure 2: Reciprocal multi-quadric RBF.

3 Radial basis functions

In this study, the following radial basis functions are employed for the valuation of the options.

Multi-quadric RBF: The function is defined as follows. The distribution of $c = 1$ is shown in Fig. 1.

$$\phi(R, R_j) = \sqrt{c^2 + \|R - R_j\|^2} \tag{14}$$

Reciprocal multi-quadric RBF: The function is defined as follows. The distribution of $c = 1$ is shown in Fig. 2.

$$\phi(R, R_j) = \frac{1}{\sqrt{c^2 + \|R - R_j\|^2}} \tag{15}$$

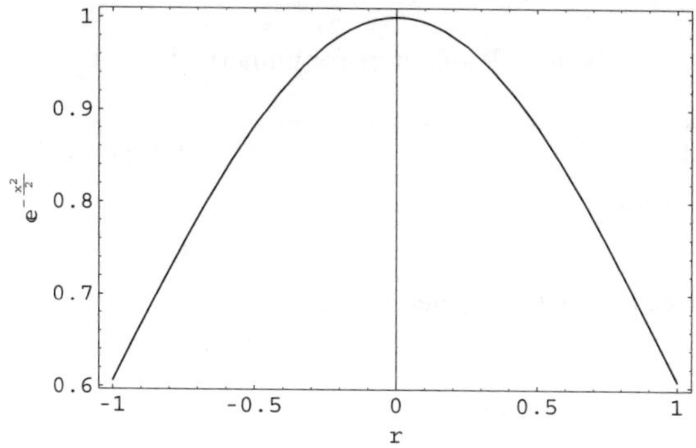

Figure 3: Gaussian RBF.

Gaussian RBF: The function is defined as follows. The distribution of $c - 1$ is shown in Fig. 3.

$$\phi(R, R_j) = \exp\left(-\frac{\|R - R_j\|^2}{2c^2}\right) \tag{16}$$

3.1 Algorithm

The algorithm of the solution procedure is defined as

1. Distribute N data points on $0 \leq R < R_{max}$ and discretize $0 \leq t \leq T$ with T/M.
2. Solve equation (12) to evaluate H^T on the expiration date $t = T$.
3. Approximate H^T by equation (12) to evaluate λ_n^T on the expiration date $t = T$.
4. $t \leftarrow T - \Delta t$.
5. Solve equation (13) to estimate λ_n^t.
6. $t \leftarrow t - \Delta t$.
7. IF $t \neq 0$, go to step 5.
8. Evaluate H^0 from equation (12) and λ_n^0 on the date $t = 0$.

4 Numerical example

The simulation parameters are defined as shown in Table 1. The total number of the data points are 101. They are distributed uniformly in the range of $0 \leq R \leq 1.0$. For comparison with the finite difference solutions, the time-step size is taken as $\Delta t = 0.0005$; the number of the time-step is $M = 1000$. The computations are performed with a personal computer of Intel Core Solo 1.06 GHz and 1 Gb-main memory and Mathematica version 5.2.

First, CPU times are compared in Table 2. The RMQ-RBF computation is fastest among them and followed with, in turn, MQ-RBF, FDM and Gaussian. Especially, the CPU times of MQ and RMQ are smaller than the FDM.

Table 1: Parameters for numerical result.

Expiration date	$T = 0.5$ [year]
Risk free interest rate	$r = 0.1$
Volatility	$\sigma = 0.4$
Crank-Nicholson θ-weighted method	$\theta = 0.5$
Maximum R	$R_{max} = 1.0$
Number of timestep(timestep size)	$M = 1000$ ($\Delta t = 0.0005$)
Number of stock data points	$N = 101$

Table 2: CPU Time.

Function	CPU Time (s)
Multi-quadric	13.453
Reciprocal Multi-quadric	11.046
Gaussian	1158.26
FDM	14.593

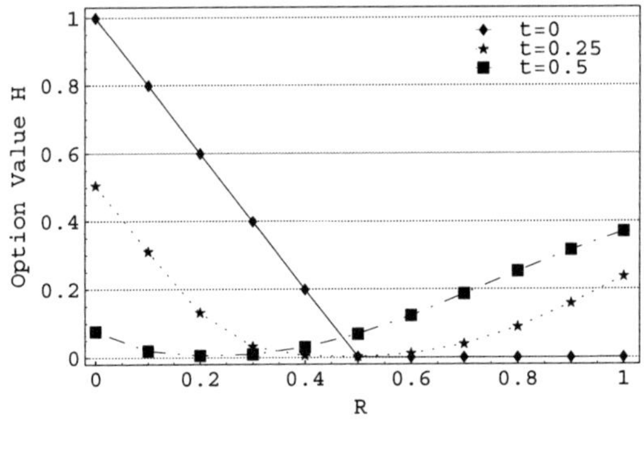

Figure 4: Multiquadrics RBF, $c = 0.04$.

Next, the computational accuracy is compared. The parameter c of the MQ-RBF (14) is taken as $c = 0.04$, which was determined from the condition number of the coefficient matrix [3]. Figure 4 shows the option value H at $t = 0$C$t = \frac{T}{2} = 0.25$C$t = T = 0.5$. The abscissa and the ordinate denote R and H, respectively. The results by finite difference method are shown in Fig. 5. We notice that the MQ-RBF solution error increases for larger value of R.

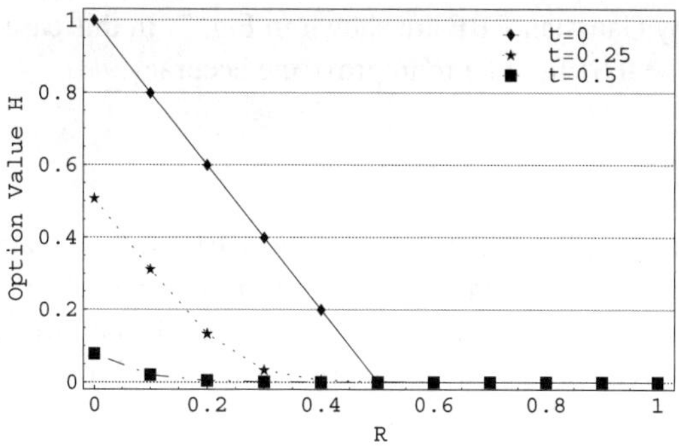

Figure 5: FDM.

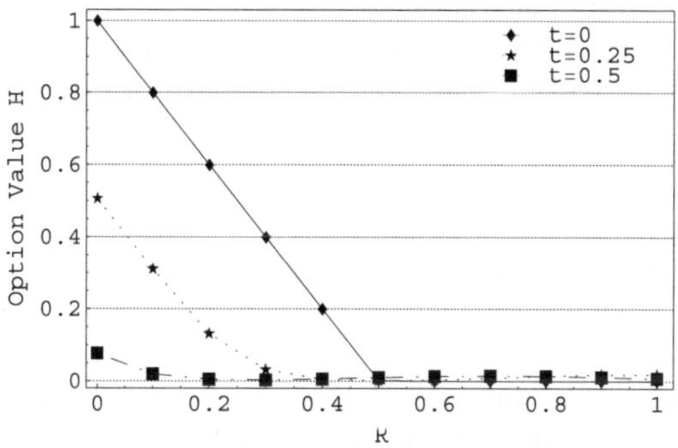

Figure 6: Reciprocal Multiquadric RBF, $c = 0.04$.

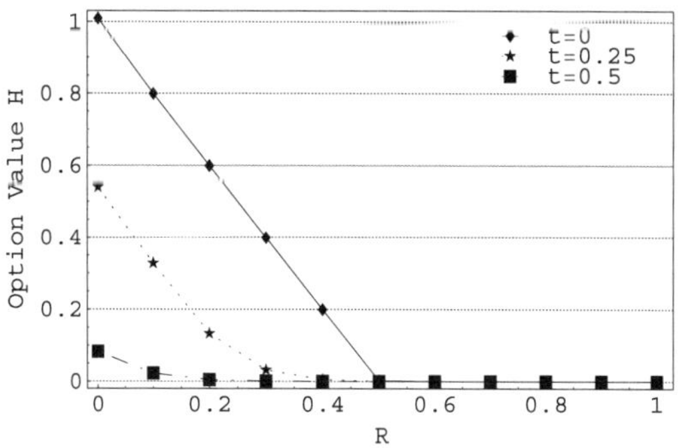

Figure 7: Gaussian RBF, $c = 0.01$.

Next, the results by RMQ-RBF are discussed. The RMQ-RBF solutions are shown in Fig. 6. When comparing 6 with Fig. 4, we notice that the RMQ-RBF solution is improved for larger value of R.

The results by Gaussian RBF are shown in Fig. 7. In this case, the parameter c is changed to $c = 0.01$ in order to improve the accuracy.

5 Conclusions

This paper described the comparison of the RBFs for evaluation of the Asian option. The formulation for evaluating the option was described and the numerical results were shown. Multi-quadric (MQ), reciprocal multi-quadric (RMQ), and Gaussian RBFs were compared in numerical examples. From the viewpoint of the CPU time, the RMQ-RBF computation was faster than the others and followed with, in turn, MQ-RBF, FDM and Gaussian. Especially, the CPU times of MQ and RMQ are smaller than the FDM. When the RBF solutions is compared with FDM one, the solutions by RMQ and Gaussian RBFs agree well with the FDM one. However, the MQ solution has relatively large error. Finally, the authors can conclude that the RMQ-RBF is more adequate for valuation of Asian option from the both sides of the CPU time and the computational accuracy.

References

[1] G. Courtadon. A more accurate finite difference approximation for valuation of options. *Journal of Financial and Quantitative Analysis*, Vol. 17, pp. 697–703, 1982.

[2] P. Wilmott, J. Dewynne, and S. Howison. *Option Pricing: Mathematical Models and Computation*. Oxford Financial Press, 1993.

[3] E. Kita, Y. Goto, F. Zhai, and K. Shen. Evaluation of Asian option by using rbf approximation. In C. A. Brebbia and J. T. Katsikadelis, editors, *Boundary Elements XXVIII* (*Skiathos, Greece, 2006*), pp. 33–40, 2006.

Inmost singularities and appropriate quadrature rules in the boundary element method

E. E. Theotokoglou & G. Tsamasphyros
Faculty of Applied Sciences, Department of Mechanics,
Laboratory of Strength of Materials,
The National Technical University of Athens, Greece

Abstract

The solution of singular integral equations (SIEs), taking into consideration the particular behaviour of its regular kernel and its right hand side function, is investigated in this paper. In particular, the problems appearing in the solution of singular integral equations in the boundary element method are verified. It is shown that the behaviour of singular integral equations does not depend only on the behaviour of the regular kernel but on the behaviour of the unknown function.
Keywords: singular integral equation, singularities, nearby poles, elasticity, quadrature formula, numerical integration, boundary element method.

1 Introduction

Many problems in the boundary element method can be reduced to a singular integral equation or to systems of singular integral equations [1]. For the solution of these equations where the known functions are Hölder continuous, approximate solutions through numerical techniques preserving the correct nature of singularities of the unknown function, have been developed [1–10].

Problems in the numerical solution of singular integral equations appear because of the Cauchy singularities, either from the unknown function or from of the poles of the known functions. In addition the regular Kernels of the singular integral equation may have either nearby poles, or branch points, or jump discontinuities.

WIT Transactions on Modelling and Simulation, Vol 44, © 2007 WIT Press
www.witpress.com, ISSN 1743-355X (on-line)
doi:10.2495/BE070141

These problems are created in the case of bodies with very narrow boundaries, in the case of the interaction of boundaries (e.g. for a crack approaching a boundary or in the case of parallel cracks when the distance between them becomes very small), and in the case of loading discontinuities. Especially, problems in the solution may appear in one of the following cases:

(i) Singular loadings which creates singularities at the unknown function.
(ii) Collinear or parallel cracks, where the distance between them is very small. There is a singular behaviour because both of them and also a nearby singularity because of the distance. In the case that one of the cracks is eliminated, its influence creates nearby singularities in the regular kernel of the integral equations and it is also influenced the unknown function.
(iii) In the case of a very narrow body. The integral equation created in one of the boundaries, is influenced by the collocation points lying on the second one and vice versa. Thus, a nearby singularity is created inside the kernel of the second one. In the case that one of the boundaries is eliminated, the unknown function is influenced and also a nearby singularity is created at the regular kernel.

In order to confront the above problems modified quadrature rules have to be used. The purpose of this paper is to show and to interpret the above problems, that appear in the solution of singular integral equation in the BEM, and to propose a quadrature rule that may solve successfully the singular integral equations.

2 The influence of poles of the regular kernel

The behavior in the solution of singular integral equations in terms of their regular kernel may appear in the solution of systems of singular integral equations when an integral equation is eliminated and the problem is reduced to a system with modified kernels and/or in the case when there is interaction of boundaries and a complex pole appears in the regular kernel.

A simple example of the above case is the antiplane shear crack under constant loading equal to 1 and approaching perpendicularly either the interface of a bimaterial plane, or the straight boundary of an elastic half-plane. If ε and a are the distances of the transverse-crack tips from the longitudinal straight boundary, the problem is governed by a system of SIEs, one along the crack and the other along the interface. Eliminating the integral along the infinite interface the problem is reduced to the following Cauchy-type singular integral equation:

$$\frac{1}{\pi}\int_{\varepsilon}^{\alpha}(t-\varepsilon)^{-1/2}(\alpha-t)^{-1/2}\left[\frac{1}{t-x}+H_1(x,t)\right]q(t)dt=1, \qquad (1)$$
$$H_1(x,t)=\lambda/(t+x); \quad \varepsilon<x<\alpha, \quad \lambda=1,$$

together with the condition:

$$\int_{\varepsilon}^{\alpha}(t-\varepsilon)^{-1/2}(\alpha-t)^{-1/2}q(t)dt=0. \tag{2}$$

It is observed the influence of the boundary in the kernel, and a **nearby** singularity appeared in (1) and (2). The dimensionless stress intensity factors at the crack tips $E(t=\varepsilon)$ and $A(t=a)$, are given by

$$K_E=-q(\varepsilon)\left(\frac{a-\varepsilon}{2}\right)^{-1/2},\quad K_A=-q(a)\left(\frac{a-\varepsilon}{2}\right)^{-1/2}. \tag{3}$$

The stress intensity factors may also be obtained [3] from the following closed-form expressions:

$$K_E=\frac{1}{a-\varepsilon}\left(\frac{2}{a+\varepsilon}\right)^{1/2}\left[a^2\frac{E(k)}{K(k)}-\varepsilon^2\right]\varepsilon^{-1/2} \tag{4}$$

$$K_A=\frac{1}{a-\varepsilon}\left(\frac{2a^3}{a+\varepsilon}\right)^{1/2}\left[1-\frac{E(k)}{K(k)}\right],\qquad k=\left(1-\frac{\varepsilon^2}{a^2}\right)^{1/2}, \tag{5}$$

where $E(k)$ and $K(k)$ elliptic integrals.

As the distance ε tends to zero, K_E tends to infinity like $\varepsilon^{-1/2}/\ell og\varepsilon$, while K_A tends to the value of the stress intensity factor of an equivalent edge crack. For the same problem the asymptotic expressions for the stress intensity factors at $E(t=\varepsilon)$ and $A(t=a=1)$, are given by [4]:

$$K_E=(1-\lambda)^{1/2}\left[\varepsilon^{1/2-s}(1-s)2^{2s-1/2}-\varepsilon^{1/2+s}(1+s)2^{-2s-1/2}+O\left(\varepsilon^{2.5-3s}\right)\right],$$
$$K_A=\sqrt{2}(1-\lambda)^{1/2}\left[8-(1-s)\varepsilon^{2-2s}2^{4s-3}+O\left(\varepsilon^{4-4s}\right)\right], \tag{6}$$

where the quantity s is given in terms of the shear modulus μ_1 and μ_2 by:

$$\cos\pi s=\lambda,\quad 0<s<1. \tag{7}$$

Another interesting problem is that of a crack approaching the interface of a bimaterial plane under plane stress or plane strain conditions. Eliminating the singular integral equation along the interface, it is obtained [5]

$$\frac{1}{\pi}\int_{\varepsilon}^{a}(t-\varepsilon)^{-1/2}(a-t)^{-1/2}\left\{\frac{1}{t-x}+H_2(x,t)\right\}q(t)dt=1, \tag{8}$$

with

$$H_2(x,t)=\frac{1}{\alpha}\left\{\frac{\gamma+\beta}{t+x}-3\beta\frac{x}{(t+x)^2}+2\beta\frac{x^2}{(t+x)^3}\right\}, \tag{9}$$

where

$$\begin{aligned}\alpha&=(m+\kappa_2)(1+m\kappa_1), \quad \beta=-4(m+\kappa_2)(1-m)\\ \gamma&=(1-m)(m+\kappa_2)+\alpha+\beta/4, \quad m=\mu_2/\mu_1,\end{aligned} \tag{10}$$

$$\kappa_i=\begin{cases}(3-\nu_i)/(1+\nu_i) & \text{for plane stress}\\ (3-4\nu_i) & \text{for plane strain}\end{cases} \tag{11}$$

In relations (8) and (9), $\kappa_i \ (i=1,2)$ are Muskhelishvili's elastic constants, expressed in terms of Poisson's ratio $\nu_i \,(i=1,2)$ of the corresponding half plane. It is also observed the influence of the eliminating boundary and a *nearby* singularity is appeared in (8). The asymptotic expression of the stress intensity factor $K_E\,(t=\varepsilon)$ obtained by Atkinson [4], is given by:

$$K_E \sim 2^{1/2}\varepsilon^{1/2}\sum_{k=1}^{N}A_k\varepsilon^{-s_k}, \quad s_k<1, \tag{12}$$

where A_k are constants independent of ε, and s_k are the first N-real roots of the equation:

$$2\alpha\cos\pi s+\beta(s-1)^2+\gamma=0. \tag{13}$$

It must be noted that the above equation gives also the eigenvalues of an infinite crack going perpendicularly to an interface as before. In the case where the material 2 is the air $(\mu_2=0)$, the asymptotic expressions of K_E and K_A take the form:

$$K_E \sim \widetilde{K}_A\varepsilon^{-1/2}/\ell og\varepsilon, \quad K_A=\widetilde{K}_A(1+1/\ell og\varepsilon), \tag{14}$$

where $\tilde{K}_A$ being the stress intensity factor of an edge crack in a half plane subjected to the same loading.

As a particular example let us consider the case of two collinear cracks $(-a,-\varepsilon)$ and $(0,b)$, i.e. the distance between the nearest tips of the collinear cracks is ε. If we represent by different symbols, ω_1 and ω_2 the quantity $(\Phi^+ - \Phi^-)$, (i.e. the distributions of dislocations) in either crack, it is obtained the following system of singular integral equations:

$$\begin{aligned} &\frac{1}{\pi}\int_{-a}^{-\varepsilon}\frac{\omega_1(x)}{x-x_1}dx+\frac{1}{\pi}\int_{0}^{b}\frac{\omega_2(x)}{x-x_1}dx=p_1, \quad -a<x_1<-\varepsilon, \\ &\frac{1}{\pi}\int_{-a}^{-\varepsilon}\frac{\omega_1(x)}{x-x_2}dx+\frac{1}{\pi}\int_{0}^{b}\frac{\omega_2(x)}{x-x_2}dx=p_2, \quad 0<x_2<b, \end{aligned} \tag{15}$$

where for the sake of simplicity we have assumed that uniformly distributed loads $p_i(i=1,2)$ are applied to either crack. Furthermore, the following equations, ensuring the single-valuedness of ω_1 and ω_2, must be taken into account:

$$\int_{-a}^{-\varepsilon}\omega_1(x)dx=0, \quad \int_{0}^{b}\omega_2(x)dx=0. \tag{16}$$

Solving the first integral of the first of equations (15) with respect to ω_1 and substituting it into the second of the second of equations (15), we obtain, after some algebra, the following singular integral equation:

$$\begin{aligned} &\frac{1}{2\pi}\int_{0}^{b}\frac{\omega_2(x)}{x-x_2}dx+\frac{1}{2\pi}\int_{0}^{b}\left[\frac{(x+a)(x+\varepsilon)}{(x_2+a)(x_2+\varepsilon)}-1\right]^{1/2}\frac{\omega_2(x)}{x-x_2}dx \\ &=p_2-p_1\left\{\frac{2x_2+a+\varepsilon}{\sqrt{(x_2+a)(x_2+\varepsilon)}}-1\right\}, \quad 0\le x_2<b. \end{aligned} \tag{17}$$

This integral equation should be supplemented by the second of equations (16). It is worthwhile noting that the kernel of the second integral in equation (17) is a regular one, possessing a nearby pole at $x_2=-\varepsilon$. Consequently, the equation may be solved by the conventional method, i.e. by taking $\omega_2(x)$ in the following form:

$$\omega_2(x)=x^{-1/2}(x-b)^{-1/2}q(x), \quad 0\le x\le b, \tag{18}$$

where $q(x)$ is a regular function. Such a procedure may give reliable results if ε is sufficiently large, but if ε tends to zero the method fails to yield reliable results. This is due to the fact that the functions $\omega_1(x)$ and $\omega_2(x)$ are equal to the difference of limiting values $(\Phi^+ - \Phi^-)$ of Muskhelishvili's complex potential $\Phi(z)$, which for this particular problem is of the form:

$$\begin{aligned}
&\Phi(z)=\frac{1}{2\pi i}X(z)\left[p_1\int_{-a}^{-\varepsilon}\frac{X(t)}{t-z}dt+p_2\int_0^b\frac{X(t)}{t-z}dt+P_2(z)\right]\\
&X(z)=(z+a)^{-1/2}(z+\varepsilon)^{-1/2}z^{-1/2}(z-b)^{-1/2}, \qquad (19)\\
&P_2(z)=c_1z^2+c_2(z)+c_3,
\end{aligned}$$

where the constants c_1, c_2, c_3 could be determined by the conditions of single-valuedness (16). Consequently, the function $q(x)$ must be written, as follows:

$$q(x)=(x+\varepsilon)^{-1/2}(x+a)^{-1/2}q_1(x), \quad 0\le x\le b, \qquad (20)$$

with $q_1(x)$, a fully regular function.

It is obvious from the last relation, that $q(x)$, for small values of ε, takes significant values in the vicinity of $x=0$. Especially, the stress intensity factor $K(0)$ corresponding to the tip with $x=0$ which is proportional to $q(0)$, has a factor $\varepsilon^{-1/2}$.

3 Applications

As application consider the singular integral equation

$$⨍_{-1}^{1}\frac{g(t)}{x-t}dt+\int_{-1}^{1}\frac{g(t)}{(t+ic)}dt=\frac{\pi}{\sqrt{c}\sqrt{1+c}}; \quad g(t)=(1-t^2)^{-1/2}q(t) \qquad (21)$$

with

$$\int_{-1}^{1}g(t)dt=\pi \qquad (22)$$

The Gauss-Chebyshev integration rule of the first kind [6] does not converge for $c<<1$. In order to have a good convergence, the proposed interpolatory formulas [10] in the case of Gauss-Chebyshev of the second kind, is applied to the regular term

$$\int_{-1}^{1}\frac{g(t)}{t^2+c}dt \cong \sum_{k=1}^{n}\frac{q(t_k)}{t_k^2+c}\pi\left\{\frac{1}{n}+\frac{1}{T_n'(t_k)}\frac{\left(\sqrt{c}-\sqrt{c+1}\right)^n}{\sqrt{c+1}}\right.$$

$$\left.\times\left(\frac{t_k}{\sqrt{c}}\sin\frac{(n-1)\pi}{2}+\cos\frac{(n-1)\pi}{2}\right)\right\};\quad g(t)=\left(1-t^2\right)^{-1/2}q(t) \tag{23}$$

Substituting (23) into (21), we have finally

$$\sum_{k=1}^{n}\frac{\pi}{n}\frac{q(t_k)}{x-t_k}+\sum_{k=1}^{n}\frac{q(t_k)}{t_k^{\,2}+c}\pi\left\{\frac{1}{n}+\frac{1}{T_n'(t_k)}\frac{\left(\sqrt{c}-\sqrt{c+1}\right)^n}{\sqrt{c+1}}\right.$$

$$\left.\times\left(\frac{t_k}{\sqrt{c}}\sin\frac{(n-1)\pi}{2}+\cos\frac{(n-1)\pi}{2}\right)\right\}=-\ell n\left|\frac{1-x}{1+x}\right|+\frac{2}{\sqrt{c}}\tan^{-1}\frac{1}{\sqrt{c}};\tag{24}$$

$$t_k=\cos\frac{2k-1}{2n}\pi,\quad \forall x:U_{n-1}(x)=0,$$

where $T_n(t)$ and $U_{n-1}(t)$ are the Chebyshev polynomials of the first and second kind [6], respectively.

Taking into consideration (24) and (22) a linear system results, whose solution gives the exact value $q(t)(\equiv 1)$ with machine accuracy for $n\geq 2$.

4 Conclusions

From the previous study it is deduced that, in many problems frequently encountered in the Boundary Element Method, the solution g of the singular integral equation is influenced by other (and perhaps weaker) singularities than the already known singularities, existing at the ends of the integration interval.

A difficult problem that may appear in the solution of singular integral equations arises from the singularities that the regular kernel may possess. The poles of the regular kernel are due to the interaction of the boundaries of a body with a very small distance between them. The above problem continues to exist when a boundary is eliminated (see relation (9)).

Problems where the singular integral equations have regular kernels with poles very close to the integration interval appear in the case of a crack parallel and near to a boundary, in the case of the antiplane shear crack, etc. (Section 2).

In the case that the regular kernel has poles very close to the integration interval $(c \ll 1)$, the classical Gauss integration rule is impossible to approximate the correct result for a few integration points (Section 3). It is observed that increasing the number of integration points the results for the error of the regular kernel do not improve. This is due to the resulted functional equation which does not represent with a "tolerant" precision the singular integral equation. The convergence is succeeded if only the modified weight quadrature rule which introduces modified weights, is used [10]. The proposed quadrature may be applied for any degree of polynomial because it has been proved [10] that it converges uniformly to the exact value of the integral with the nearby singularity.

References

[1] Brebia, C.A., Telles, J.C.F. and Wrobel, L.S., *Boundary Element Techniques*, Springer-Verlag: Berlin and New York, 1984.

[2] Erdogan, F. and Cook, T.S., Antiplane shear crack terminating at and going through a bimaterial interface. *International Journal of Fracture*, 10, pp. 227-240, 1974.

[3] Tada, H., Paris, P.C. and Irwin, G.R., *The Stress Analysis of Cracks Handbook*, 10.1. Del Research Corp., Hellertown, Pennsylvania, 1973.

[4] Atkinson, C., On the stress intensity factors associated with cracks interacting with an interface between two elastic media. *Int. J. Engineering Science*, 13, pp.489-504, 1975.

[5] Cook, T.S. and Erdogan, F., Stresses in bonded materials with a crack perpendicular to the interface. *Int. J. Engineering Science*, 10, pp.667-697, 1972.

[6] Stroud, A. and Secret, D., *Gaussian quadrature formulas*, Prentice-Hall, Englewood Cliffs, N.J, 1966.

[7] Tsamasphyros, G. and Theocaris, P.S., Equivalence and convergence of direct and indirect methods for the numerical solution of singular integral equations, *Computing*, 31, pp.71-80, 1985.

[8] Theotokoglou, E.E. and Tsamasphyros, G., An integral equation solution for cracked half-planes bonded together and containing debondings along their interface, *International Journal of Fracture*, 55(1), pp. 1-16, 1992.

[9] Lether, F.G., Subtracting out complex singularities in numerical integration. *Math. Comp.* 31(2), pp.223-229, 1997.

[10] Tsamasphyros, G. and Theotokoglou, E. E., A modified formula for integrals with nearby singularities, *International Journal for Numerical Methods in Engineering*, 67, pp. 1082-1093, 2006.

Parallelized iterative domain decomposition boundary element method for thermoelasticity

B. Gámez[1,4], D. Ojeda[1,4], E. Divo[2], A. Kassab[3] & M. Cerrolaza[4]
[1] *Departamento de Diseno Mecánico y Automatización, Universidad de Carabobo, Venezuela*
[2] *Department of Engineering Technology, University of Central Florida, USA*
[3] *Department of Mechanical, Materials, and Aerospace Engineering, University of Central Florida, USA*
[4] *National Institute of Bioengineering, Central University of Venezuela, Venezuela*

Abstract

The boundary element method (BEM) requires only a surface mesh to solve thermoelasticity problems, however, the resulting matrix is fully populated and non-diagonally dominant. This poses serious challenges for large-scale problems due to storage requirements and the solution of large sets of non-symmetric systems of equations. In this article, an effective and efficient domain decomposition, or artificial sub-sectioning technique, along with a region-by-region iteration algorithm particularly tailored for parallel computation to address these issues are developed. The domain decomposition approach effectively reduces the condition number of the resulting algebraic systems, while increasing efficiency of the solution process and decreasing memory requirements. The iterative process converges very efficiently while offering substantial savings in memory. The iterative domain decomposition technique is ideally suited for parallel computation. Results demonstrate the validity of the approach by providing solutions that compare closely to single-region BEM solutions and benchmark analytical solutions.
Keywords: domain decomposition, thermoelasticity, parallel computation, boundary element method.

WIT Transactions on Modelling and Simulation, Vol 44,
www.witpress.com, ISSN 1743-355X (on-line)
doi:10.2495/BE070151

1 Introduction

The boundary element method (BEM) is ideally suited to solve problems involving thermoelastic effects as the coupled problem can formally be solved entirely using a boundary discretization. This type of problem is encountered when a solid is subjected to heating conditions that give rise to a temperature distribution throughout its volume. This temperature distribution produces thermal expansions in the object under consideration. In an isotropic material, at a uniform reference temperature, T_{ref}, a small uniform increase in temperature can produce a pure volumetric expansion if the object is not constrained against such movement. This expansion can be expressed as a so-called thermal strain (e_{ij}^T), according to the equation:

$$e_{ij}^T = \delta_{ij}\alpha\Delta T = \delta_{ij}\alpha(T - T_{ref}) \tag{1}$$

where α is the thermal expansion coefficient. The expression above reveals that this thermal expansion can occur with absolutely no stresses present in the solid, Kane [1], Díaz et al [2]. In the standard BEM, the coefficient matrix is full and practical issues of storage and computation arise in large-scale modeling, particularly in 3D. Domain decomposition with explicit solution of the banded coefficient matrix and multipole methods have been used to successfully mitigate these problems. In this article, we propose an effective and efficient domain decomposition, or artificial sub-sectioning technique, along with a region-by-region iteration algorithm particularly tailored for parallel computation. The domain decomposition effectively reduces the condition numbers of the resulting algebraic systems, while increasing solution process efficiency and decreasing memory requirements. The iterative process converges very efficiently while offering substantial savings in memory and much promise for efficient solution of 3D thermoelasticity problems using the BEM and it is ideally suited for parallel computation.

2 BEM in thermoelasticity

The BEM can be utilized to resolve tractions, displacements and stresses on the boundary Γ and in the internal points of a domain Ω, Brebbia and Domínguez [3], based on a displacement boundary integral formulation for thermoelasticity. The thermoelastic problem is governed by the equilibrium equation:

$$\frac{\partial \sigma_{ij}}{\partial x_j} = -b_i \tag{2}$$

and the Hook's constitutive relation:

$$\sigma_{ij} = \frac{2\mu\nu}{(1-\nu)}\delta_{ij}e_{ii} + \mu e_{ij} \tag{3}$$

where σ_{ij} is the stress tensor, b_i is the body force vector, $e_{ij} = \frac{1}{2}\left(\frac{\partial u_i}{\partial x_j} + \frac{\partial u_j}{\partial x_i}\right)$ is the strain tensor, u_i is the displacement vector, δ_{ij} is the Kronecker delta, μ

WIT Transactions on Modelling and Simulation, Vol 44, © 2007 WIT Press
www.witpress.com, ISSN 1743-355X (on-line)

is the shear modulus, and ν is the Poisson ratio. Combining Eqs. (2) and (3) and introducing the strain tensor in terms of displacements yields the Navier's equation as:

$$\mu\frac{\partial u_i}{\partial x_j \partial x_j}+\frac{\mu}{(1-2\nu)}\frac{\partial u_j}{\partial x_i \partial x_j}=-b_i \tag{4}$$

On each part of the boundary, Γ, either the displacement $u_i=\overline{u}_i$ on boundary Γ_u or the traction $t_i=\bar{t}_i$ on boundary Γ_t, is prescribed in a well-posed problem. So that, Γ=$\Gamma_t \cup \Gamma_u$ is the boundary of the domain Ω. Using the Somigilana identity, an integral relation between the displacements u_i^p in a collocation point "p" and the displacements u_i and the tractions t_i on all boundary Γ is readily obtained with the body forces b_i appearing formally as a domain integral:

$$c_{ij}^p u_i^p+\oint_\Gamma H_{ij}u_i d\Gamma=\oint_\Gamma G_{ij}t_i d\Gamma+\int_\Omega G_{ij}b_i d\Omega \tag{5}$$

where G_{ij} and H_{ij} are fundamental solutions in terms of displacement and traction respectively, and $t_i=\sigma_{ij}n_j$ is the traction vector and n_j is the boundary outward normal vector; see Brebbia and Domínguez [3] and Cheng et al [4]. For thermoelasticity, the field stresses is:

$$\sigma_{ij}=\sigma_{ij}^e+\sigma_{ij}^T \tag{6}$$

where the first term σ_{ij}^e represents the contribution to the stress components due to the actual straining of the material, while the last term σ_{ij}^T represents the thermal expansion effect, Kane [1], which is given by:

$$\sigma_{ij}^T=-m\delta_{ij}(T-T_{ref}) \tag{7}$$

Therefore the body forces in the Navier's equation will have the form:

$$b_i=-m\frac{\partial T}{\partial x_i} \tag{8}$$

where m is the thermoelastic constant of the form:

$$m=\frac{2\mu\alpha(1+\nu)}{(1-2\nu)} \tag{9}$$

The domain integral can be expanded using Green's second identity and other transformations to finally obtain the boundary integral equation for displacements as:

$$c_{ij}^p u_i^p+\oint_\Gamma H_{ij}u_i d\Gamma=\oint_\Gamma G_{ij}t_i d\Gamma+\frac{m}{k}\left[\oint_\Gamma E_j q d\Gamma-\oint_\Gamma F_j(T-T_{ref})d\Gamma\right] \tag{10}$$

where k is the material thermal conductivity. Moreover, assuming thermal equilibrium, $\partial^2 T/\partial x_i \partial x_i=0$, the temperature a the collocation point p can also be related to the temperature and heat flux at the boundary by means of the following boundary integral equation:

$$c^p T^p + \oint_\Gamma G_T q d\Gamma = \oint_\Gamma H_T T d\Gamma \tag{11}$$

where c^p is a geometrical constant that possesses similar properties as c^p_{ij}, and G_T and H_T are the fundamental solutions for temperature and heat flux. The normal heat flux is defined as: $q = -k\left(\partial T/\partial x_i\right) n_i$. Additionally, one can obtain a BIE that relates the stresses to boundary displacements, tractions, temperatures and heat fluxes as:

$$\begin{aligned} c^p_{ij}\sigma^p_{ik} + \oint_\Gamma S_{ijk} u_i d\Gamma = & \oint_\Gamma D_{ijk} t_i d\Gamma \\ & + \frac{m}{k}\left[\oint_\Gamma A_{jk} q d\Gamma - \oint_\Gamma B_{jk}(T - T_{ref}) d\Gamma\right] \\ & - c^p_{jk} m (T - T_{ref})^p \end{aligned} \tag{12}$$

where S_{ijk} and D_{ijk} are the fundamental solutions of stresses. The coefficients E_j, F_j, A_{jk} and B_{jk} can be derived directly of the fundamental solution G_{ij}, Brebbia et al[3]. The discretized displacement BIE can be formulated as:

$$\begin{aligned} c^p_{ij} u^p_i = & \sum_{l=1}^{NE}\sum_{n=1}^{NN} t_i^{l,n} \int_{-1}^{1} G^l_{ij}(\eta) M^n(\eta) J^l(\eta) d\eta \\ & - \sum_{l=1}^{NE}\sum_{n=1}^{NN} u_i^{l,n} \int_{-1}^{1} H^l_{ij}(\eta) M^n(\eta) J^l(\eta) d\eta \\ & + \frac{m}{k}\sum_{l=1}^{NE}\sum_{n=1}^{NN} q^{l,n} \int_{-1}^{1} E^l_{j}(\eta) M^n(\eta) J^l(\eta) d\eta \\ & - \frac{m}{k}\sum_{l=1}^{NE}\sum_{n=1}^{NN} \left(T^{l,n} - T_{ref}\right) \int_{-1}^{1} F^l_{j}(\eta) M^n(\eta) J^l(\eta) d\eta \end{aligned} \tag{13}$$

where NE is the number of elements and NN is the number of degrees of freedom of the field variables in each element. For all examples presented in this paper, we utilize discontinuous quadratic elements with three ($NN = 3$) independent nodes for the field variables in each element with $M^n(\eta)$ denoting the respective quadratic shape functions. Also, η denotes the homogeneous parametrization variable(s) of the element geometry, and $J^l(\eta)$ is the Jacobian of the element $\Delta\Gamma_l$. To form an algebraic system, the point p is located at each of the NN nodes of all the elements NE. This generates independent equations of the form:

$$[H]\{u\} = [G]\{t\} + \{s\} \tag{14}$$

The matrices $[H]$ and $[G]$ have dimensions $N \times N$ where $N = d \times NE \times NN$ and d is the number of spatial dimensions (2 or 3). The vector $\{s\}$ contains

www.witpress.com, ISSN 1743-355X (on-line)

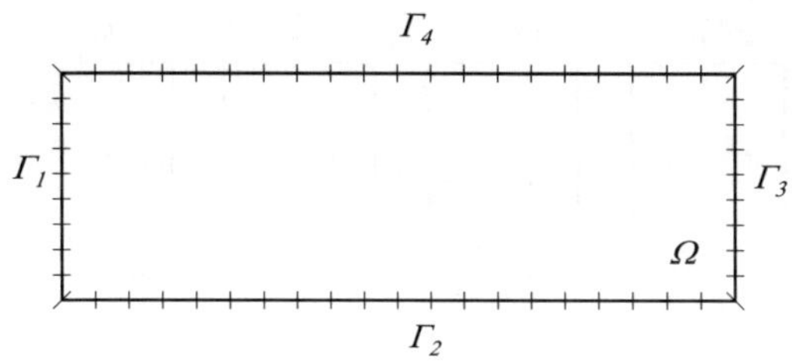

Figure 1: BEM single region discretization.

the integrated information of the thermal effects on the elastic field. Finally, the boundary conditions $\bar{t}_i$ or $\bar{u}_i$ are introduced in this algebraic system to arrive at the standard from: $[A]\{x\} = \{b\}$ where the unknown vector $\{x\}$ contains the strains and tractions that they were not specified as boundary conditions. Once the system is solved, all the field variables at the boundary are known and can be employed to determine strains and internal stresses using the appropriate boundary integral equations, Eqs. (10) and (12). It is important to emphasize that the same procedure is initially used to determine the temperature and heat fields using the temperature boundary integral equation.

3 Iterative domain decomposition (IDD)

To determine the IDD efficiency it is necessary first to indicate the requirements of memory for the resolution of the domain problem Ω in a single region, see Fig. 1, with the corresponding boundary conditions and the characteristic discretization of the boundary element method.

If a discretization of NE elements with NN independent nodes per each element is generated in a single region, the resulting algebraic system has dimensions $N \times N$, where $N = d \times NE \times NN$ and d is the number of spatial dimensions (2 or 3). So:

$$\Omega \Rightarrow [A]_{NxN}\{x\}_{Nx1} = \{b\}_{Nx1} \tag{15}$$

where the vector $\{x\}$ represents the unknown values of the tractions and displacements around the boundary. In this case the number of floating point operations required to arrive at the algebraic system above is proportional to N^2 as well as direct memory allocation also proportional to N^2. The solution of the algebraic system can be performed using a direct solution method such as LU decomposition requiring floating point operations proportional to N^3 or an indirect method such as Bi-conjugate Gradient or Generalized Minimum Residual (GMRes), which, in general, requires floating point operations proportional to N^2 to achieve convergence.

On the other hand, if a multi-region solution process is adopted instead, the original domain Ω is divided into K sub-domains $\Omega_l \therefore l = 1...K$ separated by interfaces artificially created, and each one is independently discretized, as shown

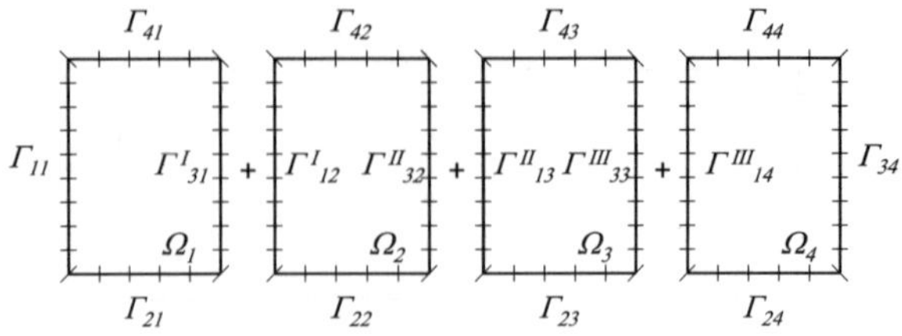

Figure 2: BEM domain decomposition and discretization.

in Fig. 2 for the case where $K = 4$. Successively, the solution in each sub-domain can be obtained independently through a standard process, as long as the boundary conditions in the artificial interfaces between the sub-domain are imposed.

For example, the first sub-domain Ω_1 in figure 2 is independently analyzed. The application of the boundary element method in this sub-domain is used to generate an algebraic system as follows:

$$[A]_{nxn}\{x\}_{nx1} = \{b\}_{nx1} \tag{16}$$

where the dimension $n = d \times ne \times NN$ is obviously a fraction of the original dimension N, Divo et al [7] provides a thorough explanation.

Naturally, the boundary conditions at the artificial interfaces between the sub-domain are originally unknown, and, therefore a scheme must be devised to guarantee the continuity and equilibrium of the field variables between the sub-domain, namely that:

$$\begin{aligned} u_i^a &= u_i^b \\ t_i^a &= -t_i^b \end{aligned} \tag{17}$$

where the superscripts a and b indicate each side of the interface at issue. In order to guarantee these conditions at each iteration including at the initial guess, a preliminary discretization of very low resolution is carried out providing a simplified model for the problem. This is solved by BEM to generate physically meaningful initial values at the interfaces. The latter are updated utilizing a refined discretization until a solution is achieved that satisfies the interfacial equilibrium and continuity conditions within a set tolerance.

The progression of the iterative process involves two stages. In the first stage, each interface is individually imposed with conditions of first kind prescribing displacements $\overline{u}_i$ and the tractions t_i are solved using the standard boundary element method in each sub-domain. These intermediate computed tractions do not agree on each side of the interfaces, thus it is necessary to alter these tractions to force them to satisfy the equilibrium conditions, and this is accomplished using the following:

$$\begin{aligned} \overline{t}_i^a &= t_i^a - \frac{t_i^a + t_i^b}{2} \\ \overline{t}_i^b &= t_i^b - \frac{t_i^a + t_i^b}{2} \end{aligned} \tag{18}$$

This guarantees that the updated tractions, $\overline{t}_i^a$ and $\overline{t}_i^b$, have the same magnitude but opposite signs satisfying the equilibrium condition. Once these tractions are updated, the second stage of the iterative procedure utilizes these tractions as the imposed interfaces conditions for each sub-domain. A new displacement field in the interfaces is obtained, and, again, the displacements do not agree on both sides of each interface. They are updated by a simple average to ensure continuous displacement, so that:

$$\begin{aligned} \overline{u}_i^a &= \frac{u_i^a + u_i^b}{2} \\ \overline{u}_i^b &= \frac{u_i^a + u_i^b}{2} \end{aligned} \tag{19}$$

The iterative norm is defined as the root mean square measure between the updated displacements and the previous ones provides is defined by:

$$L_2 = \sqrt{\frac{1}{NNI} \sum_{l=1}^{NNI} (\overline{u}_i - u_i)^2} \tag{20}$$

where NNI is the number of nodes in the interfaces. The iteration is stopped when ϵ reaches a preset value.

If a direct approach such as LU factorization is employed for all sub-domains, the LU factors of the coefficient matrices for all sub-domains can be computed only once at the first iteration and stored on disk or in RAM for later use during the iteration process. Subsequently, only a forward and a backward substitution will be required. This feature provides a significant reduction in the computational burden for the overall BEM solution.

4 Parallel implementation

The domain decomposition BEM formulation detailed above is ideally suited to parallel computing, Erhart et al [5]. The algorithm has been coded using MPICH and implemented on a multiprocessor cluster comprised of dual 64 bit Xeon 3.2 GHz nodes equipped with 6 GB RDRAM running under the Fedora core 5 operating system. A static load-balancing routine is utilized to optimally distribute the computation over the nodes. This optimization is performed using a discrete Genetic Algorithm as described in Divo and Kassab [6]. A key step in the domain decomposition is to keep each sub-domain discretization to a number of elements that allows the problem to be stored in available RAM memory to avoid disk paging, Divo et al [7].

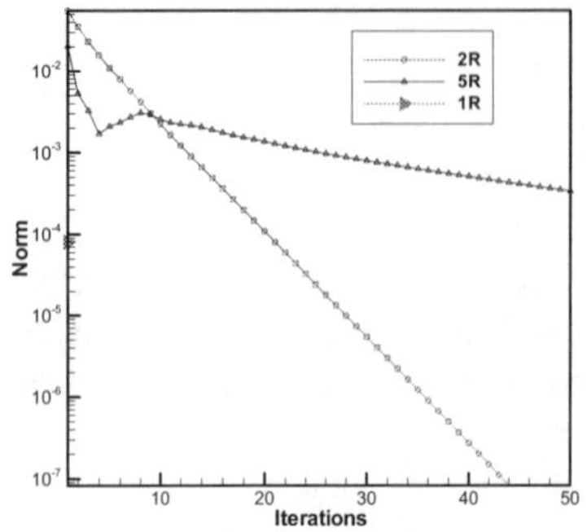

Figure 3: Elastic field convergence.

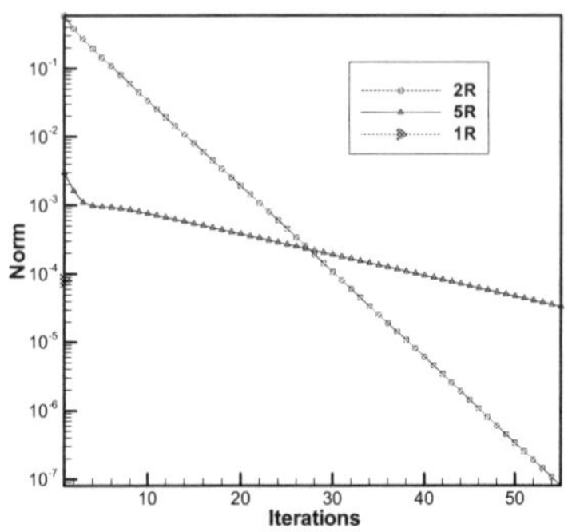

Figure 4: Thermal field convergence.

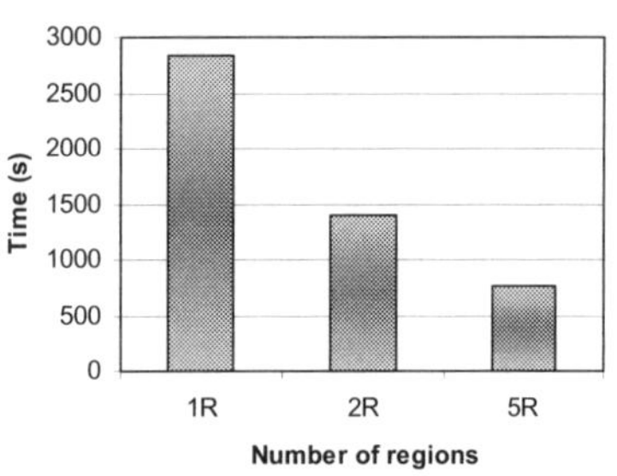

Figure 5: Time vs. number of regions.

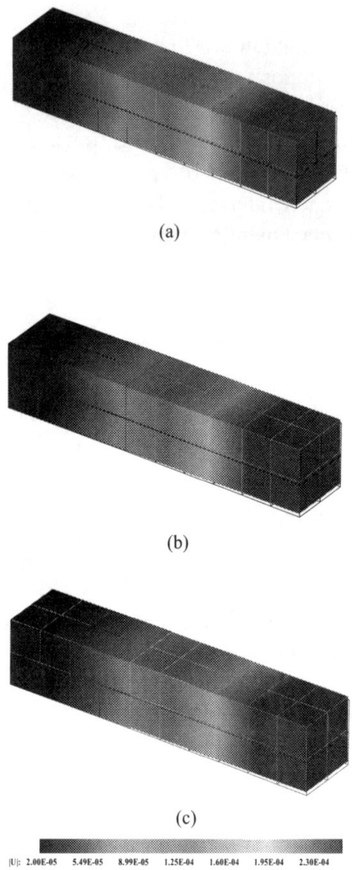

Figure 6: Contour plots of the displacement for: (a) one-region, (b) two-region, (c) five-region cases.

5 Numerical validation and examples

The example presented in this paper, consists of a square cross-section cantilever beam whose height and width are $h = w = 0.5m$ and length $L = 0.25m$. This example is provided to verify the convergence and accuracy of the domain decomposition approach in 3-D. The beam is discretized with 88 biquadratic isoparametric discontinuous elements for a total number of degrees of freedom for the single-region case of $N = 2112$ $(3\times88\times8)$. The thermal conductivity is taken as $k = 14.9W/mK$ and the thermal expansion coefficient as $\alpha = 13\cdot10^{-6}K^{-1}$. The shear modulus is set to $\mu = 60GPa$ and the Poisson ratio is set to $\nu = 0.3$. The model is solved in a single region and it is also decomposed into 2 and 5 sub-domains in the longitudinal direction. The reference temperature is $T_{ref} = 373K$. The left longitudinal surface of the beam is clamped and imposed with a temperature $T = 373K$. The rest of the surfaces are thermally insulated except for the top perimetral surface which is imposed with a heat flux $q = -1000W/m^2$. An uniform load of $P = 1MPa$ is imposed on the top perimetral surface.

Plots of the norm as a function of iterations for the elastic and thermal fields are shown in Fig. 3-4 respectively. The norm decays rapidly for both multi-region cases. The convergence criterion used was $\epsilon = 10^{-7}$ and was reached in less than 50 iterations for the two-region case and in less than 100 iterations for the five-region case. Contour plots of the displacements of the cantilever beam are shown for the one-region, two-region, and five-region cases in Fig. 6 (a)-(c).

It should be noted, in Fig. 5, that the two-region model corresponds to a computational time reduction of 50.81% when compared to a single region cantilever beam, while the five-region case provides a 73.07% reduction. It is noted that the total time to solution reported here includes load balancing, generation of H and G matrices, resolution of elastic and thermal field, and iteration to convergence.

6 Conclusions

In this article, we propose an effective and efficient BEM iterative domain decomposition (BEM-IDD) algorithm to solve large-scale 3-D BEM thermoelastic problems. In order to tackle large problems, the original domain is decomposed into a number of sub-domains. Results indicate that the proposed BEM-IDD technique is well-suited for parallel computation, converging efficiently and offering substantial savings in memory and computational time over traditional BEM formulation.

Acknowledgements

The work undertaken in this project was carried out under the institutional and financial support provided by the University of Central Florida (USA), University of Carabobo (Venezuela), and FONACIT (Venezuela).

References

[1] Kane, J., Boundary Element Analysis in Engineering Continuum Mechanics, Prentice-Hall, New Jersey 1994, pp. 123, 378.

[2] Díaz, F., Yates, J.R., and Patterson E.A., "Some improvements in the analysis of fatigue cracks using thermoelasticity," International Journal of Fatigue, Vol. 26, 2004, pp. 365-376.

[3] Brebbia, C.A. and Domínguez, J. J., Boundary Element: An Introductory Course, Computational Mechanics Publ. Boston co-publisher McGraw Hill, New York, 1989, pp. 134-250.

[4] Cheng, A.H., Chen, C.S., and Golberg, M.A., Rashed, Y.F., "BEM for thermoelasticity and elasticity with body force-a revisit," Engineering Analysis with Boundary Elements, Vol. 25, 2001, pp. 377-387.

[5] Erhart, K., Divo, E., and Kassab, A., "A parallel domain decomposition boundary element method approach for the solution of large scale transient heat conduction problems," Engineering Analysis with Boundary Elements, Vol. 30, 2006, pp. 553-563.

[6] Divo, E. and Kassab, A., "A meshless method for conjugate heat transfer problems," Engineering Analysis with Boundary Elements, Vol. 29, 2005, pp. 136-149.

[7] Divo, E., Kassab, A., and Rodríguez, F.: "Parallel domain decomposition approach for large-scale three-dimensional boundary-element models in linear and nonlinear heat conduction," Numerical Heat Transfer, Part B., Vol. 44, 2003, pp. 417-437.

Section 5
Fluid mechanics applications

Flow over a square cylinder by BEM

L. Škerget & J. Ravnik
University of Maribor, Faculty of Mechanical Engineering, Maribor, Slovenia

Abstract

Flow of an incompressible viscous fluid is considered. The velocity-vorticity formulation of the Navier–Stokes equations is used. The kinematics equation is solved for boundary vorticity values using the Boit-Savart law. Solution of the kinetics equation for the domain values is obtained by employing a macro element approach. Using macro elements enables simulations on dense meshes, since it substantially reduces the algorithm's memory requirements.

The developed numerical algorithm has been used to simulate laminar flow over a square cylinder in channel. Low Reynolds number steady state flow simulation as well as transient simulation at higher Reynolds numbers has been investigated. The results have been analysed in terms of velocity, vorticity and pressure field distributions in the wake of the cylinder.

1 Introduction

Flows past bluff bodies are an interesting topic amongst engineers and researchers. A square cylinder is a basic example of such flows, its industrial applicability ranging from wind induced motion to turbulent sound generation. In general flows past bluff bodies exhibit complex phenomena such as separation, reattachment or vortex shedding.

The numerical benchmark for laminar incompressible flow over a square cylinder was made by Breuer et al. [1] using lattice Boltzmann and finite volume methods. Recently turbulent flow over a square cylinder was considered by Sohankar [2] using the Large eddy simulation approach. Flow as well as heat transfer was considered by De and Dalal [3] for a study of a natural convection around a headed square cylinder in an enclosure. Ozgoren [4] studied experimentally the flow structure in the downstream of square and circular cylinders.

WIT Transactions on Modelling and Simulation, Vol 44, © 2007 WIT Press
www.witpress.com, ISSN 1743-355X (on-line)
doi:10.2495/BE070161

We have used the incompressible Navier–Stokes equations written in velocity - vorticity formulation to solve flow past a square cylinder. In order to decrease the computer storage requirements of BEM, wavelet compression technique was used by Ravnik et al. [5]. In this paper we present results of an alternative approach - the equations were solved by the macro element Boundary Element Method (BEM). The derived equations, the numerical procedure and the results are shown below.

2 The Navier–Stokes equations

The analytical description of motion of a continuous incompressible and isothermal fluid medium is based on conservation of mass and momentum. The primitive field functions of interest are velocity vector field $\vec{v}(\vec{r},t)$ and scalar pressure field $p(\vec{r},t)$. The dynamics of a viscous fluid flow may be partitioned into its kinematic and kinetic aspect through the use of derived vector vorticity field $\vec{w}(\vec{r},t)$, obtained as a curl of the compatible velocity field, as follows

$$\omega_i = e_{ijk}\frac{\partial v_k}{\partial x_j}, \qquad \frac{\partial \omega_j}{\partial x_j} = 0, \tag{1}$$

which is solenoidal vector by the definition, and e_{ijk} is the permutation unit tensor. By applying a curl to the vorticity definition (1) and using the solenoidal constraint for the velocity vector $\vec{\nabla}\cdot\vec{v} = 0$, the following vector elliptic Poisson's equation for the velocity vector is obtained

$$\triangle\vec{v} + \vec{\nabla}\times\vec{\omega} = 0. \tag{2}$$

The equation (2) represents the kinematics of an incompressible fluid motion, expressing the compatibility and restriction conditions between velocity and vorticity field functions.

The kinetic aspect is governed by the parabolic diffusion convection vorticity equation, obtained by applying the curl differential operator to the momentum equation. For the two-dimensional plane flow, the vorticity vector has just one component, which is perpendicular to the plane of the fluid motion. Thus we obtain a scalar transport equation for vorticity

$$\frac{D\omega}{Dt} = \nu\triangle\omega\,, \tag{3}$$

being ν constant kinematic viscosity and D/Dt Stokes substantial derivative.

The vorticity transport equation as itself is highly non-linear partial differential equation due to the inherent non-linearity caused by the compatibility and restriction conditions among velocity and vorticity fields, and due to the product of velocity and vorticity field functions in the convective term.

To derive the pressure equation, depending on known velocity and vorticity field functions, the divergence of momentum equation should be considered, resulting in the elliptic Poisson's pressure equation

$$\Delta p - \vec{\nabla} \cdot \vec{f_p} = 0, \tag{4}$$

where the pressure force term $\vec{f_p}$ is for the planar flow cases

$$\frac{\partial p}{\partial x_i} = f_{pi} = -\eta e_{ij} \frac{\partial \omega}{\partial x_j} - \rho a_i + \rho g_i, \tag{5}$$

in which the dynamic viscosity is $\eta = \rho\nu$, and $\vec{a} = D\vec{v}/Dt$ and $\vec{g}$ are the inertia and gravitational acceleration vectors, respectively. The Neumann boundary conditions for the pressure equation may be determined for the whole solution boundary Γ and the following relation is valid

$$\frac{\partial p}{\partial n} = \vec{f_p} \cdot \vec{n}. \tag{6}$$

3 Boundary-domain integral representations

To apply the boundary element method, we must rewrite the governing differential equations in integral form. The singular boundary-domain integral representation for the velocity vector can be formulated by using the Green theorems for scalar functions or weighting residuals technique rendering the following vector integral formulation, e.g. the plane two-dimensional kinematics is given by two scalar equations as follows [4] (Škerget et al. [6])

$$c(\xi)\, v_i(\xi) + \int_\Gamma v_i \frac{\partial u^\star}{\partial n} d\Gamma = e_{ij} \int_\Gamma v_j \frac{\partial u^\star}{\partial t} d\Gamma - e_{ij} \int_\Omega \omega \frac{\partial u^\star}{\partial x_j} d\Omega, \tag{7}$$

where $u^\star$ is the elliptic Laplace fundamental solution, ξ the source point and Γ the boundary of the solution domain Ω.

Considering the vorticity kinetics in an integral representation one has to take into account parabolic diffusion convection character of the vorticity transport equation [3]. With the use of the linear parabolic diffusion-convective differential operator, the following integral formulation can be written (Škerget et al. [7])

$$c(\xi)\, \omega(\xi) + \int_\Gamma \omega \frac{\partial U^\star}{\partial n} d\Gamma = \int_\Gamma \frac{\partial \omega}{\partial n} U^\star d\Gamma - \frac{1}{\nu} \int_\Gamma v_n \omega U^\star d\Gamma$$
$$+ \frac{1}{\nu} \int_\Omega \widetilde{v}_j \omega \frac{\partial U^\star}{\partial x_j} d\Omega + \beta \int_\Omega \omega_{F-1} U^\star d\Omega, \tag{8}$$

where the velocity field is decomposed into an average constant vector $\overline{v}_j$ and a perturbated one $\widetilde{v}_j$, such that $v_j = \overline{v}_j + \widetilde{v}_j$, the quantity $U^\star = \nu u^\star$ with $u^\star$ is now the fundamental solution of diffusion-convective equation with first order reaction, and $\beta = 1/\nu\Delta t$.

The pressure equation (4) is an elliptic Poisson equation, and therefore employing again the linear elliptic Laplace differential operator [5] the following form of the pressure integral equation is obtained (Škerget and Samec [8])

$$c(\xi)\, p(\xi) + \int_\Gamma p \frac{\partial u^\star}{\partial n} d\Gamma = \int_\Omega f_{pi} \frac{\partial u^\star}{\partial x_i} d\Omega, \tag{9}$$

where the vector $\vec{f}_p$ is given by equation (5).

4 Numerical example: flow over a square cylinder

The developed numerical algorithm was used to simulate incompressible laminar 2D flow over a square cylinder. The cylinder was positioned in a centre of a channel. A parabolic velocity profile was prescribed at the inflow of the channel. At the channel walls and at the cylinder a no slip boundary condition was applied. At the outflow boundary normal derivatives of all field functions were set to zero. Boundary and initial conditions are shown schematically in Figure 1. The flow configuration was the same as in Camarri and Giannetti [9] with the Reynolds number valued defined by $Re = Dv_c/\nu$. The blockage ratio was $\beta = D/H = 1/8$. The length of the flow domain before the cylinder was $L_i/D = 12$ and behind the cylinder $L_o/D = 35$.

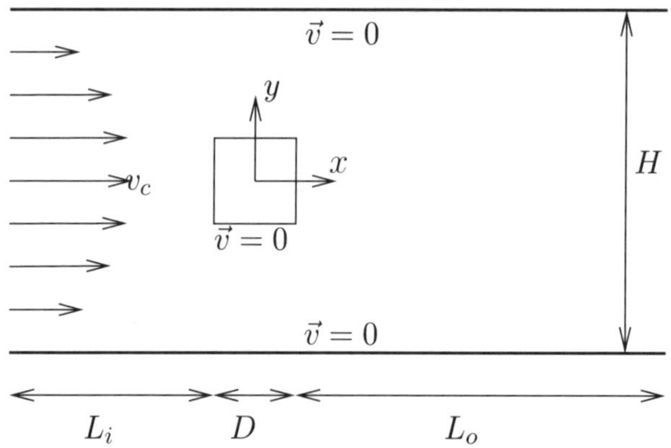

Figure 1: Flow configuration, boundary conditions and computational domain, not in scale.

The simulation was performed on a computational mesh consisting of 4700 Lagrangian 9 node domain cells. The mesh is illustrated in Figure 2.

The flow was simulated for Reynolds number values from $Re = 9$ to $Re = 90$. Flow at $Re = 9$, $Re = 18$ and $Re = 36$ is steady. Steady state simulation was used to obtained converged results for $Re = 9$ and $Re = 18$. At $Re = 36$ transient

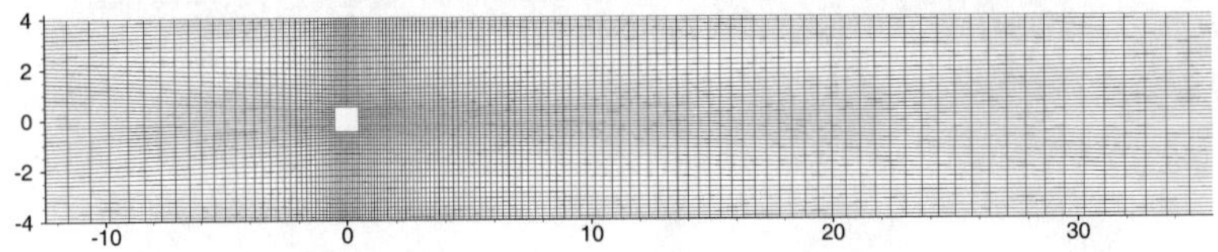

Figure 2: Computational mesh for the simulation of flow over a square cylinder. 4700 Lagrangian domain cells were used.

simulation was performed, having the steady state $Re = 18$ results for initial conditions. A time step of $\Delta t = 0.1$ was used. The simulation ran 140 time steps to achieve steady state. The contours of vertical and horizontal velocity, vorticity and pressure are shown in Figure 3 for $Re = 9$, in Figure 4 for $Re = 18$ and in in Figure 5 for $Re = 36$. One may observe that the recirculation region behind the cylinder grows with the increasing Reynolds number. The flow speeds up between the cylinder and the walls of the channel, while it slows in front of and behind the cylinder. Looking at vorticity contours we observe that the largest vorticity can be found at the lower left and upper left corner of the cylinder. Vorticity is generated at solid walls and is transported by diffusion and advection into the flow. Looking at the pressure field, we notice the high pressure zone in front of the cylinder and a low pressure zone at the top, bottom and behind the cylinder.

At Reynolds number value $Re = 90$ the flow becomes unsteady. A time step of $\Delta t = 0.1$ was used. The von Kármán vortex street is formed behind the cylinder, i.e. clockwise and counter-clockwise vortices are shed from the upper and lower sides of the cylinder. Figure 6 shows the instantaneous contours of vertical and horizontal velocity, vorticity and pressure at $t = 30$.

Breuer et al. [1] and Camarri and Giannetti [9] reported the Strouhal number ($St = fD/v_c$, f being the vortex-shedding frequency) to be $St = 0.135$ at $Re = 90$. By observing vorticity behind the cylinder at $x = 0.968, y = 0$ (Figure 7) we were able to measure the vortex shedding frequency $f = 0.2$. Using the measure shedding frequency we calculated the Strouhal number $St = 0.133$, which is in excellent agreement with the reference results.

5 Conclusions

A macro element boundary element method based method was developed for the simulation of two-dimensional laminar incompressible viscous fluid flows. The method was tested on a flow over a square cylinder confined in a channel.

Incompressible flow past a square cylinder is steady for Reynolds number values $Re = 9$, $Re = 18$, $Re = 36$ with the length of the recirculation region increasing with the increasing Reynolds number. At Reynolds number value $Re = 90$ we simulated unsteady behaviour. Vortices are shed from the upper and lower walls of the cylinder. The Strouhal number exhibited by the flow using our simulation was $St = 0.133$, which in excellent agreement with the benchmark results.

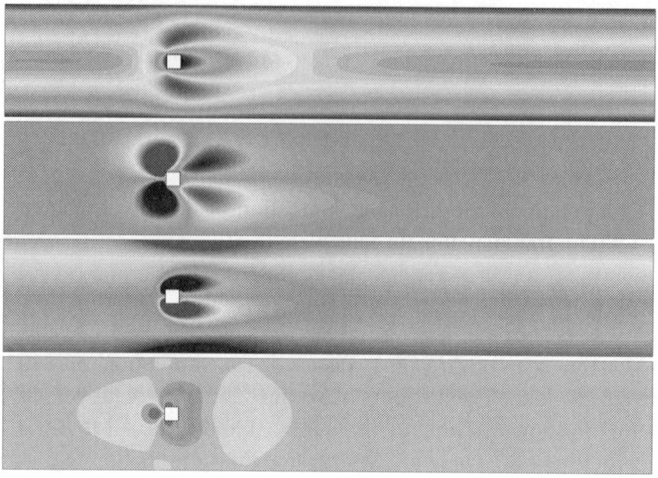

Figure 3: Steady state flow over a square cylinder at $Re = 9$. From top to bottom: horizontal velocity, vertical velocity, vorticity and pressure contours.

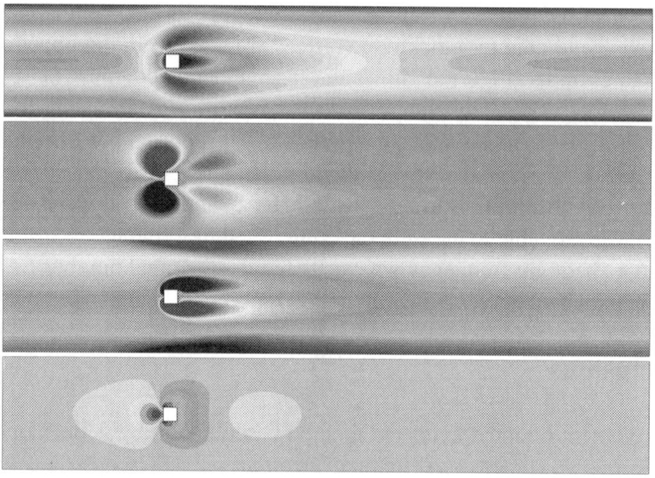

Figure 4: Steady state flow over a square cylinder at $Re = 18$. From top to bottom: horizontal velocity, vertical velocity, vorticity and pressure contours.

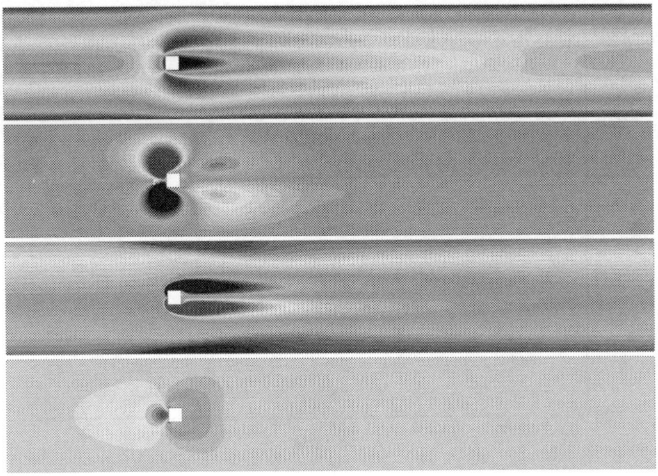

Figure 5: Steady state flow over a square cylinder at $Rc = 36$. From top to bottom: horizontal velocity, vertical velocity, vorticity and pressure contours.

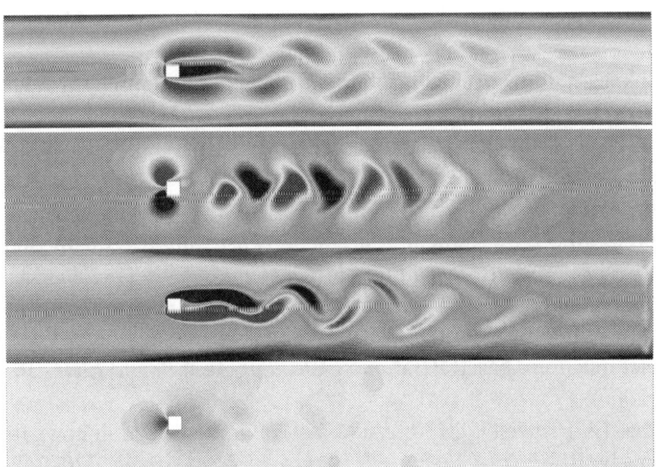

Figure 6: Unsteady flow over a square cylinder at $Re = 90$, $t = 30$. From top to bottom: horizontal velocity, vertical velocity, vorticity and pressure contours.

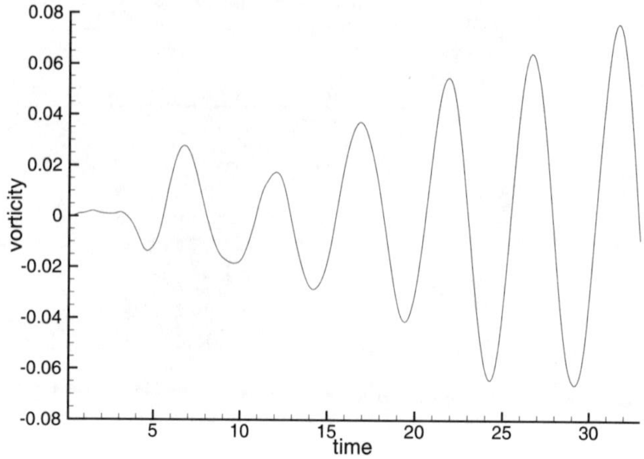

Figure 7: Flow over a square cylinder at $Re = 90$. Vorticity at $x = 0.968, y = 0$ was used to measure the Strouhal number.

References

[1] Breuer, M., Bernsdorf, J., Zeiser, T. & Durst, F., Accurate computations of the laminar flow past a square cylinder based on two different methods: lattice-Boltzmann and finite-volume. *Int J Heat and Fluid Flow*, **21**, pp. 186–196, 2000.

[2] Sohankar, A., Flow over a bluff body from moderate to high reynolds numbers using large eddy simulation. *Computers & Fluids*, **35**, pp. 1154–1168, 2006.

[3] De, A.K. & Dalal, A., A numerical study of natural convection around a square, horizontal, heated cylinder placed in an enclosure. *Int J Heat Mass Transf*, **49**, pp. 4608–4623, 2006.

[4] Ozgoren, M., Flow structure in the downstream of square and circular cylinders. *Flow Measurement and Instrumentation*, **17**, pp. 225–235, 2006.

[5] Ravnik, J., Škerget, L. & Hriberšek, M., 2D velocity vorticity based LES for the solution of natural convection in a differentially heated enclosure by wavelet transform based BEM and FEM. *Eng Anal Bound Elem*, **30**, pp. 671–686, 2006.

[6] Škerget, L., Hriberšek, M. & Žunič, Z., Natural convection flows in complex cavities by BEM. *Int J Num Meth Heat & Fluid Fl*, **13**, pp. 720–735, 2003.

[7] Škerget, L., Hriberšek, M. & Kuhn, G., Computational fluid dynamics by boundary domain integral method. *Int J Num Meth Eng*, **46**, pp. 1291–1311, 1999.

[8] Škerget, L. & Samec, N., BEM for the two-dimensional plane compressible fluid dynamics. *Eng Anal Bound Elem*, **29**, pp. 41–57, 2005.

[9] Camarri, S. & Giannetti, F., Analysis of the inversion of the von Kármán street in the wake of a confined square cylinder. *EUROMECH Fluid Dynamics Conference*, pp. 19–25, 2006.

Meshless analysis of flow and concentration in a water reservoir

M. Kanoh[1], N. Nakamura[1], K. Kai[1], T. Kuroki[2] & K. Sakamoto[3]
[1]*Department of Civil Engineering, Kyushu Sangyo University, Japan*
[2]*Baikoen 1-15-21 Dazaifu, Fukuoka 818-0124, Japan*
[3]*Environment Division, Matsue Doken Co., Ltd. Japan*

Abstract

In an earlier study, a boundary element methodology was developed to obtain numerically stable and convergent results for the concentration distribution and flow of a water reservoir. In the process to apply the boundary element method to the flow analysis around a machine that supplies dissolved oxygen (DO), we realised that the divergence and accuracy of the very delicate flow were sensitive to the mesh (domain) and boundary discretisation. In other words, it seemed difficult to determine the appropriate lengths of the mesh (domain) and boundary discretisation for obtaining stability and convergence in the computational analysis. In this paper, a new meshless method is presented, which overcomes the difficulties of the boundary element method described above. The method is based on the idea of mesh-free radial basis functions (RBFs), which is a collocation method. Referring to the velocity vectors of the water flow calculated by the weighted finite difference method (WFDM) and the finite element method (FEM) and observed in the model simulation of a water reservoir constructed in the sanitary and environmental engineering laboratory of Kyushu Sangyo University, the effect and accuracy of the alternative meshless method were estimated.
Keywords: flow and concentration in water reservoirs, meshless method, weighted finite difference method, finite element method, observed velocity in model simulation of water reservoir.

1 Introduction

The poor-oxygen layer, which is short of or lacking in dissolved oxygen (DO), sometimes results in pollution of the water in a reservoir. An attempt was made

WIT Transactions on Modelling and Simulation, Vol 44, © 2007 WIT Press
www.witpress.com, ISSN 1743-355X (on-line)
doi:10.2495/BE070171

to ameliorate the concentration of oxygen in the lower layer of the reservoir by using a machine that supplies DO (Kanoh et al. [1]). In order to numerically confirm the efficiency of the improvement in DO, we applied a meshless method to represent and calculate the slow but very delicate flow caused by the DO-supplying machine. For this purpose, two techniques were used: (1) the first is the penalty method, in which the pressure terms are eliminated in the Navier–Stokes equations for the meshless method; (2) the second is to identify the boundary conditions for the velocity and concentration against the vertical wall or the bottom and on the free surface.

2 Governing equations

The three equations, which are continuous, Navier–Stokes (N–S) and convective-diffusion equations, govern the flow in the flow domains of a water reservoir. In the vertical (x_1, x_2) plane, as illustrated in Figure 1, these equations are shown as follows:

$$\frac{\partial u_1}{\partial x_1}+\frac{\partial u_2}{\partial x_2}=0 \tag{1}$$

$$\frac{\partial u_1}{\partial t}+u_1\frac{\partial u_1}{\partial x_1}+u_2\frac{\partial u_1}{\partial x_2}-\nu\left(\frac{\partial^2 u_1}{\partial x_1^{\,2}}+\frac{\partial^2 u_1}{\partial x_2^{\,2}}\right)=-\frac{1}{\rho}\frac{\partial P}{\partial x_1} \tag{2_1}$$

$$\frac{\partial u_2}{\partial t}+u_1\frac{\partial u_2}{\partial x_1}+u_2\frac{\partial u_2}{\partial x_2}-\nu\left(\frac{\partial^2 u_2}{\partial x_1^{\,2}}+\frac{\partial^2 u_2}{\partial x_2^{\,2}}\right)=-\frac{1}{\rho}\frac{\partial P}{\partial x_2}+g \tag{2_2}$$

$$\frac{\partial T}{\partial t}+u_1\frac{\partial T}{\partial x_1}+u_2\frac{\partial T}{\partial x_2}-D\left(\frac{\partial^2 T}{\partial x_1^{\,2}}+\frac{\partial^2 T}{\partial x_2^{\,2}}\right)=0 \tag{3}$$

$$\rho = aT + b \tag{4}$$

where x_1 and x_2 are the horizontal and vertical directions, u_1 and u_2 describe the velocities of the x_1 and x_2 directions, P is the pressure, g is the gravity acceleration, ν is the kinematic viscosity, T is the water temperature, and D is the diffusion coefficient. Here, the density ρ is connected to the water temperature T, as written in Equation (4), with the coefficients a and b. The water temperature T is compatible to the concentration of dissolved oxygen (DO) in case it is necessary to calculate the DO or other values.

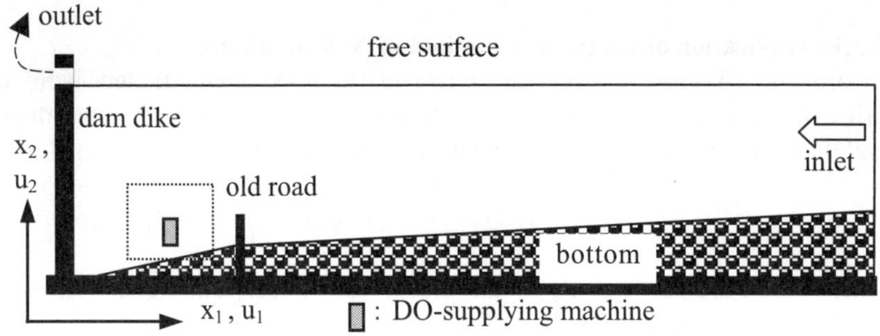

Figure 1: The entire domain of a water reservoir.

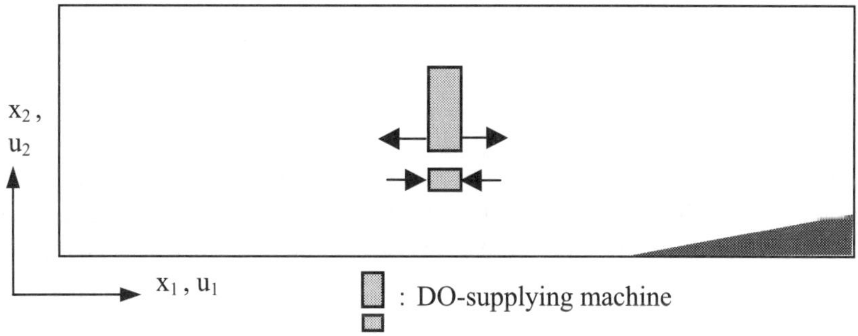

Figure 2: The domain around a DO-supplying machine in a reservoir.

3 Application of the meshless method

3.1 Meshless method for flow and concentration analysis

We deal with the flow and concentration analysis in the water reservoir as shown in Figures 1 and 2, and try to apply the meshless method to the problem. The meshless method is based on the idea of the mesh-free RBF collocation method (e.g., Divo et al. [2]). The penalty method was used so that the pressure terms would be eliminated in the N–S equations and the difficulty of the pressure boundary conditions would be avoided in the meshless method.

3.2 Meshless method formulation

3.2.1 Application of the penalty method to N–S equations

In order to eliminate the pressure terms in the N–S equations and avoid the difficulty of the pressure boundary conditions, the penalty method is introduced, and the following equations are obtained as described below

$$u_j \cdot u_{i,j} - \lambda(u_{1,2} + u_{2,2})_{,j} \cdot \delta_{i,j} - \nu(u_{1,2} + u_{2,2})_{,j} = 0 \qquad \text{i, j} = 1, 2 \tag{5}$$

where λ means $R_e \cdot K/\rho$, R_e is the Reynolds number, and K describes the coefficient of the penalty method (Kanoh et al. [1]).

3.2.2 Simultaneous equations for the meshless method

Substituting $\underline{u}_1$ and $\underline{u}_2$ at time (t-Δt) into the above Equation (5), the following expression is obtained:

$$\underline{u}_1 \cdot u_{1,1} + \underline{u}_2 \cdot u_{1,2} - \lambda(u_{1,11} + u_{2,21}) - \nu(u_{1,22} + u_{2,12} + 2u_{1,11}) = 0 \tag{6_1}$$

$$\underline{u}_1 \cdot u_{2,1} + \underline{u}_2 \cdot u_{2,2} - \lambda(u_{1,12} + u_{2,22}) - \nu(u_{2,11} + u_{1,21} + 2u_{2,22}) = 0 \tag{6_2}$$

The global expansion function Xj (= $(r^2+c^2)^{-1/2}$) is employed here so that the unknowns (u_1, u_2, and T) at time (t) can be developed for the mesh-free RBF collocation method, where r equals $\{(x-x_j)+(y-y_j)\}^{1/2}$ and c is the constant. The unknown values u_1, u_2, and T are expressed as Equation (7) using the global expansion function Xj.

$$u_1 = \alpha_j\, Xj = \alpha_j /(r^2+c^2)^{1/2} \tag{7_1}$$

$$u_2 = \beta_j\, Xj = \beta_j /(r^2+c^2)^{1/2} \tag{7_2}$$

$$T = \gamma_j\, Xj = \gamma_j /(r^2+c^2)^{1/2} \tag{7_3}$$

The simultaneous equations for calculating the above unknowns (α_j, β_j, and γ_j) in the steady state are obtained as Equation (8).

$$\left\{\left(\underline{u}_1 \frac{\partial X_j}{\partial x_1} + \underline{u}_2 \frac{\partial X_j}{\partial x_2}\right) - (\nu+\lambda)\frac{\partial^2 X_j}{\partial x_1^2} - \nu\left(\frac{\partial^2 X_j}{\partial x_1^2} + \frac{\partial^2 X_j}{\partial x_2^2}\right)\right\}\alpha_j - (\nu+\lambda)\frac{\partial^2 X_j}{\partial x_1 \partial x_2}\beta_j = 0 \tag{8_1}$$

$$-(\nu+\lambda)\frac{\partial^2 X_j}{\partial x_1 \partial x_2}\alpha_j + \left\{\left(\underline{u}_1 \frac{\partial X_j}{\partial x_1} + \underline{u}_2 \frac{\partial X_j}{\partial x_2}\right) - (\nu+\lambda)\frac{\partial^2 X_j}{\partial x_2^2} - \nu\left(\frac{\partial^2 X_j}{\partial x_1^2} + \frac{\partial^2 X_j}{\partial x_2^2}\right)\right\}\beta_j = 0 \tag{8_2}$$

$$\left\{\left(\underline{u}_1\frac{\partial X_j}{\partial x_1}+\underline{u}_2\frac{\partial X_j}{\partial x_2}\right)-D\left(\frac{\partial^2 X_j}{\partial x_1^{\ 2}}+\frac{\partial^2 X_j}{\partial x_2^{\ 2}}\right)\right\}\gamma_j=0 \qquad (8_3)$$

Substituting the obtained values of α_j, β_j, and γ_j into Equation (7), the values of three unknowns (u_1, u_2, and T) in the steady state can be calculated using the meshless method.

3.3 Boundary conditions

The boundary conditions for the free surface, the bottom and the vertical wall boundary have been previously proposed for BEM and WFDM in the flow region (Kanoh et al. [1]). Regarding the boundary conditions for flow analysis by our meshless method, the outline is as follows: (1) the velocities that exist in the normal and tangential directions at the wall or on the bottom are zero; (2) the pressure on the free surface is defined as zero. The pressure that is defined on the inside point neighbouring the wall or bottom is calculated as shown below:

$$P_{inside} = P_{outside} - \rho\cdot\nu\cdot u_{n-1}/(\Delta n)\,, \qquad (9)$$

where P_{inside} and $P_{outside}$ are the pressures on the inside and outside points, respectively, and u_{n-1} is the velocity value on the point that exists in the length of Δn to the boundary. This pressure boundary condition requires that the meshless method have the points existing inside the wall or bottom.

4 Results and discussion

The numerical results of the meshless method, the finite element method (FEM), and the weighted finite difference method (WFDM) are compared and discussed in this section. The two kinds of flow analysis calculated by these three methods are described here. Namely, the first analysis is the flow around a DO-supplying machine in a water reservoir, and the second is the flow analysis of an entire domain of the water reservoir.

4.1 Flow analysis around a DO-supplying machine in a water reservoir

As described above, in the process to apply the boundary element method to the flow analysis around the DO-supplying machine, we realised that the divergence and accuracy of the very delicate flow were sensitive to the mesh (domain) and boundary discretisation. In other words, it is difficult to determine the appropriate lengths of the mesh (domain) and boundary discretisation for obtaining convergence and stability in the computational analysis. The numerical results of the flow around the DO-supplying machine certified that our new meshless method overcame the difficulties of the boundary element method

Figure 3: Velocity vectors using meshless method around a DO-supplying machine.

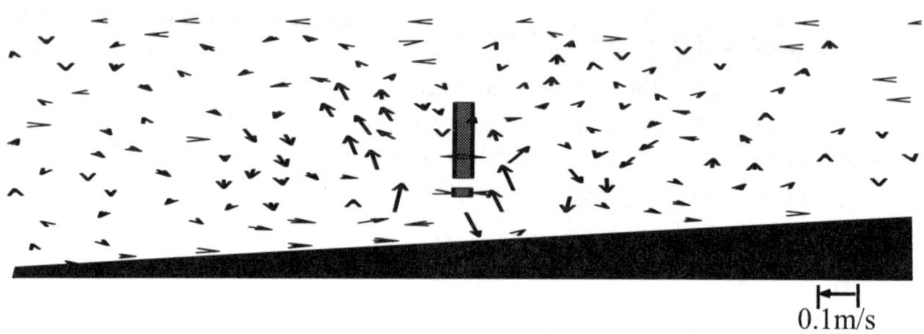

Figure 4: Velocity vectors using FEM around a DO-supplying machine.

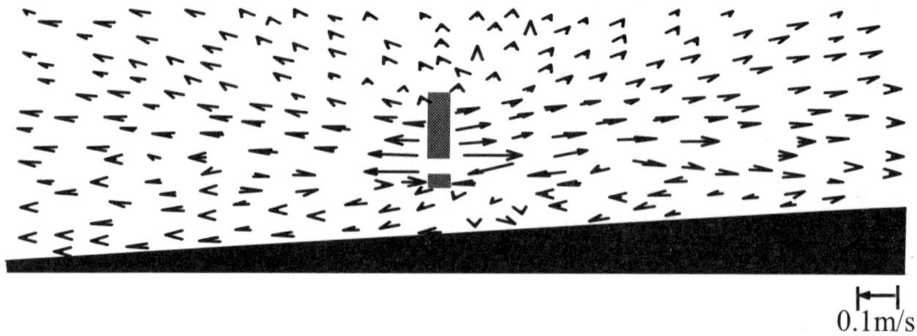

Figure 5: Velocity vectors using WFDM around a DO-supplying machine.

described above. It was reported in our previous work that the poor oxygen concentration frequently yielded a bad odour and dissolved heavy metals in the lower water layer of the B water reservoir. We were successful at improving the poor oxygen concentration in the lower layer of the B water reservoir using the DO-supplying machine. In order to economically improve the poor oxygen concentration of other water reservoirs, it seemed necessary to appropriately determine the capacity and the number of the DO-supplying machines required for improving the poor oxygen. For that purpose, we calculated the flow around a DO-supplying machine in a water reservoir by using the three methods described in this paper.

4.1.1 Meshless method calculation of the flow around the machine

Figure 3 illustrates the flow velocity vector distribution around a DO-supplying machine in a water reservoir calculated by the meshless method, in which the number of the points in the meshless method is 651 and the three values of λ, C, and ν are 1000.0, 1.0, and 0.001, respectively. The boundary element method could not yield stable and convergent results to the flow analysis around the machine to supply DO; on the other hand, the stability and convergence of the flow analysis using the meshless method seemed satisfactory.

4.1.2 FEM calculation of the flow around the machine

Figure 4 illustrates the flow velocity vector distribution around a DO-supplying machine in a water reservoir calculated by the finite element method, in which the number of the elements in the FEM is 300 and the two values of λ and ν are 1000.0 and 0.001, respectively. The penalty method was also introduced so that the pressure terms would be eliminated in the Navier–Stokes equations and the difficulty of the pressure boundary conditions could be avoided in the finite element method as well as in the meshless method. The kinds of flow velocity vector distributions around the DO-supplying machine calculated by both the finite element method and the meshless method closely resemble each other. The stability and convergence of the flow analysis using the finite element method also seemed satisfactory.

4.1.3 WFDM calculation of the flow around the machine

Figure 5 illustrates the flow velocity vector distribution around a DO-supplying machine in a water reservoir calculated by the weighted finite difference method, in which the number of meshes in the WFDM is 4750 and the value of ν is 0.0000015. We consider that the flow velocity vectors around the DO-supplying machine calculated by the WFDM are reasonable, since the WFDM yielded very similar solutions to the true results of several flow problems that were observed in the simulation models constructed in the sanitary and environmental engineering laboratory of Kyushu Sangyo University.

4.1.4 Domain discretisation of the four methods around a DO-supplying machine

Figure 6 shows the domain discretisation of the four methods around a DO-supplying machine. Comparing the node distribution of the meshless method with the boundary element distribution of the BEM around the DO-supplying machine, we can see that both the number of the points of the meshless method and the number of the boundary elements of the BEM closely resemble each other. As described above, the BEM could not give stability to the flow analysis around the DO-supplying machine; on the other hand, the stability of our meshless method was satisfactory. Namely, our meshless method overcame the difficulty of the BEM to determine the appropriate combinations of the lengths of the domain and boundary discretisation for obtaining the stability in the computational analysis. Comparing the node distribution of the meshless method with those of the FEM and WFDM in Figure 6(a), (c), and (d), we can see that the three sets of the number of the points of the meshless method, the finite elements of the FEM, and the meshes of the WFDM are almost identical. Moreover, all three methods can give stability and convergence to the flow analysis around the DO-supplying machine.

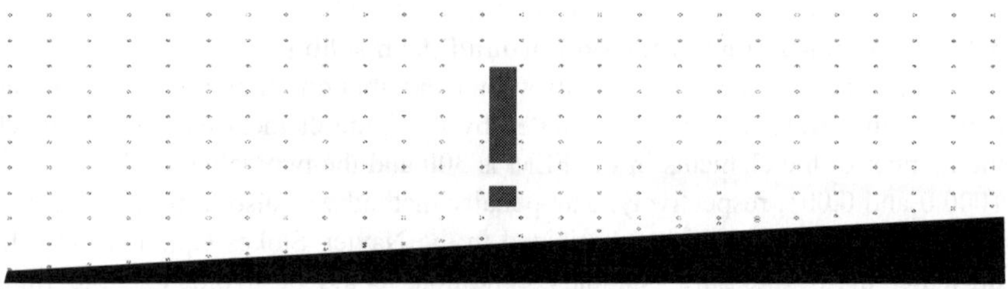

(a) Node distribution of the meshless method around a DO-supplying machine.

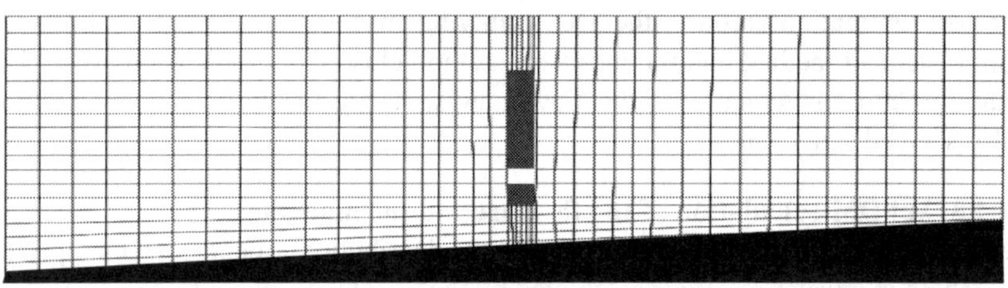

(b) Boundary element distribution of BEM around a DO-supplying machine.

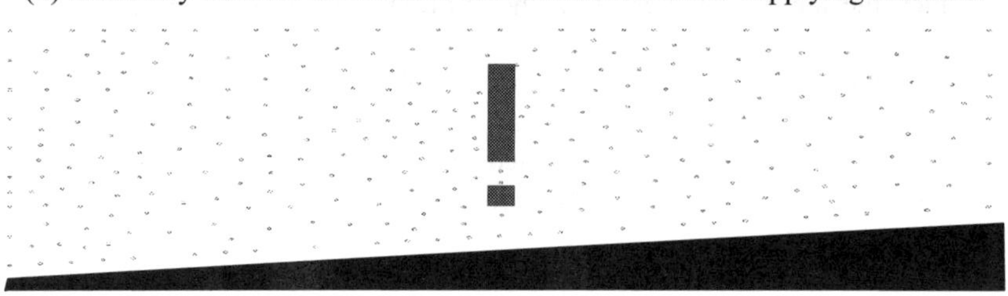

(c) Element distribution of FEM around a DO-supplying machine.

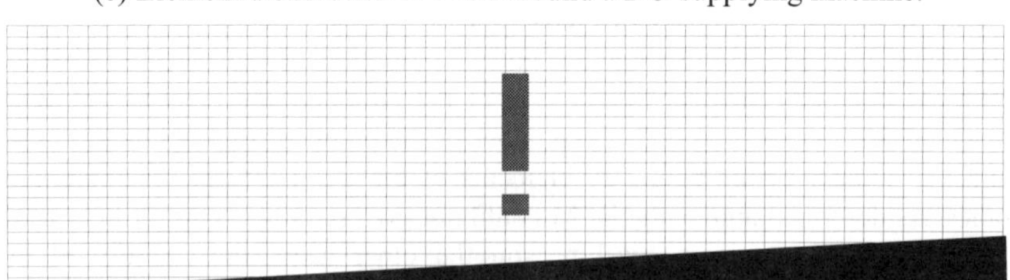

(d) Mesh distribution of WFDM around a DO-supplying machine.

Figure 6: Domain discretisation of the four methods around a DO-supplying machine.

4.2 Flow analysis of the entire domain of a water reservoir

4.2.1 Meshless method calculation of the entire domain

Figure 7 illustrates the velocity vectors of the entire domain in the B water reservoir calculated using the meshless method, in which the number of the points in the meshless method is 801 and the three values of λ, C, and ν are 1000.0, 0.1, and 0.001, respectively. We consider that the stability and convergence of the meshless method for this problem are satisfactory.

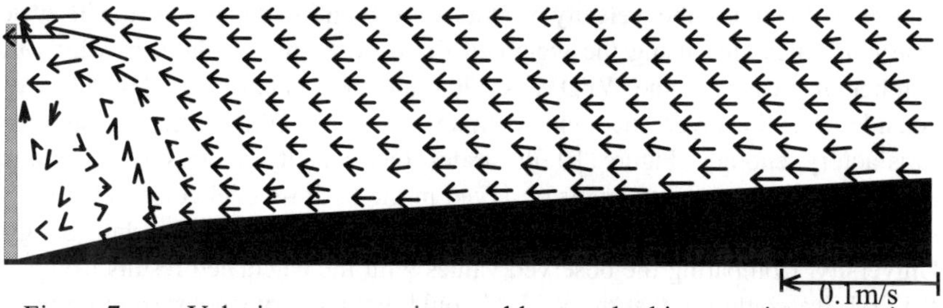

Figure 7: Velocity vectors using meshless method in an entire reservoir.

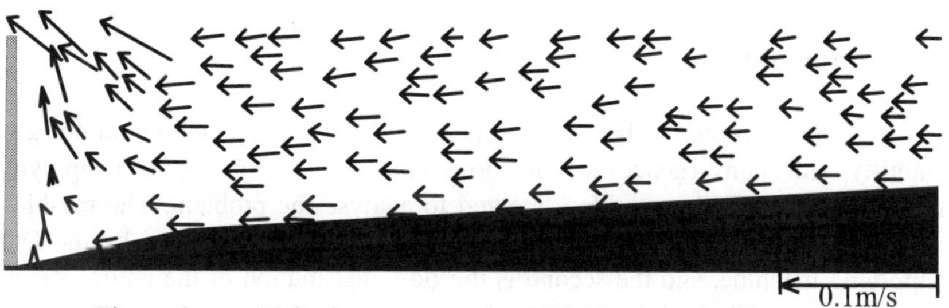

Figure 8: Velocity vectors using FEM in an entire reservoir.

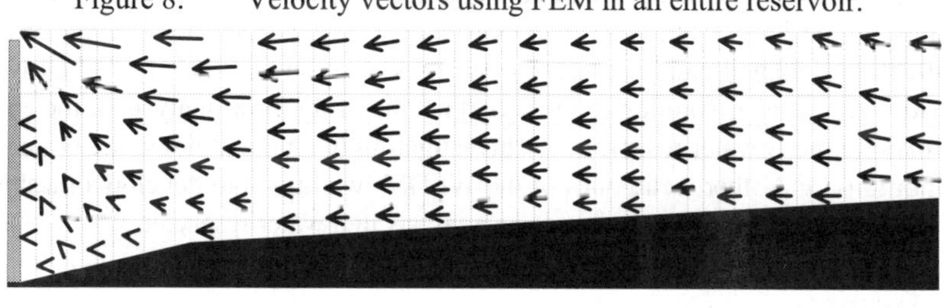

Figure 9: Velocity vectors using WFDM in an entire reservoir.

Figure 10: Observed velocity vectors in a model simulation of a reservoir.

4.2.2 Finite element method calculation of the whole domain

Figure 8 shows the velocity vectors of the entire domain in the B water reservoir calculated using the finite element method, in which the number of the elements in the FEM is 800 and the two values of λ and ν are 1000.0 and 0.001, respectively. We consider that the stability and convergence of the FEM for this problem are also satisfactory.

4.2.3 Weighted finite difference method calculation of the whole domain

Figure 9 illustrates the velocity vectors of the entire domain in the B water reservoir calculated using the weighted finite difference method, in which the number of meshes in the WFDM is 2905 and the value of ν is 0.000001. We consider that the divergence and accuracy of the WFDM for this problem are satisfactory enough. Figure 10 illustrates the velocity vectors of the entire domain in the B water reservoir observed in the hydro-model constructed in the sanitary and environmental engineering laboratory of Kyushu Sangyo University. Comparing the observed values with the calculated results using the WFDM, we could see that the WFDM could yield a very similar solution to the actual results of the problem.

5 Conclusion

In order to overcome the fact that the boundary element method could not give stability and convergence to the flow analysis around the DO-supplying machine, we applied a meshless method to analyse the problem. The meshless method could calculate two kinds of flow; the first is that caused by the DO-supplying machine, and the second is the flow distribution of the entire domain in the B water. The stability and convergence of the two kinds of flow analysis using the meshless method seemed satisfactory. Comparing the observed values with the calculated results using the WFDM, we can see that the WFDM can yield a very similar solution to the actual results of the flow analysis of the entire domain in the B water reservoir. By investigating the methodology, the boundary conditions, and other techniques of the WFDM, we intend to develop a meshless method for the flow and concentration analysis in a water reservoir.

References

[1] Kanoh, M., Nakamura, N., and Kuroki T., Boundary element method for the analysis of flow and concentration in a water reservoir, *Proc. of the 28th World Conf. on Boundary Elements and Other Mesh Reduction Methods*, ed. C.A. Brebbia, WIT Press, Skiathos, Greece, pp. 231-240, 2006.

[2] Divo E., Kassab A., and Zahab El., Parallel domain decomposition meshless modeling of dilute convection-diffusion of species, *Proc. of 27th World Conf. on Boundary Elements and Other Mesh Reduction Methods*, ed. C.A. Brebbia, WIT Press, Florida, USA, pp.79-89, 2005.

Numerical analysis of compressible fluid flow in a channel with sharp contractions

L. Škerget & J. Ravnik
University of Maribor, Faculty of Mechanical Engineering, Maribor, Slovenia

Abstract

The problem of unsteady fluid flow in a channel with a sharp contraction is studied numerically. An incompressible and full compressible Navier–Stokes set of equations is considered. The thermal energy equation is written in its most general form including the Rayleigh and reversible expansion rate terms. Flows for different Reynolds number values are studied in the context of unsteadiness of the flow. The influence of the additional nonlinearity due to compressibility of the fluid, dissipation and reversible rate of work are analyzed. Also, their influence on the stability of the flow is considered.

The boundary element numerical model is used, with the velocity vorticity formulation of the Navier–Stokes equations. The pressure field is evaluated from the pressure Poisson equation. Material properties are taken to be for the ideal fluid (air), and assumed to be pressure and temperature dependent.

1 Introduction

There are a large variety of forced-convection processes for gases flowing in closed conduits which can be reasonably approximated as constant pressure processes. In this study we examine the thermal energy equation for a flow which is neither constant density nor constant pressure. Here we consider the unsteadiness of compressible viscous flow in channels with sharp contractions. Due to sharp discontinuity in the channel geometry the thermal energy equation is written in its expanded form with the terms such as the rate of reversible work and the rate of irreversible or dissipation work. The coupled momentum and thermal energy transport equations, specially due to mentioned rate of work terms, drastically increased

WIT Transactions on Modelling and Simulation, Vol 44, © 2007 WIT Press
www.witpress.com, ISSN 1743-355X (on-line)
doi:10.2495/BE070181

the nonlinearity of the governing set of equations, and as a consequence decreased the stability of the numerical algorithm.

The unsteadiness of force convection flow are considered in a channel of aspect ratio $A = H/L = 8$, with height H and length L, with one and two sharp contractions. The flow behavior is computed for the Reynolds number values $Re = 20, 200$ and 2000. For the Reynolds number value $Re = 20$ the steady flow exists, while for the Reynolds number values $Re = 200$ and 2000 the flow is unsteady and oscillating with the transition to turbulent flow situation.

2 Governing equations for the primitive variables formulation

The field functions of interest are velocity vector field v_i, scalar pressure field p, temperature field T and the field of mass density ρ, so that the mass, momentum and energy equations are given by the following set of nonlinear equations

$$\frac{\partial v_j}{\partial x_j} = \mathcal{D} = -\frac{1}{\rho}\frac{D\rho}{Dt}, \tag{1}$$

$$\rho\frac{Dv_i}{Dt} = -e_{ijk}\frac{\partial \eta\omega_k}{\partial x_j} + 2e_{ijk}\frac{\partial \eta}{\partial x_j}\omega_k + 2\frac{\partial \eta}{\partial x_j}\frac{\partial v_i}{\partial x_j} + \frac{4}{3}\frac{\partial \eta\mathcal{D}}{\partial x_i} - 2\mathcal{D}\frac{\partial \eta}{\partial x_i} - \frac{\partial p}{\partial x_i} + \rho g_i, \tag{2}$$

$$c\frac{DT}{Dt} = \frac{\partial}{\partial x_j}\left(k\frac{\partial T}{\partial x_j}\right) + \beta T\frac{Dp}{Dt} + \Phi, \tag{3}$$

in the Cartesian frame x_i, where c denotes changeable isobaric specific heat capacity per unit volume, $c = c_p\,\rho$, t is time, g_i is gravitational acceleration vector, while β is a volume coefficient of thermal expansion and Φ is the Rayleigh viscous dissipation function. Because of the analytical reasons to develop the velocity-vorticity formulation of the governing equations, the momentum equation is given in the second extended form. The following forms of the linear constitutive models for compressible viscous shear fluid are considered, such as the Newton and Fourier law of momentum and thermal energy diffusion

$$\tau_{ij} = 2\eta\dot{\varepsilon}_{ij} - \frac{2}{3}\eta\,\mathcal{D}\delta_{ij}, \qquad q_i = -k\frac{\partial T}{\partial x_i}, \tag{4}$$

where $\mathcal{D} = div\,\vec{v} = \dot{\varepsilon}_{ii}$ represents the divergence of the velocity field, and the Rayleigh dissipation function may be stated as

$$\Phi = \tau_{ij}\frac{\partial v_i}{\partial x_j} = \eta\left(\frac{\partial v_i}{\partial x_j}\frac{\partial v_i}{\partial x_j} + \frac{\partial v_j}{\partial x_i}\frac{\partial v_j}{\partial x_i}\right) - \frac{2}{3}\eta\,\mathcal{D}^2. \tag{5}$$

Representing the material properties of the fluid the dynamic viscosity η, heat conductivity k, the specific heat per unit volume c, and the mass density ρ, respectively, as a sum of a constant and variable part, e.g. $\eta = \eta_o + \widetilde{\eta}$, $k = k_o + \widetilde{k}$, $c = c_o + \widetilde{c}$, and $\rho = \rho_o + \widetilde{\rho}$, the momentum and energy eqs. (2) and (3) may be written in analogy to the basic conservation equations formulated for the constant material properties

$$\frac{D\vec{v}}{Dt} = -\nu_o \vec{\nabla} \times \vec{\omega} - \frac{1}{\rho_o}\vec{\nabla} p + \frac{\rho}{\rho_o}\vec{g} + \frac{1}{\rho_o}\vec{f}^m, \tag{6}$$

$$\frac{DT}{Dt} = a_o \triangle T + \frac{S^m}{c_o}, \tag{7}$$

where the pseudo body force term $\vec{f}^m$ and pseudo heat source term S^m are introduced into the momentum eq. (6) and into energy eq. (7) respectively, capturing the variable material property effects, and the nonlinear effects due to rate of reversible and irreversible work, and given by expressions, e.g. for plane flow problems

$$f_i^m = -e_{ij}\frac{\partial \widetilde{\eta}\omega}{\partial x_j} + 2e_{ij}\frac{\partial \widetilde{\eta}}{\partial x_j}\omega + 2\frac{\partial \eta}{\partial x_j}\frac{\partial v_i}{\partial x_j} + \frac{4}{3}\frac{\partial \eta \mathcal{D}}{\partial x_i} - 2\mathcal{D}\frac{\partial \eta}{\partial x_i} - \widetilde{\rho} a_i, \tag{8}$$

while the pseudo heat source term is given by an expression

$$S^m = \vec{\nabla}\left(\widetilde{k}\,\vec{\nabla} T\right) - \widetilde{c}\frac{DT}{Dt} + \beta T\frac{Dp}{Dt} + \Phi, \tag{9}$$

in which the kinematic viscosity is $\nu_o = \eta_o/\rho_o$, the heat diffusivity $a_o = k_o/c_o$ and the inertia acceleration vector is $\vec{a} = D\vec{v}/Dt$.

3 Governing equations for the velocity-vorticity formulation

The kinematics of the flow motion may be obtained by applying the curl operator to vorticity definition [2]:

$$\vec{\nabla} \times \vec{\omega} = \vec{\nabla} \times (\vec{\nabla} \times \vec{v}) = \vec{\nabla}(\vec{\nabla} \cdot \vec{v}) - \Delta \vec{v}, \tag{10}$$

and by using the continuity eq. (1), the following vector elliptic Poisson equation for the velocity vector is derived

$$\Delta \vec{v} + \vec{\nabla} \times \vec{\omega} - \vec{\nabla}\mathcal{D} = 0. \tag{11}$$

The kinetics of the flow representing by the vorticity transport equation is obtained by applying the curl differential operator to the both sides of eq. (6), rendering the following statement for the two-dimensional plane flow written in Cartesian tensor notation form as

$$\frac{\partial \omega}{\partial t} + \frac{\partial v_j \omega}{\partial x_j} = \nu_o \frac{\partial^2 \omega}{\partial x_j \partial x_j} - \frac{1}{\rho_o} e_{ij}\frac{\partial \rho g_i}{\partial x_j} - \frac{1}{\rho_o} e_{ij}\frac{\partial f_i^m}{\partial x_j}. \tag{12}$$

To derive the pressure equation, depending on known velocity field, vorticity field and material functions, the divergence of momentum equation should be calculated, resulting in the elliptic Poisson pressure equation

$$\Delta p - \vec{\nabla} \cdot \vec{f}_p = 0, \tag{13}$$

where the pressure force term $\vec{f}_p$ is for the planar flow cases

$$\frac{\partial p}{\partial x_i} = f_{pi} = -\eta_o e_{ij} \frac{\partial \omega}{\partial x_j} - \rho_o a_i + \rho g_i + f_i^m. \tag{14}$$

The Neumann boundary conditions for pressure equation may be determined for the whole solution domain and the following relation is valid

$$\frac{\partial p}{\partial n} = \vec{f}_p \cdot \vec{n} \quad on \quad \Gamma. \tag{15}$$

Due to the variable material property terms, and rate of reversible and irreversible work acting as additional temperature, pressure and velocity field dependent source terms, the vorticity, thermal energy, and pressure equations are coupled, making the numerical solution procedure of this highly nonlinear coupled set of equations very severe. Already, the vorticity transport equation as itself is highly nonlinear partial differential equation due to the inherent nonlinearity caused by the compatibility and restriction conditions among velocity, vorticity and dilatation fields. The dilatation and the vortical part of the flow, $\mathcal{D}$ and $\vec{\omega}$ field functions respectively, and all other nonlinear terms have to be underrelaxed to achieve convergence of the numerical solution procedure.

4 Boundary-domain integral equations

The singular boundary-domain integral representation for the velocity vector can be formulated by using the Green theorems for scalar functions or weighting residuals technique rendering the following vector integral formulation [1–10]), e.g. the plane two-dimensional kinematics is given by two scalar equations as follows

$$c(\xi)\, v_i(\xi) + \int_\Gamma v_i q^\star d\Gamma = e_{ij} \int_\Gamma v_j q_t^\star d\Gamma - e_{ij} \int_\Omega \omega q_j^\star d\Omega + \int_\Omega \mathcal{D} q_i^\star d\Omega, \tag{16}$$

or in the form of integral vector formulation for the general flow situation

$$c(\xi)\, \vec{v}(\xi) + \int_\Gamma \vec{v} q^\star d\Gamma = \int_\Gamma (\vec{q}^\star \times \vec{n}) \times \vec{v} d\Gamma + \int_\Omega \vec{\omega} \times \vec{q}^\star d\Omega + \int_\Omega \mathcal{D} \vec{q}^\star d\Omega. \tag{17}$$

Considering the vorticity kinetics in an integral representation one has to take into account parabolic diffusion convection character of the vorticity transport equation. With the use of the linear parabolic diffusion differential operator, the following integral formulation can be written

$$c(\xi)\, \omega(\xi, t_F) + \int_\Gamma \omega Q^\star d\Gamma = \frac{1}{\eta_o} \int_\Gamma \left(\eta_o \frac{\partial \omega}{\partial n} - \rho_o v_n \omega + \rho g_t + f_t^m \right) U^\star d\Gamma$$

$$+ \frac{1}{\eta_o} \int_\Omega (\rho_o v_j \omega + \rho e_{ij} g_i + e_{ij} f_i^m)\, Q_j^\star d\Omega + \int_\Omega \omega_{F-1} u_{F-1}^\star d\Omega, \tag{18}$$

in which a constant variation of all field and material functions within the individual time increment $\Delta t = t_F - t_{F-1}$ is assumed [11], e.g. the values at $t = t_F$ are considered for each time step, and where v_n, g_t and f_t^m are the normal velocity, and the tangential gravity and nonlinear material source components, respectively, e.g. $v_n = \vec{v}\cdot\vec{n}$, $g_t = \vec{g}\cdot\vec{t} = -e_{ij}g_i n_j$ and $f_t^m = \vec{f}^m\cdot\vec{t} = -e_{ij}f_i^m n_j$.

The pressure eq. (13) is an elliptic Poisson equation, and therefore employing again the linear elliptic Laplace differential operator the following form of the pressure integral equation is obtained

$$c(\xi)\,p(\xi) + \int_\Gamma p q^\star d\Gamma = \int_\Omega f_{pi} q_i^\star d\Omega, \tag{19}$$

where the vector $\vec{f}_p$ is given by eq. (14).

The integral representation of the nonlinear heat energy transport equation is derived considering the linear parabolic diffusion differential operator, therefore the following integral representation for the thermal energy kinetics can be evaluated

$$\begin{aligned} c(\xi)\,T(\xi,t_F) + \int_\Gamma T Q^\star d\Gamma &= \frac{1}{k_o}\int_\Gamma \left(k\frac{\partial T}{\partial n} - c v_n T\right) U^\star d\Gamma \\ &-\frac{1}{k_o}\int_\Omega \left(\tilde{k}\frac{\partial T}{\partial x_j} - c v_j T\right) Q_j^\star d\Omega \end{aligned} \tag{20}$$

$$+\frac{1}{k_o}\int_\Omega \left(T v_j \frac{\partial c}{\partial x_j} + cTD - \tilde{c}\frac{\partial T}{\partial t} + \beta T\frac{Dp}{Dt} + \Phi\right) U^\star d\Omega + \int_\Omega T_{F-1} u_{F-1}^\star d\Omega.$$

5 Numerical example

Incompressible and compressible fluid flow circumstances in channels with one or two contractions are studied. Very coarse mesh is applied first, consisting of 168 quadratic boundary elements and 888 quadratic internal cells with 3721 total nodes, thus the numerical simulation results are more or less of only qualitative value to show the applicability of the presented numerical algorithm. The solution domain is shown in Figure 1.

The ideal gas (air) is chosen as a working fluid with the inflow temperature $T_o = 600K$ and pressure $p_o = 101325Pa$, the conduit walls are adiabatic, while at the outflow the convective temperature boundary conditions are prescribed. The flow field at Reynolds number values $Re = 20, 200$, and 2000 are simulated. In presented cases the Pr number value is assumed to remain constant (0.71), while the temperature dependence of the viscosity is given by the Sutherland's model

$$\frac{\eta(T)}{\eta^*} = \left(\frac{T}{T^*}\right)^{\frac{3}{2}} \frac{T^* + S}{T + S}, \tag{21}$$

and the heat conductivity is expressed as

$$k(T) = \frac{\eta(T)c_p}{Pr} \tag{22}$$

with $T^* = 273K$, $S = 110.5K$, $\eta^* = 1.68\cdot 10^{-5} kg/m/s$, $c_p = \kappa R/(\kappa - 1)$, $\kappa = 1.4$ and $R = 287.0\ J/kgK$. The influence of temperature on c_p is neglected.

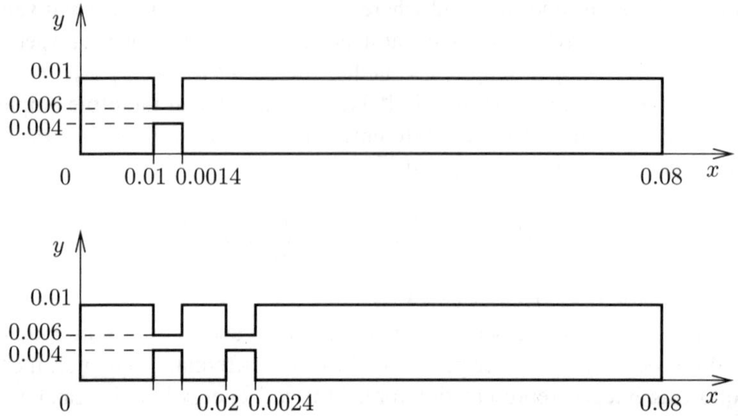

Figure 1: Numerical simulation flow domain: channel with one and two sharp contractions.

The steady state numerical simulation results for the velocity vector, pressure and temperature contours are presented in Figure 2 for $Re = 20$ for the compressible coupled momentum energy transport.

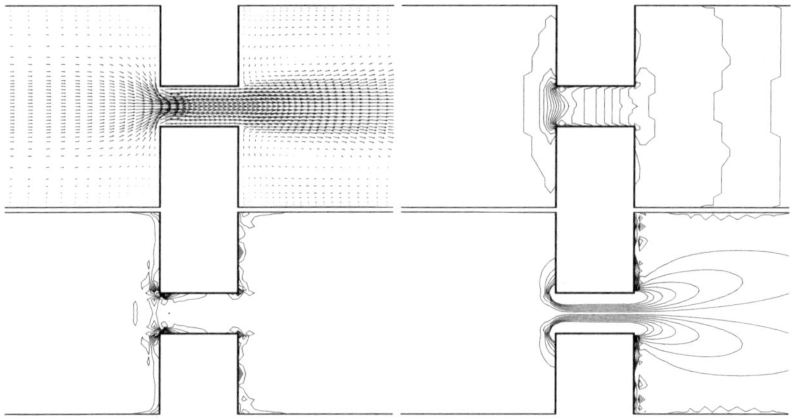

Figure 2: Numerical simulation results for $Re = 20$ for the compressible coupled momentum energy transport: velocity vector, pressure, temperature and vorticity contours.

It is evident that the flow is basically incompressible and that all nonlinear effects, such as the irreversible and reversible rate of work, are negligible and localized only on the motion around sharp edges of the contraction. For the qualitative purposes, for this very course mesh, $T_{min} = 584K$ and $T_{max} = 612K$.

The numerical solution results for the compressible and incompressible cases in general coincide.

For the $Re = 200$ flow case the steady state do not exist and the flow oscillates with strong separation of the boundary layer. The unsteady state numerical simulation results for the compressible coupled momentum energy transport for the velocity vector, pressure and temperature contours are presented in Figure 3 for $Re = 200$.

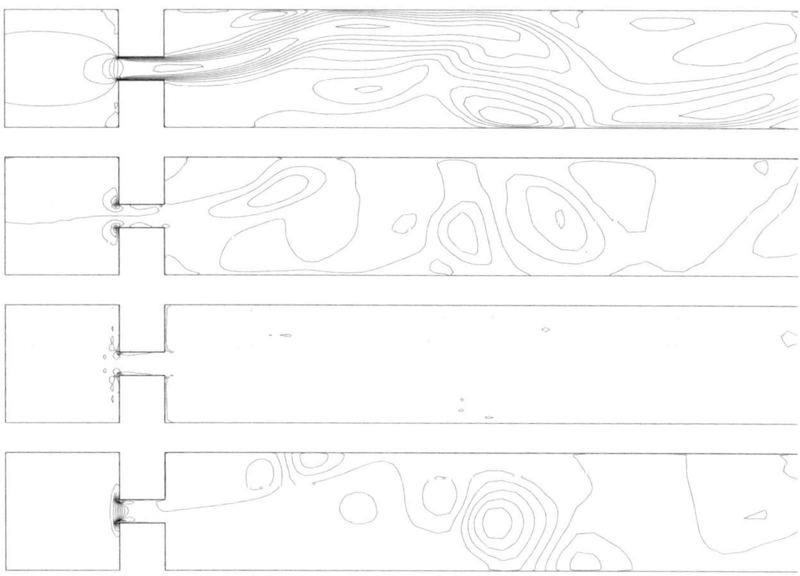

Figure 3: Numerical simulation results for $Re = 200$ for the compressible coupled momentum energy transport: velocity vector v_x, v_y, temperature and pressure contours.

Again the compressibility and all other nonlinear effects in momentum and thermal energy processes are negligible in fluid flow domain and only localized on the areas around sharp edges of the contraction. For the qualitative purposes $T_{min} = 379K$ and $T_{max} = 666K$, thus the nonlinear effects are more severe but still very localized. The increased nonlinearity decrease the stability of the numerical procedure, very small time increment $\triangle t = 10^{-4}$ has to be applied.

The flow field at Reynolds number value $Re = 2000$ was simulated next with the time increment $\triangle t = 10^{-5}$. Figure 4 shows the velocity contours, vorticity magnitude and pressure contours at a chosen time instant for the incompressible uncoupled transport case. Time traces of velocity components, vorticity and pressure a location on the centerline of the channel is plotted in Figure 5.

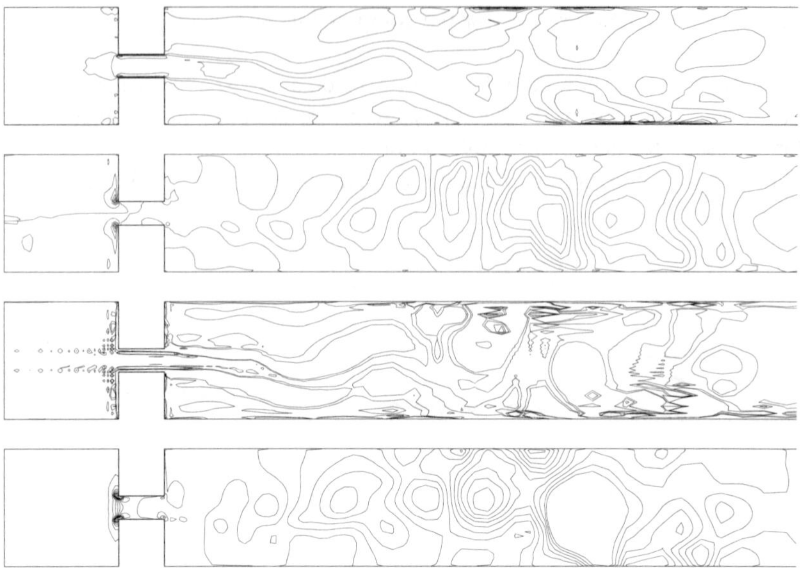

Figure 4: Numerical simulation results for $Re = 2000$ for the incompressible uncoupled flow case: horizontal velocity, vertical velocity, vorticity and pressure contours a $t = 4 \cdot 10^{-3}$.

6 Conclusions

In this work the boundary element integral approach to the solution of incompressible and compressible fluid motion in channels with sharp contractions is presented. The derived numerical model is characterized by decomposition of flow into its kinematic, vorticity kinetics, thermal energy kinetics, and pressure formulation a result of the velocity-vorticity formulation of the Navier–Stokes equation for a compressible fluid. The described numerical scheme leads to strong coupling between velocity, vorticity, temperature, pressure and mass density fields. The application of the elliptic Laplace and parabolic diffusion fundamental solutions in the derivation of integral representations ensures an accurate computation of the flow field variables.

The computed test examples confirm the applicability of BEM based numerical scheme also for a highly nonlinear transport phenomena, what compressible coupled momentum and thermal energy convection flow certainly is. The transient simulation results show development of the flow field with time caused by the compressible and viscous effects, as represented by the local expansion and vorticity field functions, respectively. The BEM mesh applied in the paper is appropriate to describe the compressible coupled momentum and thermal energy transport at Reynolds number values $Re = 20$ and $Re = 200$, but to coarse for transport phenomena at $Re = 2000$. The nonlinearities in thermal energy equation are to severe,

www.witpress.com, ISSN 1743-355X (on-line)

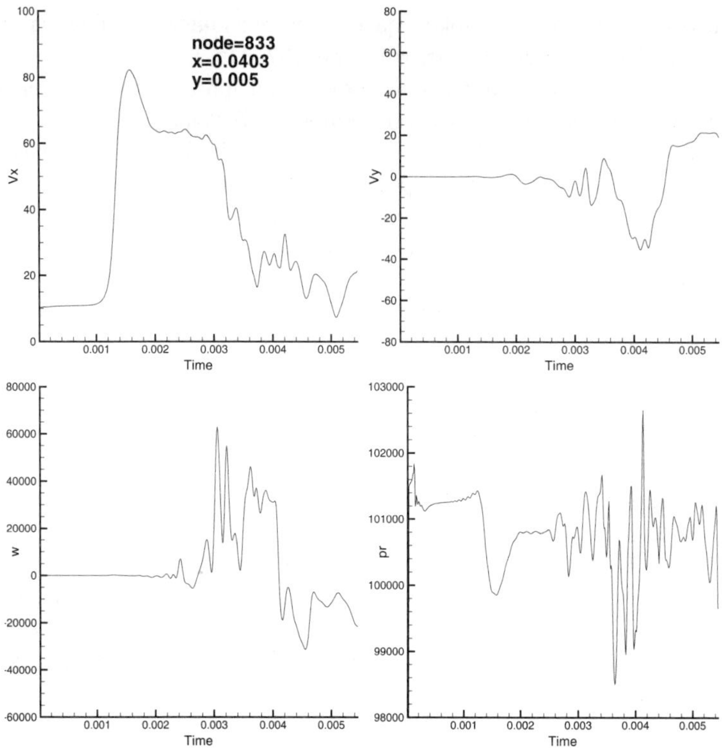

Figure 5. Numerical simulation results for $Re = 2000$ for the incompressible uncoupled flow case: time traces of horizontal velocity, vertical velocity, vorticity and pressure at $x = 0.0403, y = 0.005$.

for the mesh applied, and only compressible and incompressible uncoupled cases are simulated successfully at $Re = 2000$.

Results for the case of two contractions as well as further numerical analyses of the transition from periodic to turbulent flow circumstances will be presented at the conference.

References

[1] Škerget, L., Hriberšek, M., Kuhn, G. (1999) : Computational fluid dynamics by boundary-domain integral method. Int. J. Numer. Meth. Engng., Vol. 46, pp. 1291–1311.

[2] Škerget, L., Hriberšek, M., Žunič, Z. (2003) : Natural convection flows in complex cavities by BEM. International Journal of Numerical Methods for Heat & Fluid Flow, Vol. 13, No. 6.

[3] Wu, J.C., Thompson, J.F. (1973) : Numerical solution of time dependent

incompressible Navier–Stokes equations using an integro-differential formulation. *Computers and Fluids*, Vol. 1, pp. 197–215.

[4] Rizk, Y.M. (1980) : An integral representation approach for time dependent viscous flow. *PhD Thesis, Georgia Institute of Technology*.

[5] Wu, J.C. (1982) : Problems of general viscous fluid flow. *Developments in BEM*, Vol. 2, Ch. 2, Elsevier Appl. Sci. Publ., London and N.Y.

[6] Wu, J.C., Rizk, Y.M., Sankar, N.L. (1984) : Problems of time-dependent Navier–Stokes flow. *Developments in BEM*, Vol. 3, Ch. 6, Elsevier Appl. Sci. Publ., London and N.Y.

[7] Skerget, L., Alujevic, A., Brebbia, C.A., Kuhn, G. (1989) : Natural and forced convection simulation using the velocity-vorticity approach. *Topics in Boundary element Research*, Vol. 5, pp. 49–86, Springer-Verlag, Berlin.

[8] Škerget, L., Hriberšek, M., Kuhn, G. (1999) : Computational Fluid Dynamics by Boundary Domain Integral Method; *Int. J. Num. Meth. Eng.*, Vol. 46, pp. 1291–1311.

[9] Wrobel. L.C. (2002) : *The boundary element method*. Vol. 1, Applications in Thermo-fluids and acoustics. Wiley.

[10] Skerget, L., Jecl, R. (2004) : Compressible fluid dynamics in porous media by the boundary element method. *Emerging Technologies and Techniques in Porous Media*, Eds. Ingham, D.B., Bejan, A., Mamut, E.. Pop, I. NATO Science Series, Vol. 134, Ch. 6, Kluwer Academic Publ., Dordrecht, Boston, London.

[11] Škerget, L., Samec, N. (2005) : BEM for the two-dimensional plane compressible fluid dynamics; *Eng. Analysis with Boundary Elements*, Vol. 29, pp. 41–57.

Section 6
Heat and mass transfer

Multiscale simulation coupled DRBEM with FVM for the two-phase flow with phase change process of micrometer scale particles

W.-Q. Lu & K. Xu
College of Physical Science, Graduate University of Chinese Academy of Sciences, Beijing, People's Republic of China

Abstract

A multiscale computational method couple dual reciprocity boundary element method (DRBEM) with finite volume method (FVM) is developed, and used to numerically simulate the hydrothermal gasification and evaporation of micrometer scale particles in two-phase flow. The engineering applied background of the problems include: hydrogen production from biomass in supercritical water, water mist fire suppression, etc. Numerical results shows the multiscale computational method coupled DRBEM with FDM is credible and effective. This paper also presents some valuable numerical results for these engineering problems.
Keywords: multiscale computational method, DRBEM, FVM, hydrogen production by biomass, water mist fire suppression

1 Introduction

The two-phase flow with evaporation and gasification process of micrometer scale particles occurs in some engineering and scientific problems. The hydrogen production process by biomass particles is implemented in the column reactor with electric heating wall. The uniform mixture of supercritical water and biomass particles continuously flows into the reactor. Hydrogen and other flammable gases are produced by the hydrothermal gasification of biomass particles. There exist complex process of two phase flow, heat transfer and hydrothermal gasification of biomass particles. The process of water mist fire suppression is that micrometer scale water droplets are jetted into high

WIT Transactions on Modelling and Simulation, Vol 44,
www.witpress.com, ISSN 1743-355X (on-line)
doi:10.2495/BE070191

temperature environment. The moving water droplets continuously absorb heat from the environment and are vaporized, so that depresses the environment temperature and wins fire suppression.

The main object of this paper is to explore the multiscale numerical method to solving these problems, and open out some valuable results for these engineering practical problems.

2 Physical models

Figure 1 is the sketch map of hydrogen production from biomass particles. The reactor is a cylindrical tube with the length 650 mm and the radius 3 mm. The wall temperature keeps 650°C. The volume fraction and radius of biomass particles at inlet are 0.03 and 0.15~0.35 mm, respectively. Hydrothermal gasification of biomass particles occurs in supercritical water. The products of gas flow out at right end of the tube.

Figure 2 is the sketch map of water mist fire suppression. In the domain of the high 3 m with infinite length and width, spray nozzle is fixed at 2.5 m high over ground. The micro water droplets and air simultaneously are ejected. The initial radius and velocity of the droplets are 250 μm and 1m/s, respectively. The mass ratio between air and droplets is 0.5. The air velocity is 0.01 m/s at spray nozzle. The top wall temperature keeps 400 K and the bottom temperature keeps 350 K.

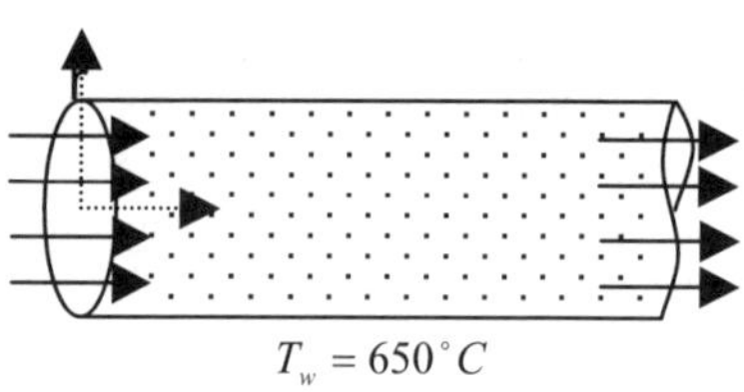

Figure 1: Schematic hydrogen production.

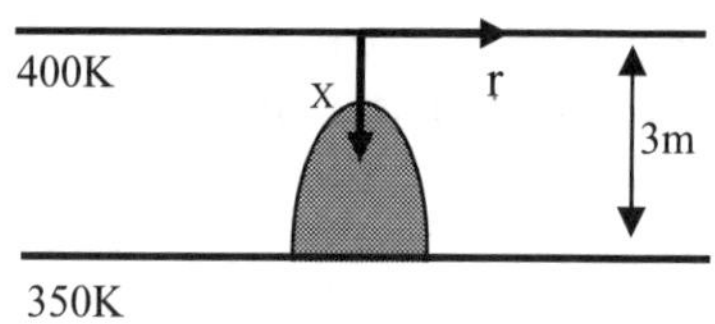

Figure 2: Schematic water mist fire suppression.

In both processes, all micro particles are assumed as spherical particles. In numerical method, it needs to consider the numerical models of particles, fluids and their coupling. In order to simplify computation, the two flow fields are considered as axial symmetrical and steady.

2.1 The gasification model of biomass particles

Hydrothermal gasification of biomass particles in supercritical water are chemical reaction processes which simultaneously occur at its surface and inside. It is assumed that the gasification of single particle is the process to gradually increase its porosity and decrease its density.

Many researches on the mechanism of the chemical reaction about gasification of biomass particles in supercritical water have been made up to now [1]. In this paper, some simplified assumptions are made: the main component of biomass is assumed as cellulose, it can be rapidly transform into glucose in supercritical water, then gasified. In addition, few products of multi carbon gases C_2H_4 and C_2H_6, etc can be neglected. The mechanism of glucose gasification reaction is summed up on the basis of experimental data [2, 3], it can be expressed as the following equation:

$$C_6H_{10}O_5 + 4.5H_2O \rightarrow 4.5CO_2 + 7.5H_2 + 0.5CO + CH_4 \quad (1)$$

Consider a micro control volume of particle's inner, its energy balance equation in cylindrical coordinates (see Fig. 3) can be written as the follows:

$$\frac{\partial}{\partial t}\left(\rho_s c_{p,s} T_s\right) = k_s\left(\frac{\partial^2 T_s}{\partial r^2} + \frac{1}{r}\frac{\partial T_s}{\partial r} + \frac{\partial^2 T_s}{\partial z^2}\right) + \left(-\Delta H\right)\left(-\frac{\partial \rho_s}{\partial t}\right) \quad (2a)$$

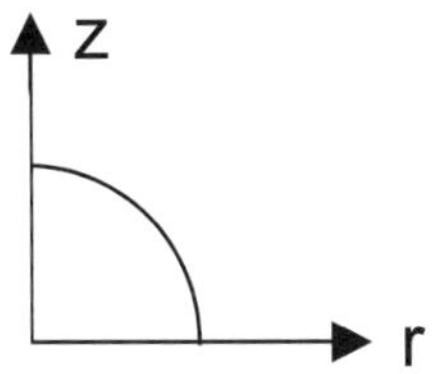

Figure 3: Computational domain of particle's quarter-sphere.

Boundary Condition:

$$-k_s\left(\frac{\partial T_s}{\partial r}\right)_{r=R} = h\left(T_s - T_f\right) \quad (2b)$$

where, ρ_s and T_s are the local density and temperature of the particle. ΔH is reaction heat. h is the convective heat transfer coefficient. T_f is fluid temperature.

2.2 The evaporation model of water droplet

Moving droplet is continually vaporized since convective heat transfer from surrounding fluid. This is a phase change process which occurs at the surface of the droplet. Its energy equation can be written as the follows:

$$\frac{\rho_l c_{pl}}{k_l}\frac{\partial T_l}{\partial t} = \frac{\partial^2 T_l}{\partial r^2} + \frac{1}{r}\frac{\partial T_l}{\partial r} + \frac{\partial^2 T_l}{\partial z^2} \quad (3a)$$

Boundary condition at the surface:

$$-k_l A\frac{dT_l}{dr} + hA\left(T_f - T_l\right) = -L\dot{m} \qquad r = r_w \quad (3b)$$

where, ρ_l and T_l are the density and temperature of the droplet. L is the latent heat of evaporation. A is the surface area of the particle. h is the convective heat transfer coefficient. $\dot{m}$ is the evaporation rate of the droplet: $\dot{m}=\rho_l 4\pi r^2 dr/dt$. The rate change of the droplet radius is [4]:

$$\frac{dr}{dt} = -\frac{\rho_f}{\rho_l}\frac{D_f}{2r} Sh \ln(1+B) \tag{4}$$

where, D_f and Sh are the coefficient of mass diffusion and Sherwood number, respectively. B is the mass transfer number.

2.3 The motion equation of a particle

$$m_d \frac{d\mathbf{v}}{dt} = C_D \rho_f (\mathbf{u}_f - \mathbf{v})|\mathbf{u}_f - \mathbf{v}|\frac{A_d}{2} + m_d \mathbf{g} \tag{5}$$

where, m_d and A_d are the mass and surface area of particle, respectively. $\mathbf{u}_f$ and $\mathbf{v}$ are the velocity of fluid and particle respectively. The drag coefficient C_D is given by the following expression [4]:

$$C_D = \begin{cases} 24\,(1+0.166\ \mathrm{Re}^{2/3})\,/\,\mathrm{Re} & \mathrm{Re} < 1000 \\ 0.424 & \mathrm{Re} > 1000 \end{cases} \tag{6}$$

where $\mathrm{Re} = \rho_f d_p |\mathbf{u}_f - \mathbf{v}| / \mu_f$.

Particle's track can be determined from:

$$\mathbf{x} = \mathbf{x}_{\mathbf{d,0}} + (\mathbf{v} + \mathbf{v}_0)\Delta t / 2 \tag{7}$$

where $\mathbf{x}_{\mathbf{d,0}}$ is the droplet position at the beginning of the time increment.

2.4 The controlling equations of fluid

The controlling equations are axial symmetric steady N-S equations:

$$\frac{\partial}{\partial x}(\rho_f u\Phi) + \frac{1}{r}\frac{\partial}{\partial r}(r\rho_f v\Phi) = \frac{\partial}{\partial x}\left(\Gamma_\Phi \frac{\partial \Phi}{\partial x}\right) + \frac{1}{r}\frac{\partial}{\partial r}\left(r\Gamma_\Phi \frac{\partial \Phi}{\partial r}\right) + S_\Phi + S_\Phi^p \tag{8}$$

where, when Φ are given as different values, equations (8) are expressed as mass, momentum, energy, turbulent kinetic energy and the dissipation rate equations, respectively. S_Φ^p is the source term produced by particles. Table 1 expresses the corresponding parameter values for different equations.

where, $G_k = \mu_{eff}\left\{2\left[\left(\frac{\partial u}{\partial x}\right)^2 + \left(\frac{\partial v}{\partial r}\right)^2 + \left(\frac{v}{r}\right)^2\right] + \left(\frac{\partial u}{\partial r} + \frac{\partial v}{\partial x}\right)^2\right\}$, μ_{eff} is effective turbulent viscosity coefficient. σ_k, σ_ε, σ_h, σ_Y, c_1 and c_2 are experience constants. S_m, S_u, S_v, S_h and S_Y are the source terms of mass, momentum, energy and species produced by particles.

In the calculation of the species in hydrogen production process, employing equation (1) to calculate gas production: CO, CO_2, H_2, CH_4. In addition, the water gas shift reaction is also considered:

$$CO + H_2O \rightarrow CO_2 + H_2 \tag{9}$$

This is a reversible reaction. However, since the concentration of water is more than that of other species, therefore the above single-direction reaction is only considered.

Table 1: Corresponding parameters for N-S equations.

Equations	Φ	Γ_Φ	S_Φ	S_Φ^p
Mass	1	0	0	S_m
Axial momentum	u	μ_{eff}	$-\frac{\partial p}{\partial x}+\frac{\partial}{\partial x}(\mu_{eff}\frac{\partial u}{\partial x})+\frac{1}{r}\frac{\partial}{\partial r}(r\mu_{eff}\frac{\partial v}{\partial x})$	S_u
Radial momentum	v	μ_{eff}	$-\frac{\partial p}{\partial r}+\frac{\partial}{\partial x}(\mu_{eff}\frac{\partial u}{\partial r})+\frac{1}{r}\frac{\partial}{\partial r}(r\mu_{eff}\frac{\partial v}{\partial r})-2\mu_{eff}\frac{v}{r^2}$	S_v
Kinetic energy	k	μ_{eff}/σ_k	$G_k-\rho\varepsilon$	0
Dissipation rate	ε	$\mu_{eff}/\sigma_\varepsilon$	$\frac{\varepsilon}{k}(c_1G_k-c_2\rho\varepsilon)$	0
Enthalpy	h	μ_{eff}/σ_h	0	S_h
Species	Y	μ_{eff}/σ_Y	0	S_Y

3 Numerical methods

Since the volume fractions of particles are small in these problems, particles are dispersed in fluids. Employing the two fluids model of two-phase flow to calculate these problems is not appropriate. The model of dispersive particles groups is considered in this calculation. In this model, it needs to calculate the behaviors of the dispersive particles groups and fluid. A multiscale computational method coupled macroscopic fluid fields with micro particle fields is constructed as the follows.

3.1 The calculation of micro particle fields

Our improved axisymmetric DRBEM (ADRBEM) [5] is used to solve equation (2) and (3) for both different problems respectively. The choice of function f is very important for the calculation by DRBEM. However, in early ADRBEM, it is impossible to arrange nodes on the symmetrical axis since the singularity of function f early used in ADRBEM. In reference [5], the singularity is avoided by integral averaging and selecting different assemblage of functions f. In this calculation, the marked B-02 and B-05 type f function [5] are used to calculate biomass particle and droplet, respectively.

B-02:

$$r \neq 0: f = 1 - \frac{1}{3}\frac{r_i}{r} + \hat{p}\left(1 - \frac{1}{4}\frac{r_i}{r}\right) + \hat{p}^2\left(1 - \frac{1}{5}\frac{r_i}{r}\right),\ \hat{p} = \sqrt{(r-r_i)^2 + (z-z_i)^2}$$

$$r = 0: f = 1 + \hat{p}\left(1 - \frac{1}{4}\frac{r_i^2}{\hat{p}^2}\right) + \hat{p}^2\left(1 - \frac{2}{5}\frac{r_i^2}{\hat{p}^2}\right),\ \hat{p} = \sqrt{r_i^2 + (z-z_i)^2} \tag{10}$$

B-05:

$r \neq 0$:

$$f = 1 - \frac{1}{3}\frac{r_i}{r} + \hat{p}\left(1 - \frac{1}{4}\frac{r_i}{r}\right) + \hat{p}^2\left(1 - \frac{1}{5}\frac{r_i}{r}\right) + \hat{p}^3\left(1 - \frac{1}{6}\frac{r_i}{r}\right) + \hat{p}^4\left(1 - \frac{1}{7}\frac{r_i}{r}\right) + \hat{p}^5\left(1 - \frac{1}{8}\frac{r_i}{r}\right),$$

$$\hat{p} = \sqrt{(r-r_i)^2 + (z-z_i)^2}$$

$r = 0$:

$$f = 1 + \hat{p}\left(1 - \frac{1}{4}\frac{r_i^2}{\hat{p}^2}\right) + \hat{p}^2\left(1 - \frac{2}{5}\frac{r_i^2}{\hat{p}^2}\right) + \hat{p}^3\left(1 - \frac{1}{2}\frac{r_i^2}{\hat{p}^2}\right) + \hat{p}^4\left(1 - \frac{4}{7}\frac{r_i^2}{\hat{p}^2}\right) + \hat{p}^5\left(1 - \frac{5}{8}\frac{r_i^2}{\hat{p}^2}\right)$$

$$,\ \hat{p} = \sqrt{r_i^2 + (z-z_i)^2} \tag{11}$$

The inner temperature distributions of two different kinds of particles are obtained respectively. The variations of biomass particle's density and droplet size are further gotten respectively. It is noticed that the differences between the calculations in two different kinds of particles exist as the follows.

In equation (2), $\partial\rho_s / \partial t = -k_g \rho_s$, where, k_g is the coefficient of the reaction [2, 3]:

$$k_g = -10^{3.09\pm0.26} \exp(-63.6 \pm 3.9 / RT) \tag{12}$$

Since equation (2) is nonlinear, Newton iterative method is used to DRBEM calculation.

In equation (3), the mass transfer number B is the function of particle's temperature, therefore $\dot{m}$ is also correlated with particle's temperature. Hence boundary condition (3b) is nonlinear. In order to simplify the calculation, boundary condition (3b) is transformed into:

$$-k_l \frac{dT_l}{dr} = h'(T_l - T_f) \qquad r = r_w \tag{13}$$

where, h' is effective coefficient of convective heat transfer, which can be calculated by iteration.

After solving equations (5-7), the velocities and tracks of particles are gotten. Finally, the source terms S_Φ^p of particle are calculated by particle- source-in cell (PSIC) method [6].

3.2 The calculation of macroscopic fluid fields

Finite volume method with SIMPLEC scheme is used to solving equations (8). All physical quantities of macroscopic fluid fields are obtained.

3.3 The couple calculation of both fields

Macroscopic fluid fields and micro particle fields are coupled by the source terms S_Φ^p. This couple computation is an iterative calculation between fluid fields and the source terms of particles.

4 Numerical results

4.1 Hydrogen production by biomass particles in supercritical water

Figure 4 shows the comparison between experimental data [7] and numerical results of main gas production under the conditions: particles radius 0.2mm, reaction time 0.5min. As shown in Fig. 4, the better agreement between both results indicates the models and methods are credible.

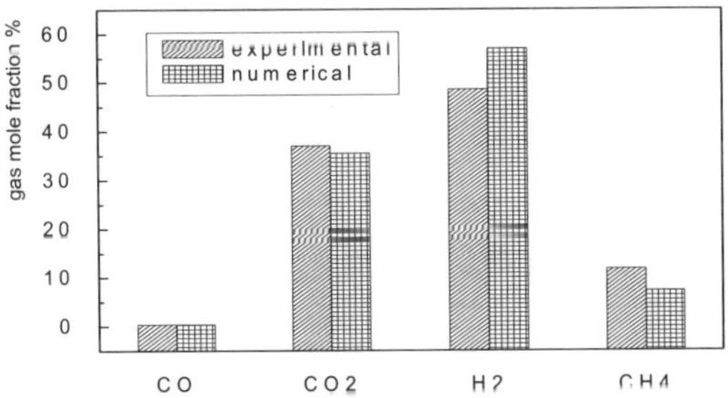

Figure 4: The numerical gas mole fraction compared with experimental data.

Figure 5 shows the effect of the wall temperature on the mole fraction of gas production in particles radius 0.2mm. As shown in the figure, H_2 increases with increasing temperature, but CO is decreased with increasing temperature. The reason is when temperature is higher, the rate of water gas shift reaction (9) between water and vapor is rapider. In the case of enough water, CO is consumed very much, and simultaneously H_2 is produced.

Figure 6 depicts the variation of hydrogen yield with particle radius under same flux, temperature and mass fraction. As shown in the figure, Hydrogen yield is decreased with increasing particle radius. This is since under same flux and mass fraction, the smaller the particle is, the more the particle's number is,

www.witpress.com, ISSN 1743-355X (on-line)

so that the interactional area between water and particles increase. This induces the hydrogen yield increases.

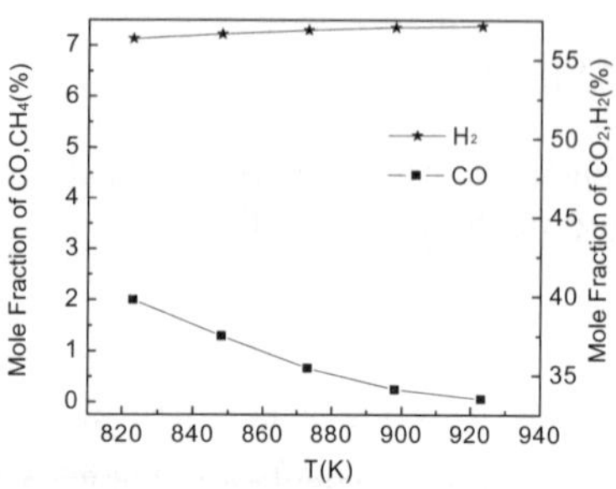

Figure 5: Effect of reactor temperature on mole fraction of gas.

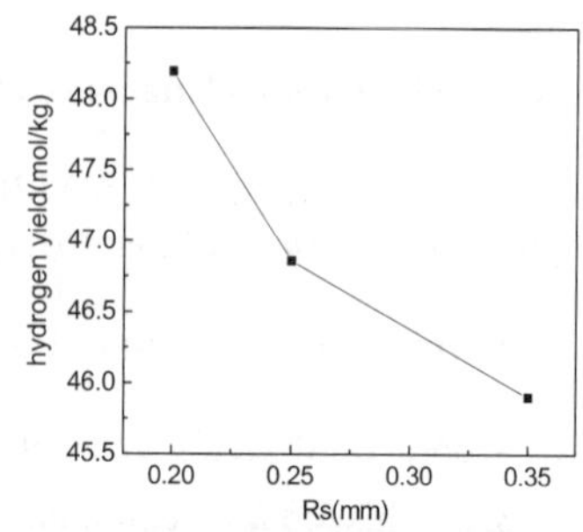

Figure 6: Effect of particle size on hydrogen yield.

4.2 Water spraying

Figure 7 shows the velocity field of the fluid. In neighborhood of the nozzle, the motion of fluid is dragged by downward moving particles, which induces to form an anticlockwise vortex. In the radial location 20−30m, the velocity field keeps invariable. Therefore the radial location over 20m can be considered as infinite beyond.

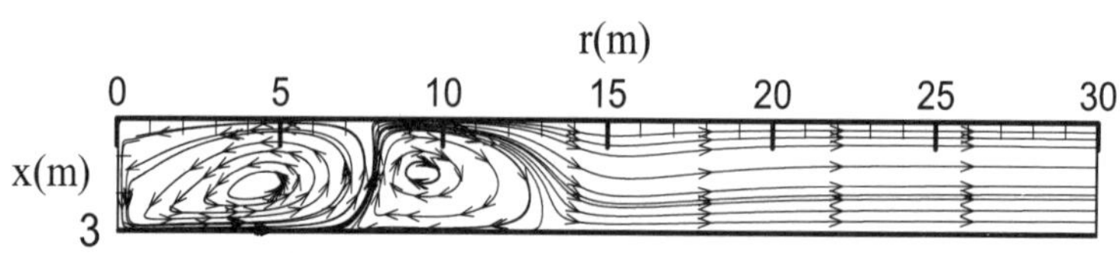

Figure 7: Velocity vector map of fluid.

Figure 8 depicts the temperature field of fluid. The low temperature domain in neighborhood of central axis is produced since here the concentration of droplet is higher, and the plentiful quantity of heat is absorbed by evaporation, that decreases local temperature.

Figure 9 shows a moving track of a particle and the flow field in neighborhood of central axis. As shown in the figure, initial moving track of the particle is near parabola, afterward its track gradually tends to local streamline of fluid. This reason is that the radius of the particle becomes gradually small since evaporation, and the effect of fluid motion on the particle is gradually enhanced.

Figure 10 depicts the variation of water drip's radius with time. As shown in the figure, the variation velocity of the drip size is slower and slower. This reason is that the speed of evaporation becomes slower and slower with the drop moving. This is also corresponded to the temperature variation of environment as shown in Fig. 8.

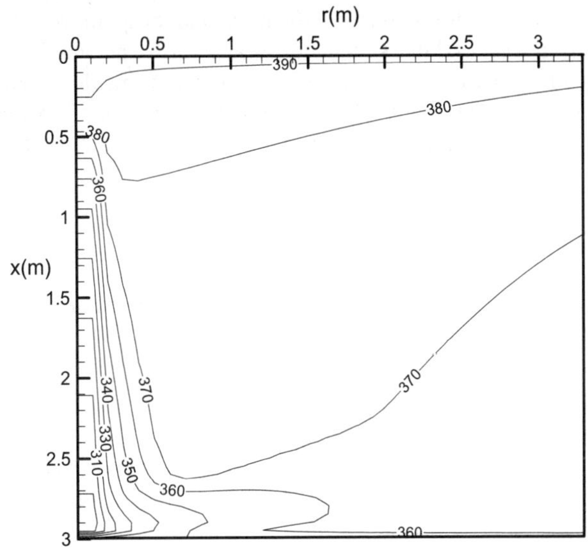

Figure 8: The temperature field of fluid.

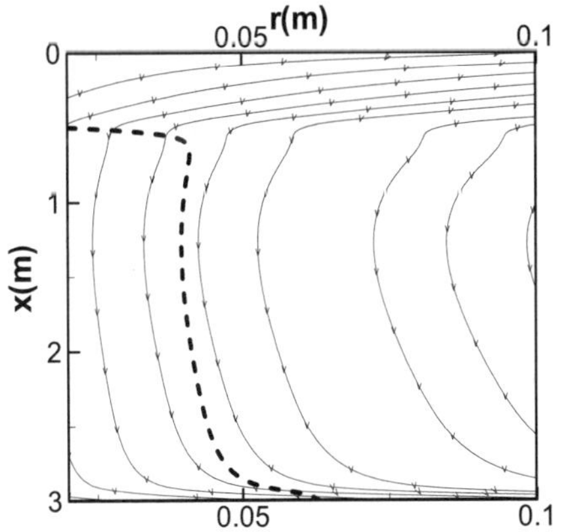

Figure 9: Moving track of particle and flow field (particle's track).

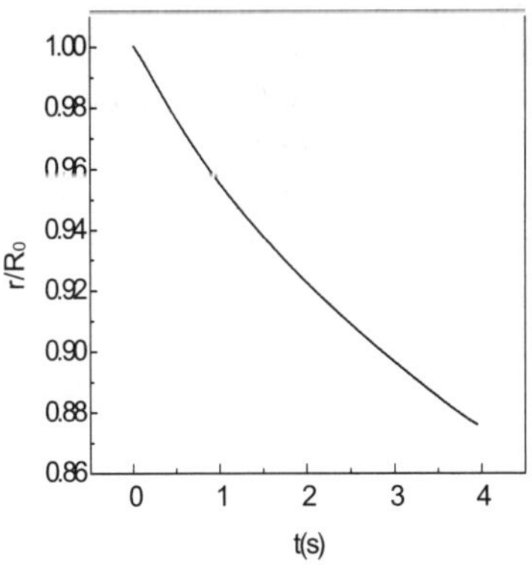

Figure 10: The variation of water droplet radius.

5 Conclusions

The multiscale computational method coupled macroscopic fluid fields by FVM with micro particles fields by DRBEM is developed and used to numerically simulate the hydrogen production by biomass particles in supercritical water and water spraying. The hydrothermal gasification and evaporation of dispersive-phase particles are successfully simulated respectively. The simulation gives the comparable with experimental data and physical reasonable results of micro particles fields and fluid fields. Present numerical examples express that the method is effective and credible. Some valuable results for these engineering problems are also given.

Acknowledgement

The project is supported by the National Natural Scientific Foundation of China (Grant 50676102 and 50536030)

References

[1] Lu, Y.J. & Guo, L.J., etc, Hydrogen production by biomass gasification in supercritical water: A parametric study. *International Journal of Hydrogen Energy* 31, pp. 822-831, 2006.

[2] Wei, F. & Hedzer, J., etc, Biomass conversions in subcritical and supercritical water: driving force, phase equilibrium, and thermodynamic analysis. *Chemical Engineering and Processing,* 43, pp. 1459–1467, 2004.

[3] Lee, I. & Kim, M.., etc, Gasification of glucose in supercritical water. *Ind. Chem. Eng. Res.* 41, pp. 1182–1188, 2002.

[4] Bracco, F.V. Modelling of engine sprays, *SAE Paper* 850394, 1985.

[5] Bai, F.W. & Lu,W.-Q., The selection and assemblage of approximation functions and disposal of its singularity in axisymmetric DRBEM for heat transfer problems. *Engineering Analysis with Boundary Elements*, **28**, pp. 955-965, 2004.

[6] Crowe, C.T. & Sharma, M.P., etc, The Particle-Source-In Cell(PSI-CELL) Model for Gas-Droplet Flows. *Journal of Fluids Engineering,* 99, pp. 325-330, 1977.

[7] Lv, Y.J. & Guo, L.J., etc, Experimental research of hydrogen production from biomass particles gasification in supercritical water, *The National Great Basic Research Development Project of China (Grant 2003CB214500) Research Paper* 3-06, 2004.

Boundary Element Method for double diffusive natural convection in a horizontal porous layer

J. Kramer[1], R. Jecl[1] & L. Škerget[2]
[1] *Faculty of Civil Engineering, University of Maribor, Slovenia*
[2] *Faculty of Mechanical Engineering, University of Maribor, Slovenia*

Abstract

A numerical study of double-diffusive natural convection in porous media using the Boundary Element Method is presented. The studied configuration is a horizontal layer filled with fluid saturated porous media, where different temperature and concentration values are applied on the horizontal walls, while the vertical walls are adiabatic and impermeable. Transport phenomena in porous media are described with the use of modified Navier–Stokes equations in the form of conservation laws for mass, momentum, energy and species. The results for different governing parameters (Rayleigh number, Darcy number, buoyancy ratio and Lewis number) are presented and compared with those in published studies.
Keywords: Boundary Element Method, porous media, double-diffusive natural convection.

1 Introduction

Transport phenomena in porous medium is a subject of intensive research in last couple decades, mainly because of wide range of applications in many engineering branches. Problems of natural convection in porous media are most commonly studied examples. Many reported studies are dealing with natural convection driven by thermally buoyancy forces. A related problem that has received less attention is the so-called double-diffusive convection, where density differences occur due to combined thermal and compositional gradients across the porous layer. Some applications, where thermal natural convection or combined double-diffusive natural convection are observed, are fibrous insulation, geothermal energy, underground spreading of contaminants, solidification processes.

In horizontal layers, where horizontal walls are maintained at different temperatures and solute concentrations, the convective flow is possible above the

WIT Transactions on Modelling and Simulation, Vol 44, © 2007 WIT Press
www.witpress.com, ISSN 1743-355X (on-line)
doi:10.2495/BE070201

critical Rayleigh number. In cases where the density differences are a result of combined temperature and concentration gradients, the critical Rayleigh number is a function of the Darcy number, Lewis number and buoyancy coefficient [1]. The flow structure in these cases becomes multi-cellular and is also called a Rayleigh–Benard flow structure [2]. Most of the studies regarding double-diffusive convection or thermohaline convection (the case where the constituent is salt) in a horizontal porous layers are focused on the problem of convective instability. There are many studies dealing with the onset of convection on the basis of linear stability theory [3, 4] or nonlinear perturbation theory [5]. In these studies the critical Rayleigh numbers for the onset of convective flows are predicted. The theoretical and numerical study of heat and mass transfer affected by a high Rayleigh number Benard convection in a porous layer heated from below is obtained in [6]. The numerical results and a scale analysis of the flow in a porous medium are presented, where the buoyancy effect is due entirely to temperature gradients.

Some further numerical results for a double-diffusive convection in a horizontal porous layer with two opposing buoyancy sources can be found in [7]. The influence of the governing parameters (Rayleigh number, Lewis number, buoyancy ratio) on the overall heat and mass transfer is discussed for the case of a square cavity. Double diffusive convection in a horizontal layer with some numerical results is also discussed in [8]. The critical values of Rayleigh numbers for the onset of convection are predicted on the basis of nonlinear parallel flow approximation.

All the above-mentioned numerical results are obtained on the basis of the Darcy flow model, which is more convenient for porous media with low permeability. The Brinkman extended Darcy model, on the other hand accounts for friction due to macroscopic shear is thus more appropriate when describing fluid flows in the porous matrix. This model was used in [1] to investigate the onset and development of double-diffusive convection in a horizontal porous layer with uniform heat and mass fluxes specified at the horizontal boundaries. The obtained analytical solutions are compared to some numerical results for different values of governing parameters.

The present study is focused on the development of the Boundary Element Method, for the problem of combined heat and mass transfer through horizontal porous layer. The Brinkman extended Darcy formulation is used to model the fluid flow in porous media, where the momentum equation is equivalent to the classical Navier–Stokes equations for pure fluid flow. The general set of equations is transformed with use of velocity-vorticity formulation, which consequently separates the numerical scheme into a kinematic and kinetic computational part [9].

2 Mathematical formulation

The configuration studied in the present paper is shown in fig. 1.

It is a horizontal layer of width D and height H, filled with homogenous nondeformable porous media, fully saturated with Newtonian fluid. The horizontal

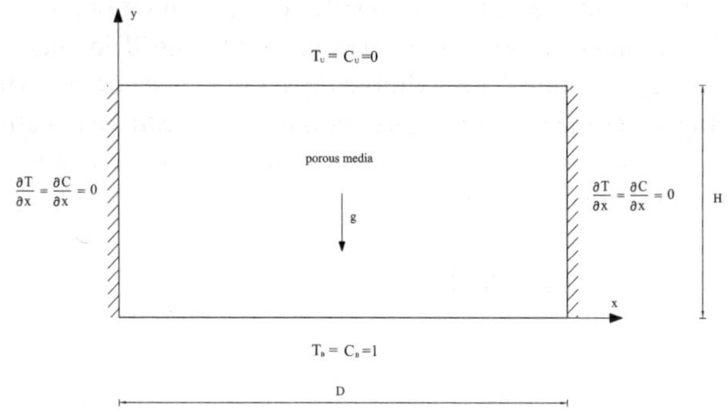

Figure 1: Geometry of a horizontal layer with boundary conditions.

walls are subjected to different temperature and concentration values (T_B, C_B at the bottom boundary and T_U, C_U at the upper boundary), while the vertical walls are adiabatic and impermeable. The fluid saturating the porous media is modelled as a Boussinesq incompressible fluid, where the density depends only on temperature and concentration variations: $\rho = \rho_0(1-\beta_T(T-T_0)-\beta_C(C-C_0))$, where the subscript 0 refers to a reference state, β_T and β_C are volumetric thermal and concentration expansion coefficients.

Transport phenomena in porous media is described using modified Navier–Stokes equations. The general set of macroscopic equations for conservation of mass, momentum, energy and species are written considering the fact that only a part of the volume, expressed with porosity (ϕ) is available for the flow of the fluid:

$$\frac{\partial v_i}{\partial x_i} = 0, \tag{1}$$

$$\frac{1}{\phi}\frac{\partial v_i}{\partial t} + \frac{1}{\phi^2}\frac{\partial v_j v_i}{\partial x_j} = -\frac{1}{\rho_0}\frac{\partial p}{\partial x_i} + F g_i - \frac{\nu}{K} v_i + \frac{\partial}{\partial x_j}\left(2\frac{\nu}{\phi}\dot{\varepsilon}_{ij}\right), \tag{2}$$

$$\frac{\partial}{\partial t}\left[\phi c_f + (1-\phi)c_s\right]T + c_f\frac{\partial v_j T}{\partial x_j} = \frac{\partial}{\partial x_j}\left(\lambda_e \frac{\partial T}{\partial x_j}\right), \tag{3}$$

$$\phi\frac{\partial C}{\partial t} + \frac{\partial v_j C}{\partial x_j} = \frac{\partial}{\partial x_j}\left(D\frac{\partial C}{\partial x_j}\right). \tag{4}$$

The parameters, used above are: v_i volume-averaged velocity, x_i the i-th coordinate, ϕ porosity, t time, ρ density, ν kinematic viscosity, $\partial p/\partial x_i$ the pressure gradient, g_i gravity and K permeability of porous media. F is the normalized density difference function and is given as: $F = (\rho - \rho_0)/\rho_0 = -\left[\beta_T(T-T_0) + \beta_C(C-C_0)\right]$. Furthermore $c_f = (\rho c)_f$ and $c_s = (\rho c)_s$ are the heat capacities for the fluid and solid phases, respectively, T is temperature, λ_e the effective thermal conductivity of the porous media given as $\lambda_e = \phi\lambda_f + (1-\phi)\lambda_s$, where λ_f and λ_s are thermal conductivities for the fluid and solid phases, respectively. In the final equation C stands for concentration, and D for mass diffusivity. The momentum equation (2) is known as the Brinkman extension of the classical Darcy equation. The additional Brinkman viscous term (fourth on the r.h.s.) is analogous to the Laplacian term in the Navier–Stokes equations for pure fluid and accounts for viscous resistance or viscous drag force exerted by the solid phase on the flowing fluid at their contact surfaces.

3 Boundary Element Method

In the present study the extension of the classical Boundary Element Method (BEM) is used, the so called Boundary Domain Integral Method (BDIM) [9, 10]. Because in the obtained set of integral equations boundary and domain integrals are present, the discretization of surface and domain is required.

To use the BDIM the above given general set of equations should first be modified. Firstly, the modified velocity $v_i' = v_i/\phi$ is introduced. The material properties, kinematic viscosity ν in the momentum equation, thermal diffusivity a_P in the energy equation and mass diffusivity D in the species equation are divided into a constant and variable part as follows: $\nu = \bar{\nu} + \tilde{\nu}$, $a_P = \bar{a}_P + \tilde{a}_P$ and $D = \bar{D} + \tilde{D}$. The momentum, energy and species equation can now be written as:

$$\frac{\partial v_i'}{\partial t} + \frac{\partial v_j' v_i'}{\partial x_j} = -\frac{1}{\rho_0}\frac{\partial p}{\partial x_i} + F g_i - \frac{\nu\phi}{K} v_i' + \bar{\nu}\frac{\partial^2 v_i'}{\partial x_j \partial x_j} + \frac{\partial}{\partial x_j}\left(2\tilde{\nu}\dot{\varepsilon}_{ij}\right), \tag{5}$$

$$\frac{\sigma}{\phi}\frac{\partial T}{\partial t} + \frac{\partial v_j' T}{\partial x_j} = \frac{\bar{a}_P}{\phi}\frac{\partial^2 T}{\partial x_j \partial x_j} + \frac{\partial}{\partial x_j}\left(\frac{\tilde{a}_P}{\phi}\frac{\partial T}{\partial x_j}\right), \tag{6}$$

$$\frac{\partial C}{\partial t} + \frac{\partial v_j' C}{\partial x_j} = \frac{\bar{D}}{\phi}\frac{\partial^2 C}{\partial x_j \partial x_j} + \frac{\partial}{\partial x_j}\left(\frac{\tilde{D}}{\phi}\frac{\partial C}{\partial x_j}\right), \tag{7}$$

where $\dot{\varepsilon}_{ij}$ represents the strain rate tensor $\dot{\varepsilon}_{ij} = 1/2(\partial v_i'/\partial x_j + \partial v_j'/\partial x_i)$ and σ is the heat capacity ratio given as $\sigma = \phi + (1+\phi)c_s/c_f$.

3.1 Velocity-vorticity formulation

In the next step the above-stated governing equations are transformed by the use of velocity-vorticity formulation (VVF), consequently the computational scheme is partitioned into its kinematic and kinetic parts [9].

In the present study a two-dimensional problem is considered, so all subsequent equations will be written for case of planar geometry.

The vorticity vector, which represents the curl of the velocity field $\omega = e_{ij}\partial v_j/\partial x_i$ is introduced, where e_{ij} is the unit permutation tensor. The kinematic part is represented by the elliptic velocity vector equation:

$$\frac{\partial^2 v_i'}{\partial x_j \partial x_j} + e_{ij}\frac{\partial \omega'}{\partial x_j} = 0, \tag{8}$$

where ω' is the modified vorticity $\omega' = \omega/\phi$. The kinetics is governed by the vorticity, energy and species transport equation. The vorticity transport equation is obtained as a curl of the Brinkman momentum equation (5):

$$\frac{\partial \omega'}{\partial t} + v_j' \frac{\partial \omega'}{\partial x_j} = \bar{\nu} \frac{\partial^2 \omega'}{\partial x_j \partial x_j} + e_{ij}\, g_j \frac{\partial F}{\partial x_j} - \frac{\nu \phi}{K} \omega' + \frac{\partial}{\partial x_j} \left(\tilde{\nu} \frac{\partial \omega'}{\partial x_j} \right) + \frac{\partial f_j}{\partial x_j}, \quad (9)$$

where f_j is the contribution arising on account of nonlinear material properties. Equations (6), (7), (8) and (9) represent the leading non-linear set of equations, to which the weighted residual technique has to be applied. Integral representation of kinematic equation is:

$$c(\xi) v_i'(\xi) + \int_\Gamma v_i' q^* d\Gamma = \int_\Gamma \frac{\partial v_i'}{\partial n} u^* d\Gamma + \int_\Omega b_i u^* d\Omega, \quad (10)$$

where the parameter $c(\xi)$ denotes coefficient related to the position of the source point. b_i stands for the pseudo-body source term and is in this case $b_i = e_{ij} \partial \omega' / \partial x_j$, u^* is the elliptic Laplace fundamental solution and q^* is its normal derivative e.g. $q^* = \partial u^* / \partial n$. The fundamental solution u^* for the case of planar geometry is given by the expression:

$$u^* = \frac{1}{2\pi} ln \left(\frac{1}{r(\xi, s)} \right), \quad (11)$$

where r is the vector from the source point ξ to the reference field point s. With further mathematical reformulations and use of Gauss-divergence theorem following integral formulation for kinematics can be written:

$$c(\xi) v_i'(\xi) + \int_\Gamma v_i' q^* d\Gamma = e_{ij} \int_\Gamma v_j'\, q_t^* d\Gamma - e_{ij} \int_\Omega \omega' q_j^* d\Omega, \quad (12)$$

where q_t^* is the tangential derivative of the fundamental solution $q_t^* = \partial u^* / \partial t$. The formulations for the vorticity, temperature and concentration can generally be written as a non-homogeneous elliptic diffusion-convective equation [10]:

$$\bar{o} \frac{\partial^2 u}{\partial x_j \partial x_j} - \frac{\partial \bar{v}_j' u}{\partial x_j} - \frac{u}{\Delta t} + b_i = 0, \quad (13)$$

where u is taken as vorticity ω', temperature T and concentration C, respectively, $\bar{o}$ is defined by considering the conservation laws and constitutive hypothesis, and b_i stands for the pseudo-body source term. Since the fundamental solution exists only for steady diffusion-convective PDE with constant coefficients, the velocity field is decomposed into an average constant vector $\bar{v}_i$ and perturbated vector $\tilde{v}_i$, such that $v_i = \bar{v}_i + \tilde{v}_i$. Thus the following integral formulation can be obtained:

$$c(\xi)u(\xi) + \bar{o}\int_\Gamma u\frac{\partial u^*}{\partial n}d\Gamma = \int_\Gamma \left(\bar{o}\frac{\partial u}{\partial n} - u\bar{v}'_n\right)d\Gamma + \int_\Omega b_i u^* d\Omega, \tag{14}$$

where u^* is the elliptic diffusion-convective fundamental solution of steady diffusion-convective PDE, in the form of:

$$u^* = \frac{1}{2\pi\bar{o}}K_0(\mu r)exp\left(\frac{\bar{v}_j r_j}{2\bar{o}}\right), \tag{15}$$

for the plane case. Parameter μ is defined as $\mu = \sqrt{(\bar{v}'/2\bar{o})^2 + \beta}$, where $\bar{v}'^2 = \overline{v'_j v'_j}$, $\beta = 1/\bar{o}\Delta t$, K_0 is the modified Bessel function of the second kind of order 0, and r is the magnitude of the vector from the source to the reference point, i.e. $r = |x_i(\xi) - x_i(s)|$. The following integral representations for vorticity, temperature and concentration kinetics are obtained according to equation (14):

$$\begin{aligned} c(\xi)\omega'(\xi) + \int_\Gamma \omega'\frac{\partial U^*_\omega}{\partial n}d\Gamma = \\ = \frac{1}{\bar{\nu}}\int_\Gamma\left(\nu\frac{\partial\omega'}{\partial n} - \omega' v'_n + e_{ij}g_j F n_j + f_j n_j\right)U^*_\omega d\Gamma + \\ + \frac{1}{\bar{\nu}}\int_\Omega\left(\omega'\tilde{v}'_j - e_{ij}g_j F - \tilde{\nu}\frac{\partial\omega'}{\partial x_j} - f_j\right)\frac{\partial U^*_\omega}{\partial x_j}d\Omega + \\ + \frac{1}{\bar{\nu}}\int_\Omega \frac{\nu\phi}{K}\omega' U^*_\omega d\Omega + \frac{1}{\bar{\nu}\Delta t}\int_\Omega \omega'_{F-1}U^*_\omega d\Omega, \end{aligned} \tag{16}$$

$$\begin{aligned} c(\xi)T(\xi) + \int_\Gamma T\frac{\partial U^*_T}{\partial n}d\Gamma = \frac{\phi}{\sigma\bar{a}_P}\int_\Gamma\left(\frac{\sigma a_P}{\phi}\frac{\partial T}{\partial n} - Tv'_n\right)U^*_T d\Gamma - \\ - \frac{\phi}{\sigma\bar{a}_P}\int_\Gamma\left(\frac{\sigma\tilde{a}_P}{\phi}\frac{\partial T}{\partial x_j} - T\tilde{v}'_j\right)\frac{\partial U^*_T}{\partial x_j}d\Gamma + \\ + \frac{\phi}{\sigma\bar{a}_P\Delta t}\int_\Omega T_{F-1}U^*_T d\Omega, \end{aligned} \tag{17}$$

$$\begin{aligned} c(\xi)C(\xi) + \int_\Gamma C\frac{\partial U^*_C}{\partial n}d\Gamma = \frac{\phi}{\bar{D}}\int_\Gamma\left(\frac{D}{\phi}\frac{\partial C}{\partial n} - Cv'_n\right)U^*_C d\Gamma - \\ - \frac{\phi}{\bar{D}}\int_\Omega\left(\frac{\tilde{D}}{\phi}\frac{\partial C}{\partial x_j} - C\tilde{v}'_j\right)\frac{\partial U^*_C}{\partial x_j}d\Omega + \\ + \frac{\phi}{\bar{D}\Delta t}\int_\Omega C_{F-1}U^*_C d\Omega, \end{aligned} \tag{18}$$

where U^*_ω, U^*_T and U^*_C are modified elliptic diffusion-convective solutions defined as $U^*_\omega = \bar{\nu}u^*$ in the momentum equation, $U^*_T = \bar{a}_P/\phi u^*$ in the energy equation and $U^*_C = \bar{D}/\phi u^*$ in the species equation.

For approximate numerical solution, the integral equations have to be written in a dicretized manner, where the integrals over boundary and domain are approximated by the sum of integrals over all boundary elements and internal cells, respectively. The variation of field functions within each boundary element or internal cell is approximated by the use of appropriate interpolation polynomials [11]. The system of discretized equations is solved by coupling those kinetic and kinematic equations and considering the corresponding boundary and initial conditions. Since the implicit set of equations is written simultaneously for all boundary and internal nodes, this procedure results in a very large and fully-populated system matrix, influenced by diffusion and convection. The consequence of this approach is a very stable and accurate numerical scheme with substantial computer time and memory demands. The subdomain technique is used to improve the economics of the computation, where the entire computational domain is partitioned into subdomains to which the same described numerical procedure can be applied [9].

4 Test example

The obtained numerical scheme is discussed on the problem described in section 2 of this paper. The governing parameters of a problem are:

- porosity, ϕ,
- Darcy number Da, given with the expression $Da = K/\phi H^2$
- aspect ratio $A = D/H$
- modified (porous) thermal Rayleigh number $Ra = Kg\beta_T \Delta T H/a_P \nu$,
- Lewis number $Le = a_F/D$,
- buoyancy ratio $N = \beta_C \Delta C/\beta_T \Delta T$.

In the above notations ΔT and ΔC are temperature and concentration differences between the upper and bottom boundaries, and D is the mass diffusivity. For aspect ratio $A = 1$, a non-uniform computational mesh 20×20 subdomains was used with a ratio between the longest and shortest elements of $r = 6$, and for $A = 2$, $A = 4$ 20×10 subdomains were used. Time-steps ranging from $\Delta t = 10^{-16}$ (steady state) to $\Delta t = 10^{-4}$ were employed, and the convergence criterion is determined as $\varepsilon = 5 \times 10^{-6}$ for all cases.

The described numerical model was tested for different values of governing parameters (Ra, Da, Le and N). It should be noted that, in the case of $N = 0$ the buoyancy effect is due entirely to temperature gradients. The mass transfer in this case is due to temperature field and concentration differences between the horizontal boundaries. In the case of positive values for buoyancy ratio ($N > 0$), the thermal and solutal buoyancy forces aid each other (aiding convection) and for negative values of buoyancy ratio ($N < 0$) the solutal and thermal effects have opposite tendencies (opposing convection).

The results for total heat and mass transfer through the horizontal layer are given by the values of Nusselt and Sherwood numbers defined as:

$$Nu = -\int_0^1 \left(\frac{\partial T}{\partial x}\right)_{x=0} dy, \qquad Sh = -\int_0^1 \left(\frac{\partial C}{\partial x}\right)_{x=0} dy. \tag{19}$$

Table 1: Comparison of results with numerical experiments reported in the literature for $A = 1$.

	Nu		Sh	
	This Study	Ref. [7]	This study	Ref. [7]
$Ra = 600, Le = 1, N = 0$	7.01	6.6	7.01	-
$Ra = 100, Le = 10, N = 0.2$	2.48	2.4	10.00	-
$Ra = 100, Le = 10, N = 0.2$	2.50	2.5	14.8	15

The validation of the numerical code was accomplished by comparison with some published numerical experiments. In table 1 some results for $A = 1$, $Da = 10^{-5}$, different values of Rayleigh, Lewis numbers and buoyancy ratio are presented.

The values of overall heat and mass transfer are compared with the published results, where the numerical calculations based on the Darcy model are obtained [7]. The first result is for the case of $Ra = 600$, $Le = 1$ and $N = 0$, which means, that only the thermal buoyancy force is present. The overall heat and mass transfer, which are presented by Nu and Sh are identical. The other two cases are for $Ra = 100$, $N = 0.2$ and $Le = 10$, $Le = 30$. In this case, both the thermal and solutal buoyancy forces are present and aid each other. The values of Sherwood numbers are now higher than those of the Nusselt numbers, which is a result of higher Lewis number. The presented result are in agreement with published ones. Table 2 presents the influence of the Darcy number on the overall heat and mass transfer. The values of the governing parameters for this case are: aspect ratio $A = 4$, Rayleigh number $Ra = 300$, Lewis number $Le = 0.1$ and buoyancy ration $N = -2$. The negative sign for buoyancy ratio indicates, that the thermal and solutal buoyancy forces oppose to each other.

Table 2: Nu and Sh numbers for different values of Da and $A = 4$, $Ra = 300$, $Le = 0.1$, $N = -2$.

Da	10^{-1}	10^{-2}	10^{-3}	10^{-4}	10^{-5}
Nu	1.00	2.05	2.82	3.12	3.50
Sh	1.00	1.02	1.04	1.05	1.06

From the obtained results it is evident, that with any decrease in the Darcy number the vaule of the Nusselt number increases. In cases of small Lewis numbers $Le \to 0$ the values of Sherwood numbers tend to unity ($Sh \to 1$), which implies, that the mass transfer is dominated by diffusion. The same conclusions are also published in [1]. In the case of $Da = 10^{-1}$ the values of Nu and Sh are equal to 1, which means that heat and mass transfer are governed by diffusion. The Rayleigh number in this case is beyond the critical value required for the beginning of convective motion. The relationship between the critical Rayleigh number and the Darcy number is given in [1] and states that the values of the critical Rayleigh number increase in line with the Darcy number.

The velocity, temperature and concentration fields for $Ra = 2500$, $Da = 10^{-2}$, $Le = 10$, $N = 0$, $A = 2$ are presented in fig. 2.

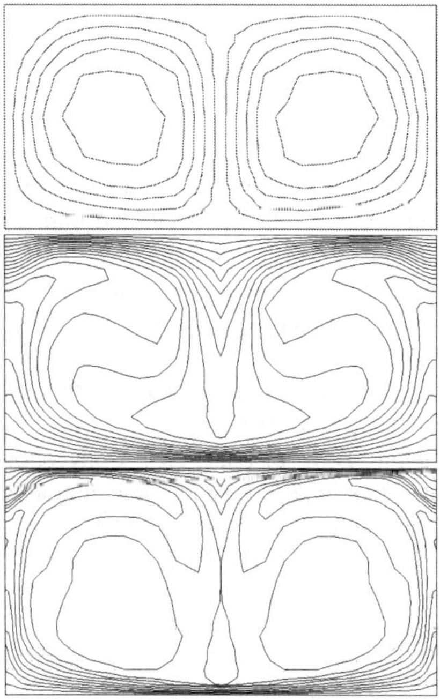

Figure 2: Streamlines, isotherms and isoconcentrations in a horizontal layer for $Ra = 2500$, $Da = 10^{-2}$, $Le = 10$, $N = 0$, $A = 2$.

The buoyancy effect in this case is due entirely to temperature gradients, so the concentration field is a result of the flow driven by the temperature gradients and the imposed concentration difference between the upper and bottom boundaries. From the fig. 2. it is evident that the flow in the horizontal layer with $A > 1$ becomes multi-cellular (in the case of $A = 2$ there are 2 cells). The flow consists of rising hot fluid in the centre of the layer and colder fluid sinking along the vertical walls. In the centre of the domain higher solute concentration is found than along the adiabatic and impermeable side walls. Thin temperature and composition boundary layers are evident at the top and bottom walls.

www.witpress.com, ISSN 1743-355X (on-line)

5 Conclusion

A numerical study of double-diffusive natural convection in a horizontal porous layer, saturated with the Newtonian and incompressible fluid is presented. For the solution of governing equations the Boundary Domain Integral Method, an extension of the classical Boundary Element Method, was used. The modified Navier–Stokes equations have been used to describe the fluid motion in porous media. The general set of equations is transformed with use of velocity-vorticity formulation, which consequently separates the computational scheme into a kinematic and kinetic part. The results for different values of governing parameters are obtained and compared to some published studies.

References

[1] Amahmid, A., Hasnaoui, M., Mamou, M. & Vasseur, P., Double-diffusive parallel flow induced in a horizontal brinkman porous layer subjected to constant heat and mass fluxes: analytical and numerical studies. *Heat and Mass Transfer*, **35**, pp. 409–421, 1999.

[2] Kladias, N. & Prasad, V., Natural convection in horizontal porous layers: Effects of darcy and prandtl numbers. *Journal of Heat Transfer*, **111**, pp. 926–935, 1989.

[3] Nield, D.A., Onset of thermohaine convection in a porous medium. *Water Resour Res*, **4**, pp. 553–560, 1968.

[4] Nield, D.A., Manole, D.M. & Lage, J.L., Convection induced by inclined thermal and solutal gradients in a shallow horizontal layer of a porous medium. *J Fluid Mech*, **257**, pp. 559–574, 1993.

[5] Rudraiah, N., Srimani, P.K. & Friedrich, R., Finite amplitude convection in a two-component fluid saturated porous layer. *Int J Heat Mass Transfer*, **25**, pp. 715–722, 1982.

[6] Trevisan, O.V. & Bejan, A., Combined heat and mass transfer by natural convection in a vertical enclosure. *Journal of Heat Transfer*, **109**, pp. 104–112, 1987.

[7] Rosenberg, N.D. & Spera, F.J., Thermohaline convection in a porous medium heated from below. *Int J Heat Mass Transfer*, **35**, pp. 1261–1273, 1992.

[8] Mamou, M., Vasseur, P. & Bilgen, E., Multiple solutions for double diffusive convection in a vertical porous enclosure. *Int J Heat Mass Transfer*, **38**, pp. 1787–1798, 1995.

[9] Škerget, L., Hriberšek, M. & Kuhn, G., Computational fluid dynamics by boundary-domain integral method. *Int J Numer Meth Engng*, **46**, pp. 1291–1311, 1999.

[10] Jecl, R. & Škerget, L., Boundary element method for natural convection in non-newtonian fluid saturated square porous cavity. *Engineering Analysis with Boundary Elements*, **27**, pp. 963–975, 2003.

[11] Brebbia, C.A. & Dominguez, J., *Boundary Elements, An Introductory Course*. McGraw-Hill Book Company, New York, 1992.

Section 7
Plates and shells

Analysis of von Kármán plates using a BEM formulation

L. Waidemam & W. S. Venturini
São Carlos School of Engineering, University of São Paulo, Brazil

Abstract

This work deals with non-linear geometrical plates in the context of von Kármán theory. The formulation is written in a way to require only boundary in-plane displacement and deflection integral equation for boundary collocations. At internal points only out of plane rotation, curvature and in-plane internal force representations are used. The non-linear system of algebraic equations to be solved is reduced to internal point collocation relations. The solution is solved by using a Newton scheme for which a consistent tangent operator was derived.
Keywords: bending plates, geometrical nonlinearities.

1 Introduction

The boundary element method (BEM) applied to solve plate-bending problems has been successfully used many times so far. An important characteristic of the boundary methods applied to plate bending is approximating all boundary values by the same shape function, avoiding therefore using higher order derivatives of displacement approximation to compute internal forces. Thus, bending and twisting moments and also shear forces are precisely evaluated. The method has already proved to be enough accurate and reliable for this kind of application.

The plate bending numerical formulation is a very important subject in engineering due to be applied to a large number of complex problems such as aircraft, ship, car, pressure vessel, off shore structures among others. Usually these complex problems require accurate plate bending models as those that take into account the geometrical non-linear effects. In this context, several BEM formulations have already been proposed so far. One of the first works treating this subject is due to Morjaria [1]. Kamiya and Sawaki [2] have proposed a BEM formulation for elastic plates governed by the Berger equation. The first BEM

WIT Transactions on Modelling and Simulation, Vol 44,
www.witpress.com, ISSN 1743-355X (on-line)
doi:10.2495/BE070211

formulation to analyze plate-bending problems within the context of von Kármán hypothesis is due to Ye and Liu [3], who have used a fictitious loading distributed over the domain to model the non-linear effects. Von Kármán hypothesis was also adopted by Tanaka et al [4] to develop a more elaborated BEM incremental formulation to deal with finite deflections of thin elastic plates. Wang et al [5] have also worked on von Kármán plates introducing the dual reciprocity approach based on global radial functions to approximate the correcting integral term.

All works reported above were proposed in the context of thin plates. Several other important works have appeared more recently pointing out the efficiency of BEM formulations to deal with shear deformable plate based on the Reissner-Mindlin hypothesis: Wen et al [6], Purbolaksono and Aliabadi [7].

In this work we came back to the BEM formulation based on the von Kármán's theory. Emphasis is given to the accurate evaluation of the domain integrals approximated by using cells and to the solution technique for which a tangent consistent operator is proposed. Examples of plate with finite deflection is analysed and the results compared with other numerical solutions.

2 Basic equations

Without loss of generality, let us consider a single thin plate region Ω with boundary Γ over which a distributed load q is applied orthogonal to the middle surface, (direction x_3), as shown in figure 1. This plate region can also be subjected to in plane forces (directions x_1 and x_2) either distributed over the domain or applied along the boundary. In order to write the field equations of this plate problems following the hypothesis can be assumed according to the von Kármán theory, for which the strains are assumed to be enough small and the final deflection of order of the plate thickness h.

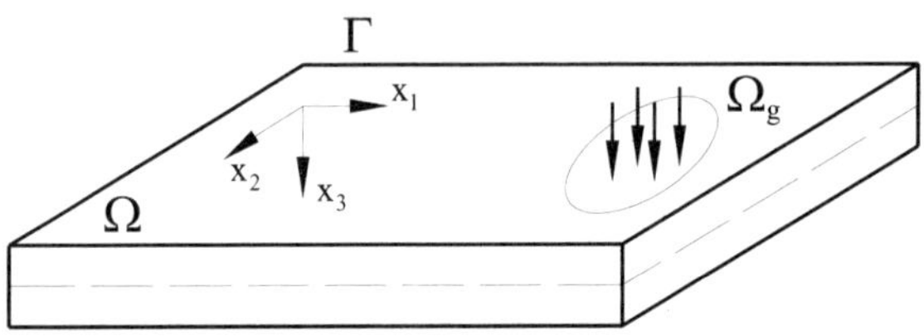

Figure 1: General plate domain.

For any point defined in Ω the following basic relationships are defined:

- Equilibrium equations for the bending problem:

$$m_{ij},_{ij} + N_{ij} w,_{ij} - b_i w,_i + m_i,_i + q = 0 \tag{1}$$

where m_{ij} are bending and twisting moments, N_{ij} is the in-plane internal forces, b_i represents the in-plane domain loads applied and m_i is the applied moment over the plate domain; the subscripts are in the range $i,j=\{1, 2\}$.

- The in-plane equilibrium equation:

$$N_{ij,j} + b_i = 0 \tag{2}$$

Assuming linear elastic material eqns (1) and (2) can be written in terms of in-plane and out of plane displacements as follows:

$$Dw_{,iijj} = g + N_{ij} w_{,ij} \qquad (i, j = 1,2) \tag{3}$$

$$\frac{1}{2(1-\nu)}(u_{j,ij} + w_{,j}\, w_{,ij}) + \frac{1}{2(1+\nu)}(u_{i,jj} + w_{,i}\, w_{,jj}) + b_i/(Eh) \tag{4}$$

where $D = Eh^3/(1-\nu^2)$ is the flexural rigidity, E and ν are the material Young modulus and Poisson's ratio.

The problem definition is then completed by assuming the following boundary conditions over Γ: $u_i = \bar{u}_i$ on Γ_1 (generalised displacements, in plane displacements, deflections and rotations) and $p_i = \bar{p}_i$ on Γ_2 (generalised tractions, in-plane tractions normal bending moment and effective shear forces), where $\Gamma_1 \cup \Gamma_2 = \Gamma$.

3 Integral representations

In this section, we are going to derive the integral equations of the plate bending and stretching problems considering geometrical non-linearities within the context previously defined. To obtain the integral equations of both problems one can apply the Betti's reciprocity to the linear parts of the stress and strain fields. Thus, for the bending problem the general reciprocity relation written for the 3D case can be integrated across the plate thickness to give:

$$\int_{\Omega_m} w^*_{,ij} m_{ij} d\Omega = \int_{\Omega_m} m^*_{ij} w_{,ij}\, d\Omega \tag{5}$$

where w^* and m^*_{ij} are the well-known fundamental solutions of the plate problem given in terms of deflection and internal moments. These fundamental values and the other resulting required values are given in the specialized literature [8].

By integrating eqn (5) by parts twice and replacing the second derivative of the internal, $m_{ij,ij}$, according to eqn (1) one has:

$$C(S)w(S)+\int_{\Gamma}\left(V_n^*(S,P)w(P)-M_n^*(S,P)w_{,n}(P)\right)d\Gamma(P)+\sum_{i=1}^{N_c}R_{ci}^*(S,P)w_{ci}(P)$$
$$=\int_{\Gamma}\left(V_n(P)w^*(S,P)-M_n(P)w_{,n}^*(S,P)\right)d\Gamma(P)+\sum_{i=1}^{N_c}R_{ci}(P)w_{ci}^*(S,P)+ \tag{6}$$
$$\int_{\Omega_g}g(p)w^*(S,p)d\Omega(p)+\int_{\Omega}N_{ij}(p)w_{,ij}(p)w^*(S,p)d\Omega$$

where V_n and M_n are effective shear forces, and moments applied along the boundary, respectively, R_{ci} represents the corner reactions, V_n^*, M_n^* and R_{ci}^* the corresponding values obtained from the fundamental solution $w*$ according to their definition.

For any internal collocation s one can differentiate eqn (6) to obtain the integral representations of rotations and curvatures, as follows:

$$w_{,i}(s)=-\int_{\Gamma}\left\{V_{n}^*{}_{,i}(s,P)w(P)-M_{n}^*{}_{,i}(s,P)w_{,n}(P)\right\}d\Gamma(P)-\sum_{k=1}^{N_c}R_{ck}^*{}_{,i}(s,P)w_{ck}(P)$$
$$+\int_{\Gamma}\left\{V_n(P)w_{,i}^*(q,P)-M_n(P)w_{,ni}^*(s,P)\right\}d\Gamma(P)+\sum_{k=1}^{N_c}R_{ck}(P)w_{ck}^*{}_{,i}(s,P)+ \tag{7}$$
$$+\int_{\Omega_g}g(p)w_{,i}^*(s,p)d\Omega(p)+\int_{\Omega}N_{jk}(p)w_{,jk}(p)w_{,i}^*(s,p)d\Omega(p)$$

$$w_{,ij}(s)=-\int_{\Gamma}\left\{V_{n}^*{}_{,ij}(s,P)w(P)-M_{n}^*{}_{,ij}(s,P)w_{,n}(P)\right\}d\Gamma(P)-\sum_{k=1}^{N_c}R_{c}^*{}_{,ij}(s,P)w_{c}(P)$$
$$+\int_{\Gamma}\left\{V_n(P)w_{,ij}^*(s,P)-M_n(P)w_{,nij}^*(s,P)\right\}d\Gamma(P)+\sum_{k=1}^{N_c}R_{c}(P)w_{c}^*{}_{,ij}(s,P)+ \tag{8}$$
$$+\int_{\Omega_g}g(p)w_{,ij}^*(s,p)d\Omega(p)+\int_{\Omega}N_{km}(p)w_{,km}(p)w_{,ij}^*(s,p)d\Omega(p)$$

Analogously, the Betti's reciprocity relation for can be in applied to the linear parts of the 2D stretching problem to give:

$$\int_{\Omega}N_{ij}^{\ell}\varepsilon_{ij}^*d\Omega=\int_{\Omega}N_{ij}^*\varepsilon_{ij}d\Omega \tag{9}$$

where N_{ij}^{ℓ} represents the linear parts of the stretching problem internal forces, therefore given by:

$$N_{ij}^{\ell}=\frac{Eh}{(1-\nu)^2}\left[\nu u_{k,k}\,\delta_{ij}+\frac{1-\nu}{2}(u_{i,j}+u_{j,i})\right] \tag{10}$$

After replacing N_{ij}^{ℓ} in eqn (13) and integrating it by parts the following integral representation is obtained for the in-plane boundary displacements:

$$C_{ij}(S)u_j(S) = -\int_\Gamma P_{ij}^*(S,P)u_j(P)d\Gamma + \int_\Gamma u_{ij}^*(S,P)p_j(P)d\Gamma + \\ + \int_\Omega u_{ij}^*(S,p)b_j(p)d\Omega - \frac{1}{2}\int_\Omega N_{ijk}^*(S,p)w,_j(p)w,_k(p)d\Omega \tag{11}$$

By differentiating eqn (11) and then applying the Hooke's law at the collocation point s accordingly, one has:

$$N_{ij}(s) = \int_\Gamma D_{ijk}(s,P)p_k(P)d\Gamma - \int_\Gamma S_{ijk}(s,P)u_k(P)d\Gamma + \int_\Omega D_{ijk}(s,p)b_k(p)d\Omega \\ -\frac{1}{2}\int_\Omega T_{ijkl}(s,p)w_{,k}(p)w_{,l}(p)d\Omega + \frac{Gh}{8(1-\bar{\upsilon})}\left(2w_{,i}(s)w_{,j}(s) + w_{,k}(s)w_{,k}(s)\delta_{ij}\right) \tag{12}$$

where $\bar{v} = v/(1+v)$ is used to simulate the plane stress conditions.

4 Algebraic equations

Before transforming the integral representations derived in the previous section to algebraic equations let us replace the rate values by the corresponding increments. Let $\Delta t = t_{n+1} - t_n$ be a typical time-step in the time discretization. Any rate quantity $\dot{x}$ integrated along the time interval Δt becomes $\Delta x = x_{n+1} - x_n$ that will replace $\dot{x}$ in all integral representations already derived.

As usual for any BEM formulation, the integral representations (6), (7), (8), (11) and (12) have to be transformed into algebraic expressions after discretizing the boundary and the domain. The plate boundary Γ is then discretized into elements, Γ_s, along which generalized displacements and tractions are approximated using continuous and discontinuous linear boundary elements. In plate stretching-bending problems discontinuities are always present, particularly at corners and traction jumps. The discontinuity is always introduced by defining the collocation along the element or at any outside point near the boundary. Before transforming the integral representations, we decided replacing the density of the domain integrals into single values. The domain value $N_{ij}(p)w_{,ij}(p)$ at a field point p will be replaced by a scalar value $T(p)$. Similarly, for the stretching problem, we replaced the domain value $w,_j(p)w,_k(p)$ by domain tensor $W_{jk}(p)$. The approximation over the cells are now applied to these new domain values $T(p)$ and $W_{jk}(p)$. Triangular internal cells with linear shape functions are used with nodes always defined at internal points. Thus, discontinuous cells are required for cells adjacent to the boundary. The cell integrals are first transformed to integral over their sides and carried out as boundary elements using either analytical or appropriate numerical integration scheme with sub-elementation [9].

Carrying out the boundary and cell integrals eqn (6) written for all boundary nodes leads to the following incremental matrix equation:

$$\boldsymbol{H}_b^w \Delta \boldsymbol{U}^b = \boldsymbol{G}_b^w \Delta \boldsymbol{P}^b + \boldsymbol{S}_b^w \Delta \boldsymbol{T}_N^\chi + \Delta \boldsymbol{B}_b^w \tag{13}$$

where $\Delta \boldsymbol{U}^b$ and $\Delta \boldsymbol{P}^b$ contains all boundary displacement and traction nodal values of the bending problem plus the additional corner displacements and reactions, $\Delta \boldsymbol{T}_N^\chi$ represents the condensate summation of the in-plane forces multiplied by curvatures, $\boldsymbol{S}_b^w$ is the corresponding matrices obtained by integrating all domain cells and $\Delta \boldsymbol{B}_b^w$ gives the domain load effects.

Similarly one can also transform the in-plane displacement integral representation, equation (7), into its incremental matrix form:

$$\boldsymbol{H}_s^u \Delta \boldsymbol{U}^s = \boldsymbol{G}_s^u \Delta \boldsymbol{P}^s + \boldsymbol{S}_s^u \Delta \boldsymbol{W}_\theta^\theta + \Delta \boldsymbol{B}_s^u \tag{14}$$

where $\Delta \boldsymbol{U}^s$ and $\Delta \boldsymbol{P}^s$ contains all incremental boundary displacement and traction nodal values of the stretching problem, $\Delta \boldsymbol{W}_\theta^\theta$ represents the increment of the rotation product defined at each domain node, $\boldsymbol{S}_s^u$ is the corresponding matrices obtained by integrating all domain cells and $\Delta \boldsymbol{B}_s^u$ gives the in-plane domain incremental load effects.

To complete the necessary algebraic relations one has to obtain the algebraic forms of the integral equations (7), (8) and (12), as follows:

$$\Delta \boldsymbol{\theta} = -\boldsymbol{H}_b^\theta \Delta \boldsymbol{U}^b + \boldsymbol{G}_b^\theta \Delta \boldsymbol{P}^b + \boldsymbol{S}_b^\theta \Delta \boldsymbol{T}_N^\chi + \Delta \boldsymbol{B}_b^\theta \tag{15}$$

$$\Delta \boldsymbol{\chi} = -\boldsymbol{H}_b^\chi \Delta \boldsymbol{U}^b + \boldsymbol{G}_b^\chi \Delta \boldsymbol{P}^b + \boldsymbol{S}_b^\chi \Delta \boldsymbol{T}_N^\chi + \Delta \boldsymbol{B}_b^\chi \tag{16}$$

$$\Delta \boldsymbol{N} = -\boldsymbol{H}_s^N \Delta \boldsymbol{U}^s + \boldsymbol{G}_s^N \Delta \boldsymbol{P}^s + \boldsymbol{S}_s^N \Delta \boldsymbol{W}_\theta^\theta + \Delta \boldsymbol{B}_s^N \tag{17}$$

where $\Delta \boldsymbol{\theta}$, $\Delta \boldsymbol{\chi}$ and $\Delta \boldsymbol{N}$ are vectors containing rotation, curvature and membrane internal force increments at the domain nodes defined by the adopted discretization.

After applying the boundary conditions equations (13), (14) and (15)–(17) become:

$$\boldsymbol{A}_b^w \Delta \boldsymbol{X}^b = \Delta \boldsymbol{F}_b^w + \boldsymbol{S}_b^w \Delta \boldsymbol{T}_N^\chi \tag{18}$$

$$\boldsymbol{A}_s^u \Delta \boldsymbol{X}^s = \Delta \boldsymbol{F}_s^u + \boldsymbol{S}_s^u \Delta \boldsymbol{W}_\theta^\theta \tag{19}$$

$$\Delta \boldsymbol{\theta} = -\boldsymbol{A}_b^\theta \Delta \boldsymbol{X}^b + \Delta \boldsymbol{F}_b^\theta + \boldsymbol{S}_b^\theta \Delta \boldsymbol{T}_N^\chi \tag{20}$$

$$\Delta \boldsymbol{\chi} = -\boldsymbol{A}_b^\chi \Delta \boldsymbol{X}^b + \Delta \boldsymbol{F}_b^\chi + \boldsymbol{S}_b^\chi \Delta \boldsymbol{T}_N^\chi \tag{21}$$

$$\Delta \boldsymbol{N} = -\boldsymbol{A}_s^N \Delta \boldsymbol{U}^b + \Delta \boldsymbol{F}_s^N + \boldsymbol{S}_s^N \Delta \boldsymbol{W}_\theta^\theta \tag{22}$$

where the matrices $\boldsymbol{A}_x^y$ are conveniently built up using columns of $\boldsymbol{H}_x^y$ and $\boldsymbol{G}_x^y$ corresponding to unknown boundary values, and $\Delta \boldsymbol{F}_x^y$ is a vector collecting all contributions of prescribed domain and boundary values.

Solving eqns (18) and (19) and replacing into eqns (20) – (22) gives:

$$\Delta\boldsymbol{\theta} = \Delta\boldsymbol{N}_b^{\theta} + \boldsymbol{Q}_b^{\theta}\,\Delta\boldsymbol{T}_N^{\chi} \tag{23}$$

$$\Delta\chi = \Delta\boldsymbol{N}_b^{\chi} + \boldsymbol{Q}_b^{\chi}\,\Delta\boldsymbol{T}_N^{\chi} \tag{24}$$

$$\Delta\boldsymbol{N} = \Delta\boldsymbol{N}_s^{N} + \boldsymbol{Q}_s^{N}\,\Delta\boldsymbol{W}_{\theta}^{\theta} \tag{25}$$

where $\Delta\boldsymbol{N}_x^y = -\boldsymbol{A}_x^y\,\Delta\boldsymbol{M}_x^z + \Delta\boldsymbol{F}_x^y$ and $\boldsymbol{Q}_x^y = -\boldsymbol{A}_x^y\boldsymbol{Q}_x^z + \boldsymbol{S}_x^y$, with $\Delta\boldsymbol{M}_x^z = \left[\boldsymbol{A}_x^z\right]^{-1}\Delta\boldsymbol{F}_x^z$ and $\boldsymbol{Q}_x^z = \left[\boldsymbol{A}_x^z\right]^{-1}\boldsymbol{S}_x^z$ and z given by the superscript of the corresponding boundary algebraic equation used to compute $\boldsymbol{Q}_x^z$, i.e. $\boldsymbol{Q}_x^w$ from eqn (18) or $\boldsymbol{Q}_x^u$ from eqn (19).

Equations (23) – (25) are the necessary relations to solve a geometrical non-linear plate problem. However, one has to treat correctly the increments $\Delta W_{jk}(p)$ and $\Delta T(p)$. We can first find the rates of the densities in eqns (6) and (11) and obtain the incremental forms of $\Delta\boldsymbol{T}$ and $\Delta\boldsymbol{W}$ for a given time interval Δt_n, as follows:

$$\Delta\boldsymbol{T} = \Delta\boldsymbol{N} \bullet \chi + \boldsymbol{N} \bullet \Delta\chi \tag{26}$$

$$\Delta\boldsymbol{W} = \Delta\boldsymbol{\theta} \otimes \boldsymbol{\theta} + \boldsymbol{\theta} \otimes \Delta\boldsymbol{\theta} \tag{27}$$

5 System solution

Equations (23) - (25) represent the non-linear system to be solved in terms of the increments $\Delta\boldsymbol{\theta}$, $\Delta\chi$ and $\Delta\boldsymbol{N}$. Replacing the increments $\Delta\boldsymbol{T}$ and $\Delta\boldsymbol{W}$ according to eqns (26) and (27) one has:

$$F_{\theta}(\Delta\boldsymbol{\theta}, \Delta\chi, \Delta\boldsymbol{N}) = -\Delta\boldsymbol{\theta} + \Delta\boldsymbol{N}_b^{\theta} + Q_b^{\theta}\,\Delta\boldsymbol{N} \bullet \chi + Q_b^{\theta}\,\boldsymbol{N} \bullet \Delta\chi = 0 \tag{28}$$

$$F_{\chi}(\Delta\boldsymbol{\theta}, \Delta\chi, \Delta\boldsymbol{N}) = -\Delta\chi + \Delta\boldsymbol{N}_b^{\chi} + Q_b^{\chi}\,\Delta\boldsymbol{N} \bullet \chi + Q_b^{\chi}\,\boldsymbol{N} \bullet \Delta\chi \tag{29}$$

$$F_N(\Delta\boldsymbol{\theta}, \Delta\chi, \Delta\boldsymbol{N}) = -\Delta\boldsymbol{N} + \Delta\boldsymbol{N}_s^{N} + Q_s^{N}\,\Delta\boldsymbol{\theta} \otimes \boldsymbol{\theta} + Q_s^{N}\,\boldsymbol{\theta} \otimes \Delta\boldsymbol{\theta} \tag{30}$$

The above non-linear system of equations is solved by applying the Newton-Raphson's scheme. Within a time increment $\Delta t_n = t_{n+1} - t_n$ an iterative process may be required to achieve the equilibrium. Then, from the solution at iteration i the next try at iteration $(i+1)$ is given by:

$$\left\{ \Delta\boldsymbol{N}_n^{i+1} \right\} = \left\{ \Delta\boldsymbol{N}_n^{i} \right\} + \left\{ \delta\Delta\boldsymbol{N}_n^{i} \right\} \tag{31}$$

$$\left\{ \Delta\boldsymbol{\theta}_n^{i+1} \right\} = \left\{ \Delta\boldsymbol{\theta}_n^{i} \right\} + \left\{ \delta\Delta\boldsymbol{\theta}_n^{i} \right\} \tag{32}$$

$$\left\{ \Delta\chi_n^{i+1} \right\} = \left\{ \Delta\chi_n^i \right\} + \left\{ \delta\Delta\chi_n^i \right\} \tag{33}$$

By linearizing eqns (28) - (30), using the first term of the Taylor's expansion, gives:

$$\begin{Bmatrix} F_\theta\left(\Delta\theta_n^i, \Delta\chi_n^i, \Delta N_n^i\right) \\ F_\chi\left(\Delta\theta_n^i, \Delta\chi_n^i, \Delta N_n^i\right) \\ F_N\left(\Delta\theta_n^i, \Delta\chi_n^i, \Delta N_n^i\right) \end{Bmatrix} + \begin{bmatrix} \frac{\partial F_\theta}{\partial \Delta\theta_n^i} & \frac{\partial F_\theta}{\partial \Delta\chi_n^i} & \frac{\partial F_\theta}{\partial \Delta N_n^i} \\ \frac{\partial F_\chi}{\partial \Delta\theta_n^i} & \frac{\partial F_\chi}{\partial \Delta\chi_n^i} & \frac{\partial F_\chi}{\partial \Delta N_n^i} \\ \frac{\partial F_N}{\partial \Delta\theta_n^i} & \frac{\partial F_N}{\partial \Delta\chi_n^i} & \frac{\partial F_N}{\partial \Delta N_n^i} \end{bmatrix} \begin{Bmatrix} \delta\Delta\theta_n^i \\ \delta\Delta\chi_n^i \\ \delta\Delta N_n^i \end{Bmatrix} + \ldots = 0 \tag{34}$$

where each term of the tangent matrix is given by:

$$[C] = \begin{bmatrix} -\boldsymbol{I} & \boldsymbol{Q}_b^\theta \boldsymbol{N}_n^i \bullet \boldsymbol{II} & \boldsymbol{Q}_b^\theta \boldsymbol{II} \bullet \chi_n^i \\ \boldsymbol{0} & -\boldsymbol{II} + \boldsymbol{Q}_b^\chi \boldsymbol{N}_n^i \bullet \boldsymbol{II} & \boldsymbol{Q}_b^\chi \boldsymbol{II} \bullet \chi_n^i \\ \boldsymbol{Q}_s^N (\boldsymbol{I} \otimes \boldsymbol{\theta}_n^i + \boldsymbol{\theta}_n^i \otimes \boldsymbol{I}) & \boldsymbol{0} & -\boldsymbol{II} \end{bmatrix} \tag{35}$$

Thus, the corrections to be cumulated during the iterations are obtained by solving eqn (34), as follows:

$$\begin{Bmatrix} \delta\Delta\theta_n^i \\ \delta\Delta\chi_n^i \\ \delta\Delta N_n^i \end{Bmatrix} = -[C]^{-1} \begin{Bmatrix} F_\theta\left(\Delta\theta_n^i, \Delta\chi_n^i, \Delta N_n^i\right) \\ F_\chi\left(\Delta\theta_n^i, \Delta\chi_n^i, \Delta N_n^i\right) \\ F_N\left(\Delta\theta_n^i, \Delta\chi_n^i, \Delta N_n^i\right) \end{Bmatrix} \tag{36}$$

6 Numerical application

To check the performance of the proposed formulation we have chosen a square plate subjected to a uniform load (figure 2). The plate side length is a, the thickness is t, with the ration t/a = 0.0.1. The Poisson's ration was assumed $\nu = 0.3$ while q is the uniform applied load.

Several boundary conditions have been analyzed. We started by assuming simple supported and clamped plate conditions. For each of this case the in-plane boundary displacements could also be prescribed equal to zero (IE) or kept free (ME).

The results obtained are shown in figures 3 and 4, for the simply supported plate and clamped plate respectively. As can be seen the results are in total agreement with the one given by Ye and Liu [3].

The obtained results, compared with other numerical and analytical solutions demonstrated that the proposed formulation is accurate. Running several other internal and boundary meshes has also demonstrated that the convergence is obtained quickly. With a rather coarse mesh, particularly to integrate the domain integrals, the results are already accurate. The presented results were obtained by using a 160 boundary elements and only 8 domain cells

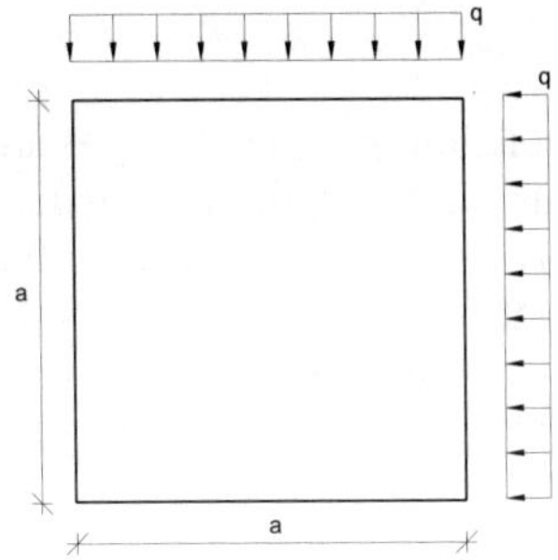

Figure 2: Square plate with uniform load.

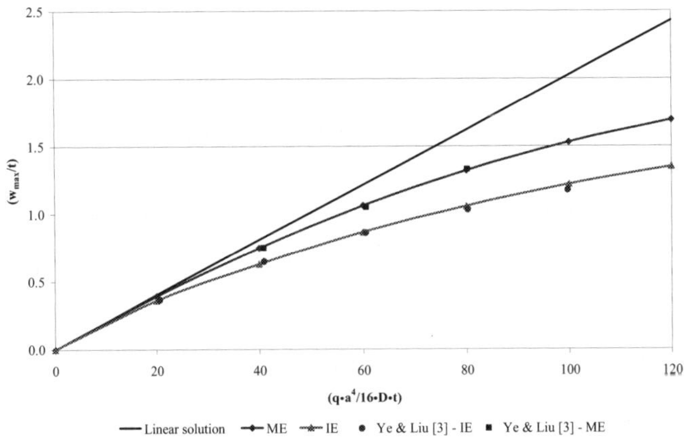

Figure 3: Simply supported plate – load displacement curve.

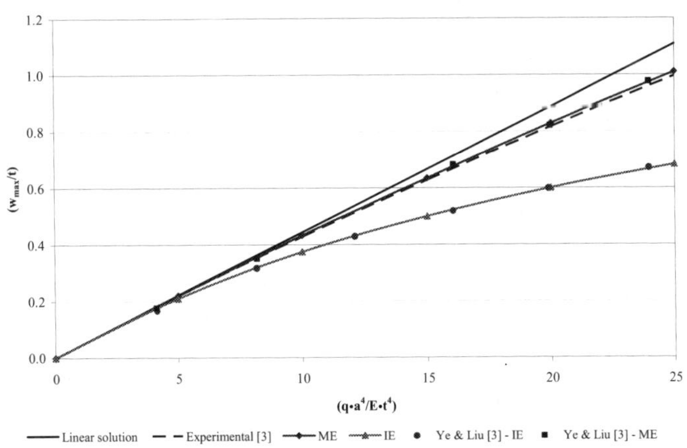

Figure 4: Clamped plate – load displacement curve.

7 Conclusions

A boundary element formulation to analyse von Kármán plates was proposed. The domain approximations were simplified by reducing the density of domain integrals into simple equivalent values reducing the computational effort to compute the matrices related with the domain values. A consistent tangent operator has been derived and the Newton process has been implement leading accurate and reliable solutions using a very small number of iterations.

Acknowledgments

To FAPESP – São Paulo State Research Foundation for the support given to this work.

References

[1] Morjaria, M., Inelastic analysis of transverse deflection of plates by the boundary element method. *Journal of Applied Mechanics-Transactions of the ASME*, 47: 291-, 1980

[2] Kamiya, N. & Sawaki, Y., An integral equation approach to finite deflection of elastic plates. *International Journal of Non-Linear Mechanics*, 17(3): 187-194, 1982.

[3] Ye, T.-Q. & Liu, Y.-J., Finite deflection analysis of elastic plate by the boundary element method. *Applied Mathematical Modelling*, 9: 183-188, 1985.

[4] Tanaka, M., Matsumoto, T. & Zheng, Z.-D. Incremental analysis of finite deflection of elastic plates via boundary-domain-element method. *Engineering Analysis with Boundary Elements*, 17: 123-131, 1996.

[5] Wang, W., Ji, X. & Tanaka, M., A dual boundary element approach for the problems of large deflection of thin elastic plates. *Computational Mechanics*, 26: 58-65, 2000.

[6] Wen, P.H., Aliabadi, M.H. & Young, A., Large deflection analysis of Reissner plate by boundary element method. *Computers & Structures,* 83 (10-11): 870-879, 2005.

[7] Purbolaksono, J. & Aliabadi, M.H., Large deformation of shear-deformable plates by the boundary element method. *Journal of Engineering Mathematics* 51 (3): 211-230, 2005.

[8] Brebbia, C.A., Telles, J.C.F & Wrobel, L.C. *Boundary Element Techniques. Theory and Applications in Engineering,* Springer-Verlag: Berlin and New York, 1984.

[9] Leite, L.G.S. & Venturini, W.S., Stiff and soft thin inclusions in two-dimensional solids by the boundary element method. *Engineering Analysis with Boundary Elements*, 29 (3): 257-267, 2005.

Linear analysis of building floor structures by a BEM formulation based on Reissner's theory

G. R. Fernandes, D. H. Konda & L. C. F. Sanches
Civil Engineering Department of São Paulo State University, UNESP, Ilha Solteira, Brazil

Abstract

In this work, the plate bending formulation of the boundary element method (BEM) based on the Reissner's hypothesis is extended to the analysis of zoned plates in order to model a building floor structure. In the proposed formulation each sub-region defines a beam or a slab and depending on the way the sub-regions are represented, one can have two different types of analysis. In the simple bending problem all sub-regions are defined by their middle surface. On the other hand, for the coupled stretching–bending problem all sub-regions are referred to a chosen reference surface, therefore eccentricity effects are taken into account. Equilibrium and compatibility conditions are automatically imposed by the integral equations, which treat this composed structure as a single body. The bending and stretching values defined on the interfaces are approximated along the beam width, reducing therefore the number of degrees of freedom. Then, in the proposed model the set of equations is written in terms of the problem values on the beam axis and on the external boundary without beams. Finally some numerical examples are presented to show the accuracy of the proposed model.
Keywords: plate bending, boundary elements, building floor structures.

1 Introduction

The boundary element method (BEM) has already proved to be a suitable numerical tool to deal with plate bending problems. The method is particularly recommended to evaluate internal force concentrations due to loads distributed over small regions that very often appear in practical problems. Moreover, the same order of errors is expected when computing deflections, slopes, moments

WIT Transactions on Modelling and Simulation, Vol 44,
www.witpress.com, ISSN 1743-355X (on-line)
doi:10.2495/BE070221

and shear forces. They are not obtained by differentiating approximation function as for other numerical techniques.

Several models to analyze plate reinforced by beams, using BEM coupled with the finite element method (FEM), have already been proposed (see Hu and Hartley [1], Tanaka. and Bercin [2], Sapountzakis and Katsikadelis [3]). In those works the BEM and FEM approximate, respectively, plate and beam elements. However, for complex floor structures the number of degrees of freedom may increase rapidly diminishing the solution accuracy.

An alternative scheme to reduce the number of degrees of freedom has been recently proposed by Fernandes and Venturini in [4] and [5] using only a BEM formulation based on Kirchhoff's hypothesis, where the building floor is modeled by a zoned plate. In the first work is proposed a formulation to perform simple bending analysis where the tractions are eliminated along the interfaces. Moreover in order to reduce the number of degrees of freedom some Kinematics assumptions were made along the beam width. In the second work this formulation is extended to take into account the membrane effects which are associated with bending due to the relative positions of the structural elements.

In this work the BEM formulation developed in [5] is modified to take into account the Reissner's hypothesis instead of the Kirchhoff's. In the proposed model the tractions related to the bending problem is no longer eliminated on the interfaces. Therefore traction and displacements related to both problems (bending and stretching) are approximated along the beam width, leading to a model where the problem values are defined only on the beams axis and on the plate boundary without beams. The accuracy of the proposed model is illustrated by numerical examples whose analytical results are known.

Note that in the Kirchhoff's theory (see Fernandes and Venturini [5], Hartmann and Zotemantel [6] and Kirchhoff [7]) are defined only four boundary values and its inaccuracy turns out to be important for thick plates, especially in the edge zone of the plate and around holes whose diameter is not larger than the plate thickness. In the Reissner's theory (see Reissner [8], Weën [9] and Palermo [10]) which can be used either for thin or thick plates, are defined six boundary values and it is more accurate because it takes into account the shear deformation effect.

2 Basic equations

Without loss of generality, let us consider the plate depicted in figure 1(a), where t_1, t_2 and t_3 are the thicknesses of the sub-regions Ω_1, Ω_2 and Ω_3, whose external boundaries are Γ_1, Γ_2 and Γ_3, respectively. The total external boundary is given by Γ while Γ_{jk} represents the interface between the adjacent sub-regions Ω_j and Ω_k.. In the simple bending analysis all sub-regions are represented by their middle surface, as shown in figure 1(c), while for the coupled stretching-bending problem the Cartesian system of co-ordinates (axes x_1, x_2 and x_3) is defined on a chosen reference surface (see figure 1(b)), whose distance to the sub-regions middle surfaces are given by c_1, c_2 and c_3. As in figure 1(b) the reference surface is adopted coincident to Ω_2 middle surface one has $c_2=0$.

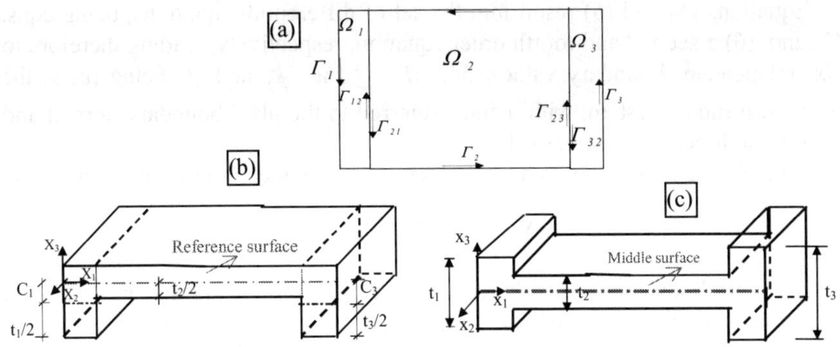

Figure 1: (a) Reinforced plate; (b) reference surface view, (c) middle surface view.

Let us consider initially, the bending problem. For a point placed at any of those plate sub-regions, the following equations are defined:

-The equilibrium equations in terms of internal forces:

$$M_{ij},_j - Q_i = 0 \qquad i, j = 1, 2 \tag{1}$$

$$Q_i,_i + g = 0 \tag{2}$$

where g is the distributed load acting on the plate middle surface, m_{ij} are bending and twisting moments and Q_i represents shear forces.

-The generalised internal forces written in terms of displacement:

$$M_{ij} = \frac{D(1-\nu)}{2}\left(\phi_i,_j + \phi_j,_i + \frac{2\nu}{1-\nu}\phi_k,_k\,\delta_{ij}\right) + \frac{\nu g}{(1-\nu)\lambda^2}\delta_{ij} \tag{3}$$

$$Q_i = \frac{D(1-\nu)}{2}\lambda^2(\phi_i + w,_i) \qquad i, j = 1, 2; \tag{4}$$

where ϕ_i is the rotation in the i direction, w the deflection, $D=Et^3/(1-\nu^2)$ the flexural rigidity, ν the Poisson's ration, λ a constant related to shear effect given by $\lambda = \sqrt{10}/t$ and δ_{ij} is the Kronecker delta.

-Finally, the plate bending differential equations given by:

$$Q_i - \frac{1}{\lambda^2}\nabla^2 Q_i + \frac{1}{(1-\nu)\lambda^2}\frac{\partial g}{\partial x_i} = -D\frac{\partial}{\partial x_i}\nabla^2 w \qquad i=1, 2 \tag{5}$$

$$\nabla^4 w = \frac{1}{D}\left[g - \frac{(2-\nu)}{(1-\nu)\lambda^2}\nabla^2 g\right] \tag{6}$$

where $w,_{iijj} = \nabla^4 w$, being ∇^4 the bi-harmonic operator; $w,_{ii} = \nabla^2 w$ being ∇^2 the bi-Laplacian operator.

Equations (5) and (6) result into the set of differential equations, being eqns. (5) and. (6) a second and fourth order equation, respectively, leading therefore to six independent boundary values: M_n, M_{ns}, Q_n, w, ϕ_n and ϕ_s being (n, s) the local co-ordinate system, with n and s referred to the plate boundary normal and tangential directions, respectively.

Considering now the stretching problem, the in-plane equilibrium equation is:

$$N_{ij},_j + b_i = 0 \tag{7}$$

where b_i are body forces distributed over the plate middle surface and N_{ij} is the membrane internal force, which, for plane stress conditions, can be written in terms of the in-plane displacements u_i derivatives as follow:

$$N_{ij} = Gt\left[\frac{2\nu}{(1-\nu)} u_k,_k \delta_{ij} + (u_i,_j + u_j,_i)\right] \tag{8}$$

The problem definition is then completed by assuming the following boundary conditions over Γ: $U_i = \overline{U}_i$ on Γ_u (generalised displacements: deflections, rotations and in-plane displacements) and $P_i = \overline{P}_i$ on Γ_p (generalised tractions: bending and twisting moments, shear forces and in-plane tractions), where $\Gamma_u \cup \Gamma_p = \Gamma$. Note that the in-plane displacements and tractions are considered only for the coupled stretching-bending problem.

3 Integral representations

For the simple bending problem the following weighted residual equation can be written for a simple plate:

$$\int_{\Omega}\left[\phi^*_{ki}(\mathrm{M}_{ij},_j - \mathrm{Q}_i) + (\mathrm{Q}_i,_i + \mathrm{g})\mathrm{w}^*_k\right]d\Omega = \int_{\Gamma_u}\left[(\overline{\phi}_i - \phi_i)\mathrm{M}^*_{ki} + (\overline{\mathrm{w}} - \mathrm{w})\mathrm{Q}^*_{kn}\right]d\Gamma$$
$$- \int_{\Gamma_P}\left[(\overline{M}_i - M_i)\phi^*_{ki} + (\overline{Q}_n - Q_n)w^*_k\right]d\Gamma \quad \text{i, j =1, 2; k=1, 2, 3} \tag{9}$$

where the superscript * refers to the fundamental problem; k is the fundamental load direction with k = 1, 2 defining unit moments applied in the x_1 and x_2 directions and k=3 is related to a unit load acting in the x_3 direction.

Integrating (9) by parts twice, considering eqns (3) and (4) and writing the values in terms of the local system of coordinates (n,s), the integral equation of the generalised displacements can be obtained:

$$c(q)U_k(q) = \int_{\Omega_g} g\left[w^*_k - \frac{\nu}{(1-\nu)\lambda^2}\phi^*_{ki},_i\right]d\Omega - \int_{\Gamma}\left[\phi_n M^*_{kn} + \phi_s M^*_{kns} + wQ^*_{kn}\right]d\Gamma +$$
$$+ \int_{\Gamma}\left[M_n\phi^*_{kn} + M_{ns}\phi^*_{ks} + Q_n w^*_k\right]d\Gamma \qquad \text{k=m, l, 3;} \qquad \text{i=1, 2} \tag{10}$$

where q is the collocation point, Ω_g the area where the load g is distributed, $c(q)$ is the free term given by: $c(q) = 0$, $c(q) = 1$ and $c(Q) = 1/2$, respectively, for external, internal and boundary points; $U_m = \phi_m$ $U_l = \phi_l$ and $U_3 = w$, being m and l the local system (n, s) for boundary points or any direction for internal points.

For a zoned plate, as the one depicted in the figure 1, eqn. (10) is valid to each sub-region separately. Then, taking into account the equilibrium and compatibility conditions, writing eqn. (10) to all sub-regions and summing them one can write the integral representation for the simple bending problem:

$$U_k(q) = \sum_{j=1}^{N_s} \int_{\Omega_g} g\left[w_k^{*j} - \frac{\nu}{(1-\nu)\lambda^2}\phi_{ki}^{*j},_i \right] d\Omega - \sum_{j=1}^{N_s} \int_{\Gamma_1} \left[\phi_n M_{kn}^{*j} + \phi_s M_{kns}^{*j} + wQ_{kn}^{*j} \right] d\Gamma$$

$$- \sum_{j=1}^{N_{\mathrm{int}}} \int_{\Gamma_{ja}} \left\{ \phi_n \left[M_{kn}^{*j} - M_{kn}^{*a} \right] + \phi_s \left[M_{kns}^{*j} - M_{kns}^{*a} \right] + w\left[Q_{kn}^{*j} - Q_{kn}^{*a} \right] \right\} d\Gamma$$

$$+ \sum_{j=1}^{N_s} \int_{\Gamma_1} \left[M_n \phi_{kn}^{*j} + M_{ns} \phi_{ks}^{*j} + Q_n w_k^{*j} \right] d\Gamma$$

$$+ \sum_{j=1}^{N_{\mathrm{int}}} \int_{\Gamma_{ja}} \left\{ M_n \left[\phi_{kn}^{*j} - \phi_{kn}^{*a} \right] + M_{ns} \left[\phi_{ks}^{*j} - \phi_{ks}^{*a} \right] + Q_n \left[w_k^{*l} - w_k^{*a} \right] \right\} d\Gamma \quad (11)$$

where N_s and N_{int} are the sub-regions and interfaces number, Γ_{ja} represents a interface for which the subscript a denotes the adjacent sub-region to Ω_j

The bending equation for the coupled stretching-bending problem is obtained from eqn. (11) by writing the moment values on the Ω_j middle surface in terms of their values on the reference surface ($\boldsymbol{M}_n^r$ and M_{ns}^r), as follow:

$$\boldsymbol{M}_n^j = \boldsymbol{M}_n^r + \boldsymbol{p}_n \boldsymbol{c}_j \quad (12)$$

$$\boldsymbol{M}_{ns}^j = \boldsymbol{M}_{ns}^r + \boldsymbol{p}_s \boldsymbol{c}_j \quad (13)$$

where p_n and p_s are the in-plane tractions.

Then the bending integral equation for the coupled problem, where all values are referred to the reference surface, reads:

$$U_k(q) = \sum_{j=1}^{N_s} \int_{\Omega_g} g\left[w_k^{*j} - \frac{\nu}{(1-\nu)\lambda^2}\phi_{ki}^{*j},_i \right] d\Omega$$

$$- \sum_{j=1}^{N_s} \int_{\Gamma_l} \left[\phi_n M_{kn}^{*j} + \phi_s M_{kns}^{*j} + wQ_{kn}^{*j} \right] d\Gamma + \sum_{j=1}^{N_s} \int_{\Gamma_l} \left[M_n \phi_{kn}^{*j} + M_{ns} \phi_{ks}^{*j} + Q_n w_k^{*j} \right] d\Gamma$$

$$- \sum_{j=1}^{N_{int}} \int_{\Gamma_{ja}} \left\{ \phi_n \left[M_{kn}^{*j} - M_{kn}^{*a} \right] + \phi_s \left[M_{kns}^{*j} - M_{kns}^{*a} \right] + w\left[Q_{kn}^{*j} - Q_{kn}^{*a} \right] \right\} d\Gamma$$

$$+\sum_{j=1}^{N_{int}} \int_{\Gamma_{ja}} \left\{ M_n \left[\phi_{kn}^{*j} - \phi_{kn}^{*a} \right] + M_{ns} \left[\phi_{ks}^{*j} - \phi_{ks}^{*a} \right] + Q_n \left[w_k^{*1} - w_k^{*a} \right] \right\} d\Gamma$$

$$+\sum_{j=1}^{N_{int}} \int_{\Gamma_{ja}} \left\{ p_n \left[c_j \phi_{kn}^{*j} - c_a \phi_{kn}^{*a} \right] + p_s \left[c_j \phi_{ks}^{*j} - c_a \phi_{ks}^{*a} \right] \right\} d\Gamma + \sum_{j=1}^{N_s} \int_{\Gamma_I} c_j \left[p_n \phi_{kn}^{*j} + p_s \phi_{ks}^{*j} \right] d\Gamma \quad (14)$$

Let us now consider the stretching problem. For simplicity and also to eliminate the in-plane tractions along the interfaces, the fundamental value $u_{ki}^{*(j)}$ related to Ω_j will be written in terms of u_{ki}^{*} referred to the sub-region where the collocation point is placed as follow:

$$u_{ki}^{*(j)} = u_{ki}^{*} \overline{E} / \overline{E}_j \quad (15)$$

where $\overline{E}_j = E_j t_j$.

From the weighted residual method and considering eqn. (15) one can derive the integral representation of displacements for one sub-region. The integral equation for a zoned plate is obtained by summing the equations of all sub-regions and enforcing equilibrium and compatibility conditions along interfaces. Moreover for the coupled problem the in-plane displacements defined over the middle surface (u_s and u_n) have to be written in terms of their values on the reference surface ($u_i^j = u_i^r - c_j \phi_i$, with i=n,s). Then the following stretching integral equation for the coupled stretching-bending problem can be obtained:

$$\left[-c(q)\phi_k(q) + u_k(q) \right] = -\sum_{i=1}^{N_S} \frac{\overline{E}_i}{\overline{E}} \int_{\Gamma} \left[u_n p_{kn}^{*} + u_s p_{ks}^{*} \right] d\Gamma +$$

$$-\sum_{m=1}^{N_{int}} \frac{\left(\overline{E}_j - \overline{E}_a \right)}{\overline{E}} \int_{\Gamma_{ja}} \left[u_n p_{kn}^{*} + u_s p_{ks}^{*} \right] d\Gamma_{ja} + \sum_{i=1}^{N_S} \frac{\overline{E}_i}{\overline{E}} \int_{\Gamma} c_i \left[p_{kn}^{*} \phi_n + p_{ks}^{*} \phi_s \right] d\Gamma +$$

$$\sum_{m=1}^{N_{int}} \frac{\left(\overline{E}_j c_j - \overline{E}_a c_a \right)}{\overline{E}} \int_{\Gamma_{ja}} \left(p_{kn}^{*} \phi_n + p_{ks}^{*} \phi_s \right) d\Gamma + \int_{\Gamma} \left(u_{kn}^{*} p_n + u_{ks}^{*} p_s \right) d\Gamma + \int_{\Omega_i} \left(u_{kn}^{*} b_n + u_{ks}^{*} b_s \right) d\Omega \quad (16)$$

Note that in eqn (16) the in-plane tractions were eliminated from the interfaces, where the only remaining values are the displacements.

Let us now consider the beam B_3 represented in figure 2(a) by the sub-region Ω_3. In order to reduce the number of degrees of freedom, the displacements w, u_s, u_n, ϕ_s and ϕ_n will be assumed to be linear along the beam width, leading to a model where the values are defined along the beam skeleton line instead of its boundary. Thus the displacement related to the beam interfaces are translated to the skeleton line, as follows:

$$\phi_k^{\Gamma_{32}} = \phi_k + \phi_{k,n} b_3 / 2 \qquad \text{k=n,s} \quad (17a)$$

$$\phi_k^{\Gamma_{31}} = -\left[\phi_k - \phi_{k,n} b_3 / 2 \right] \quad (17b)$$

$$w^{\Gamma_{32}} = w + w,_n b_3 / 2 \tag{17c}$$

$$w^{\Gamma_{31}} = w - w,_n b_3 / 2 \tag{17d}$$

$$u_k^{\Gamma_{32}} = u_k + u_k,_n b_3 / 2 \tag{18a}$$

$$u_k^{\Gamma_{31}} = -\left[u_k - u_k,_n b_3 / 2\right] \tag{18b}$$

where b_3 is the beam width , $\phi_k^{\Gamma_{ij}}$, $u_k^{\Gamma_{ij}}$ and $w^{\Gamma_{ij}}$ are displacement components along the interface Γ_{ij}; ϕ_k , w, $\phi_k,_n$, $u_k,_n$ and $w,_n$ are skeleton line components.

Observe that adopting the approximations defined in eqns (17) and (18), new variables related to the beam axis appear in the formulation: the rotations $w,_n$ $u_s,_n$ and $u_n,_n$ and the curvatures $\phi_s,_n$ and $\phi_n,_n$ whose integral representations can be easily obtained by differentiating eqns. (11), (14) or (16).

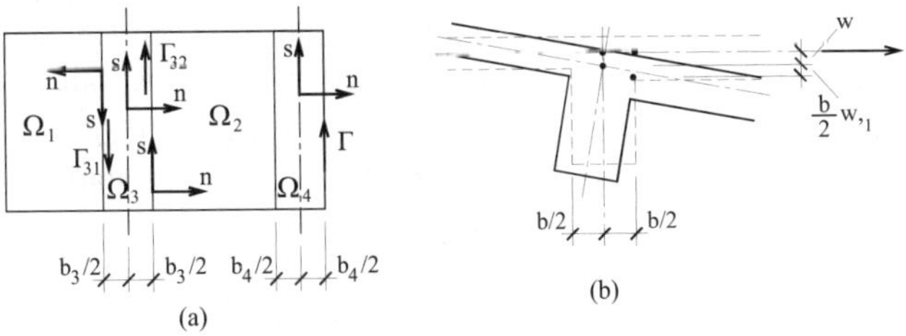

Figure 2: (a) reinforced plate view; (b) deflection approximations along interfaces.

The tractions defined on the interfaces will be written in terms of its components along the beam axis as follows:

$$Q_n^{\Gamma_{32}} = -Q_n^{\Gamma_{31}} = Q_n \tag{19}$$

$$M_n^{\Gamma_{32}} = M_n - Q_n b_3 / 2 \tag{20}$$

$$M_n^{\Gamma_{31}} = M_n + Q_n b_3 / 2 \tag{21}$$

$$M_{ns}^{\Gamma_{32}} = M_{ns}^{\Gamma_{31}} = M_{ns} \tag{22}$$

$$p_i^{\Gamma_{32}} = \frac{1}{2} p_i = -p_i^{\Gamma_{31}} \qquad \text{i=n, s} \tag{23}$$

where M_n, M_{ns}, Q_n and $\boldsymbol{p_i}$ refers to the beam axis while the tractions with superscripts Γ_{ij} are related to the local coordinate system defined on interface Γ_{ij}.

As the integrals are still performed on the interfaces and the collocation points are adopted on the beam axis, there is no problem of singularities.

4 Algebraic equations

The integral representations (11), (14) or (16) have to be transformed into algebraic expressions after discretizing the boundary and interfaces into elements. It has been adopted linear elements to approximate the problem geometry while the variables were approximated by quadratic shape functions.

Let us initially consider the simple bending analysis. Six values (w, ϕ_n, ϕ_s, Q_n, M_n and M_{ns}) are defined along the external boundary without beams, being three of them prescribed. Thus, in this case one has adopted to write eqn. (11) for an external collocation point very near to the boundary node. On the beam axis one has nine values: w, ϕ_n, ϕ_s, $\phi_{s,n}$, $\phi_{n,n}$, $w_{,n}$, Q_n, M_n and M_{ns} with collocation points adopted on the skeleton line coincident to the node or defined at element internal points when variable discontinuity is required at the element end. For external beams nodes the displacements $\phi_{s,n}$, $\phi_{n,n}$ and $w_{,n}$ are problem unknowns while three of the remaining values must be prescribed, requiring therefore six equations. In this case, one writes eqn. (11) plus the equations of $\phi_{s,n}$, $\phi_{n,n}$ and $w_{,n}$ for each collocation point. As all the nine values remain as unknowns for the internal beams nodes, in this case besides the equations adopted for the external beam nodes we also write the representations of Q_n, M_n and M_{ns}.

For the coupled stretching-bending problem, in addition to the values and equations defined previously for the simple bending problem one has to consider those related to the stretching problem. Along the external boundary without beams one has also the values u_s, u_n, p_n and p_s, being two of them prescribed. Thus in this case one has chosen to write eqns. (14) and (16) for each external collocation point. On beam nodes are also defined the following values: u_s, u_n, $u_{s,n}$, $u_{n,n}$, p_n and p_s All these values remain as unknowns in the internal beams, requiring therefore fifteen algebraic equations for each skeleton line point. In this case the adopted equations were those corresponding to the unknowns. For external beams, the displacements $u_{s,n}$ and $u_{n,n}$ are also problem unknowns while two of the four values: u_s, u_n, p_n and p_s, must be prescribed, leading to ten unknowns for each external beam node. It has been adopted to write eqns (14), (16) plus the following ones: $u_{s,n}$, $u_{n,n}$, $\phi_{s,n}$, $\phi_{n,n}$ and $w_{,n}$.

After writing the recommended algebraic relations one obtains the set of equations, which can be solved after applying the boundary conditions. For simple bending analysis and the coupled problem they are given, respectively by:

$$\underset{\sim}{H}^B \underset{\sim}{U}_B = \underset{\sim}{G}^B \underset{\sim}{P}_B + \underset{\sim}{T}^B \tag{24}$$

$$\begin{bmatrix} [H]^B & [\overline{H}]^S \\ --- & --- \\ [\overline{H}]^B & [H]^S \end{bmatrix} \begin{Bmatrix} \{U\}_B \\ \{U\}_S \end{Bmatrix} = \begin{bmatrix} [G]^B & [\overline{G}]^S \\ --- & --- \\ [0] & [G]^S \end{bmatrix} \begin{Bmatrix} \{P\}_B \\ \{P\}_S \end{Bmatrix} + \begin{Bmatrix} \{T\}^B \\ --- \\ \{T\}^S \end{Bmatrix} \tag{25}$$

where {U} and {P} are displacement and traction vectors, respectively; {T} is the independent vector due to the applied loads; [H] and [G] are matrices obtained by integrating all boundary and interfaces; B and S are related to bending and stretching problems.

In eqn. (25) the upper e bottom parts indicate, respectively, algebraic equations of the bending and stretching problems.

5 Numerical application

Let us consider the plate reinforced by two external beams depicted in figure 1, adopting $t_1=t_3=25cm$, $t_2=10.0cm$, Young's modulus $E=2.7x10^4 kN/cm^2$ and Poisson's ratio $\nu=0.0$. The two sides containing the beams are assumed free, while the other two are simply supported. The plate sides without beams as well as the beam axis were discretized by *12* quadratic elements (Figure 3), giving the total amount of 48 elements and 100 nodes. Observe that the element coincident to the beam width is automatically generated by the code.

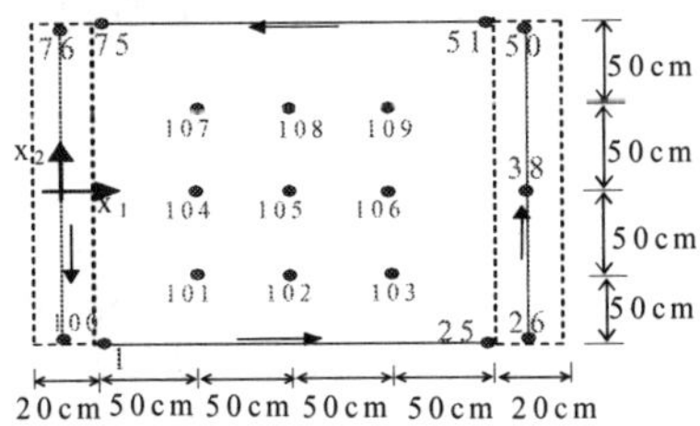

Figure 3: Discretization.

Table 1: Displacements at internal and boundary nodes.

Nodes	x_2 (cm)	w(cm) SB	w (cm) CP	ϕ_2 SB	ϕ_2 CP
1 to 25	-100	0	0.	-0.006667	0.007407
101, 102, 103, 94, 32	-50	0.25	-0.2777	-0.003333	0.003333
104, 105, 106, 38, 88	0	0.33	-0.3703	0.0	0.0

For both analyses one has prescribed appropriate boundary values to enforce constant curvatures over the entire structural element. So that displacements and internal forces would have exact solutions. For the simple bending analysis (see figure 1(c)) we have applied M_n=150 kNcm/cm and M_n= 2.34375x10^3 kNcm/cm, respectively, along the simply supported plate boundary and on the beam simply supported ends (the beam width). The prescribed loads, for the coupled stretching-bending problem (see figure 1(b)), were: M_n=-1.666667x10^5Ncm/cm on the simply supported plate boundary; p_n=3.75x10^5kN/cm and M_n=-5.416667x10^6Ncm/cm along the beam width. As expected, for both analyses the computed values are exactly the theoretical ones (see table 1, where SB and CP mean, respectively, simple bending and coupled problem).

6 Conclusions

BEM formulations based on Reissner's hypothesis for analysing plate reinforced by beams have been presented. Some approximations for displacements and tractions along the beam cross section have been considered, leading to a model where the problem values are defined on the beam axis. The performance of the proposed formulation has been confirmed by comparing the results with analytical solutions.

Acknowledgements

The authors wish to thank FAPESP (São Paulo State Foundation for Scientific Research) for the financial support.

References

[1] Hu, C. & Hartley, G.A., Elastic analysis of thin plates with beam supports. *Engineering Analysis with Boundary Elements*, **13**, pp 229-238, 1994.

[2] Tanaka, M. & Bercin, A.N., A boundary Element Method applied to the elastic bending problems of stiffened plates. In: *Boundary Element Method XIX*, Eds. C.A. Brebbia et al., CMP, Southampton, 1997.

[3] Sapountzakis, E.J. & Katsikadelis, J.T., Analysis of plates reinforced with beams. *Computational Mechanics*, 26, pp 66-74, 2000.

[4] Fernandes, G.R & Venturini, W.S., Stiffened plate bending analysis by the boundary element method. *Computational Mechanics*, 28, pp 275-281, 2002.

[5] Fernandes, G.R. & Venturini, W. S., *Building floor analysis by the Boundary element method.* Computational Mechanics, 35, pp 277-291, 2005.

[6] Hartmann, F. & Zotemantel, R., The direct boundary element method in plate bending. *International Journal for Numerical Methods in Engineering*, **23**(11), pp 2049-2069, 1986

[7] Kirchhoff, G., Uber das gleichgewicht und die bewegung einer elastischen scleibe. *J. Math.,* 40, pp 51-58, 1850.

[8] Reissner, E., On bending of elastic plates. *Quart. Appl. Math.*; 5(1), pp 55-68; 1947

[9] Weën, F. V., Application of boundary integral equation method to Reissner's plate model. Int. J. Num. Meth. Emg., 18 (1), pp. 1-10, 1982.

[10] Palermo Jr. L., Plate bending analysis using the classical or the Reissner-Mindlin models. *Engineering Analysis with Boundary Elements*, **27**, pp. 603-609, 2003.

Section 8
Wave propagation

Time and space derivatives in a BEM formulation based on the CQM with initial conditions contribution

A. I. Abreu[1], M. A. C. Ferro[2] & W. J. Mansur[1]
[1]*Department of Civil Engineering, COPPE/Federal University of Rio de Janeiro, Rio de Janeiro, Brazil*
[2]*Military Institute of Engineering, Rio de Janeiro, Brazil*

Abstract

This work is concerned with the numerical computation of time and space derivatives of the time-domain solution of scalar wave propagation problems using the boundary element method (TD-BEM). In the present formulation, the BEM based on the so-called convolution quadrature method (CQM-BEM) is employed. The CQM-BEM takes into account non-homogeneous initial conditions by means of a general procedure, known as the initial condition pseudo-force procedure (ICPF), which replaces the initial conditions by equivalent pseudo-forces. The boundary integral equation with initial conditions contribution is derived analytically and the quadrature weights of the standard ICPF-CQM-BEM formulation are transformed in order to compute time and space derivatives. Two numerical examples are presented at the end of the work illustrating the efficacy of the implemented formulation.
Keywords: wave equation, time and space derivatives, boundary element method, convolution quadrature method, initial condition pseudo-force procedure.

1 Introduction

This work presents an application of a time-domain boundary element method (TD-BEM) for the analysis of scalar wave propagation problems. The TD-BEM employs the convolution quadrature method (CQM) developed by Lubich [1, 2]. In the CQM, fundamental solutions in the Laplace transformed-domain are considered and a numerical approximation of the basic integral equations of the

WIT Transactions on Modelling and Simulation, Vol 44, © 2007 WIT Press
www.witpress.com, ISSN 1743-355X (on-line)
doi:10.2495/BE070231

TD-BEM is worked out by a quadrature formula based on a linear multi-step method. The main advantage of the CQM is that it can be applied to problems where the TD fundamental solution is not available or has a very difficult expression. The CQM-BEM was firstly applied to scalar wave propagation problems by Abreu *et al.* [3].

The CQM was employed successfully in viscoelastic and poroelastic problems by Schanz [4] and Gaul and Schanz [5]. Application of the CQM to plane frame dynamic modelling was performed by Antes *et al.* [6]. A method based on Duhamel integrals, in combination with the CQM, for the analysis of one-dimensional wave propagation in a layered medium was presented by Moser *et al.* [7], which was extended to plane strain elastodynamic BEM-FEM (finite element method) coupling later on [8]. Recently, Dobromil and Schanz [9] and Schanz *et al.* [10] applied the CQM to a poroelastic boundary elements approach shown that BEM based on CQM is very robust and suitable for such kinds of problems.

All the cited formulations were applied to problems with null initial conditions. In a recent work by Abreu *et al.* [11], a new numerical technique called Initial Condition Pseudo-Force procedure (ICPF), was presented in order to consider non-null initial conditions in a CQM-BEM context. Originally, the ICPF was employed for frequency-domain analysis with BEM and FEM formulations, as described by Mansur *et al.* [12, 13]. Basically, the ICPF consists in replacing the initial conditions by equivalent pseudo-forces. In the present work, the notation ICPF-CQM-BEM will be adopted when non-homogeneous initial conditions are considered in the CQM-BEM formulation [11]. In the following sections, the ICPF-CQM-BEM is reviewed. In the sequence, numerical procedures to compute time and space derivatives using ICPF-CQM-BEM is presented. Finally, two numerical examples illustrate the efficacy of the developed formulation.

2 The convolution quadrature method

Consider first the following equation:

$$y(t) = \int_0^t f(t-\tau)\, g(\tau)\, d\tau \tag{1}$$

In [1, 2] it has been showed that function y can be approximated at points $n\Delta t$ as:

$$y(n\Delta t) = \sum_{k=0}^{n} \omega_{n-k}^{j}(\Delta t)\, g(k\Delta t)\,, \qquad n = 0, 1, 2..., N \tag{2}$$

where N is the total number of time sampling and the weights ω_n are:

$$\omega_n(\Delta t) = \frac{1}{2\pi i}\int_{C\rho} \hat{f}\left(\frac{\gamma(z)}{\Delta t}\right) z^{-n-1}\, dz \approx \frac{\rho^{-n}}{L}\sum_{l=0}^{L-1} \hat{f}\left(\frac{\gamma(\rho\, e^{i\,l\,2\pi/L})}{\Delta t}\right) e^{-i\,n\,l\,2\pi/L} \tag{3}$$

where $\hat{f}$ is the Laplace transform of f and $C_\rho = \{z \in \mathbf{C}; |z| = \rho\}$ is the contour employed to perform the integration; i.e., ρ is the radius of a circle in the domain of analyticity of $\hat{f}(\gamma(z)/\Delta t)$. In Eq. (3) a polar coordinate system was adopted and the integral was approximated by the trapezoidal rule with L equal steps $(2\pi/L)$.

Assuming $f(0) = 0$ and applying Leibniz integral rule to Eq. (1) leads to:

$$\frac{\partial y(t)}{\partial t} = \int_0^t \frac{\partial f(t-\tau)}{\partial t} g(\tau)\, d\tau \tag{4}$$

Eq. (4) shows for analogy with Eq (1) that $\partial y/\partial t$ can be approximated at points $n\Delta t$ as:

$$\frac{\partial y(n\Delta t)}{\partial t} = \sum_{k=0}^{n} \bar{\omega}_{n-k}(\Delta t)\, g(k\Delta t) \tag{5}$$

where now $\bar{\omega}_n$ can be obtained using the Laplace transform of $\partial f/\partial t$, which, evaluated at point s gives $s\hat{f}(s)$. That is:

$$\bar{\omega}_n(\Delta t) \approx \frac{\rho^{-n}}{L} \sum_{l=0}^{L-1} s_l \hat{f}(s_l)\, e^{-i n l 2\pi/L} \tag{6}$$

and $s_l = \gamma(\rho\, e^{i l 2\pi/L})/\Delta t$.

3 Boundary integral equation with initial conditions

The time-domain integral equation corresponding to problems governed by the 2D scalar wave equation with non-homogeneous initial conditions is, for any interior point ξ [11]:

$$\begin{aligned} u(\xi,t) = & \int_{\Gamma_p} \int_0^{t^+} u^*(r,t-\tau)\, p(X,\tau)\, d\tau d\Gamma_p - \int_{\Gamma_u} \int_0^{t^+} p^*(r,t-\tau)\, u(X,\tau)\, d\tau d\Gamma_u \\ & + \int_{\Omega} \int_0^{t(\Delta t)} u^*(r,t-\tau)\, f_{v_0}(X,t)\, d\tau d\Omega - \int_{\Omega} \int_0^{t^+} u^*(r,t-\tau)\, f_{u_0}(X,t)\, d\tau d\Omega \end{aligned} \tag{7}$$

In Eq. (7) $p(X,t) = \partial u(X,t)/\partial n$ represents the normal flux, $\Gamma = \Gamma_u \cup \Gamma_p$ is the boundary of the domain Ω. The distance between the source point ξ and the field point $X = (x,y)$ is represented by $r = |X - \xi|$ where $X \in \Omega \cup \Gamma$, $u^*(r,t-\tau)$ is the fundamental solution and $p^*(r,t-\tau)$ is its normal derivative. Eq. (7) considers the fictitious domain source contributions f_{u_0} and f_{v_0} corresponding to the initial displacement and velocity fields $u_0(X)$ and $v_0(X)$, respectively. The expressions for the pseudo-forces can be calculated according to [11]:

$$f_{v_0}(X,t) = \frac{1}{c^2\, \Delta t} v_0(X)\, \delta(t-0) \quad \text{and} \quad f_{u_0}(X,t) = c^2 \nabla^2 u_0(X)\, H(t-0) \tag{8}$$

where the response due to the initial velocity field $v_0(X)$ can be obtained considering an impulsive pseudo-force as described by Eq. (8) (Δt is the time interval sampling). The contribution of the initial displacement field $u_0(X)$ can be computed subtracting from the initial displacement field itself, the response corresponding to a pseudo-force as presented in Eq. (8). In Eq. (8) $\delta(t-0)$ is the Dirac delta generalized function and $H(t-0)$ is the Heaviside function.

Once the boundary is discretized in J elements (Γ_j, $j = 1, 2, \ldots J$) and the subdomain of Ω with non-homogeneous initial condition in Ce cells (Ω_j, $j = 1, 2, \ldots Ce$), the discretized version of the integral Eq. (7) employing the CQM is, for an interior point ξ:

$$u(\xi,t_n) = \sum_{j=1}^{J}\sum_{k=0}^{n} \boldsymbol{g}_{n-k}^{j}(\xi,\Delta t)\, \boldsymbol{p}_k^{j}(X) - \sum_{j=1}^{J}\sum_{k=0}^{n} \boldsymbol{h}_{n-k}^{j}(\xi,\Delta t)\, \boldsymbol{u}_k^{j}(X) + \sum_{j=1}^{Ce} \boldsymbol{m}_n^{j}(\xi,\Delta t)\, \boldsymbol{f}_{v_0\,0}^{\;j}(X) - \sum_{j=1}^{Ce}\sum_{k=0}^{n} \boldsymbol{m}_{n-k}^{j}(\xi,\Delta t)\, \boldsymbol{f}_{u_0\,k}^{\;j}(X) \tag{9}$$

The CQM weights $\boldsymbol{g}_n$, $\boldsymbol{h}_n$ and $\boldsymbol{m}_n$ in Eq. (9) are:

$$\boldsymbol{g}_n^{j}(\xi,\Delta t) = \frac{\rho^{-n}}{L} \sum_{l=0}^{L-1} \int_{\Gamma_j} \hat{u}^*(r,s_l)\, \boldsymbol{\Phi}^{j}(X)\, d\Gamma\, e^{-i\,n\,l\,2\pi/L} \tag{10}$$

$$\boldsymbol{h}_n^{j}(\xi,\Delta t) = \frac{\rho^{-n}}{L} \sum_{l=0}^{L-1} \int_{\Gamma_j} \hat{p}^*(r,s_l)\, \boldsymbol{\Phi}^{j}(X)\, d\Gamma\, e^{-i\,n\,l\,2\pi/L} \tag{11}$$

$$\boldsymbol{m}_n^{j}(\xi,\Delta t) = \frac{\rho^{-n}}{L} \sum_{l=0}^{L-1} \int_{\Omega_j} \hat{u}^*(r,s_l)\, \boldsymbol{\Phi}_{ce}^{j}(X)\, d\Omega\, e^{-i\,n\,l\,2\pi/L} \tag{12}$$

In the above expressions, $\boldsymbol{\Phi}^{j}(X)$ and $\boldsymbol{\Phi}_{ce}^{j}(X)$ represent the interpolation functions employed in the boundary and domain discretization, respectively. The function $\gamma(z)$, used in Eq. (10) to (12), is the quotient of the characteristic polynomials generated by a linear multi-step method [1, 2].

In Eqs. (10) and (12), $\hat{u}^*(r,s)$ is the Laplace transform of the fundamental solution $u^*(r,t)$ and in Eq. (11) $\hat{p}^*(r,s)$ is the Laplace transform of $p^*(r,t)$. The expressions of these functions are given by [14]:

$$\hat{u}^*(r,s) = 2K_0(sr) \quad \text{and} \quad \hat{p}^*(r,s) = \frac{\partial \hat{u}^*(r,s)}{\partial r}\frac{\partial r}{\partial n} = -2sK_1(sr)\frac{\partial r}{\partial n} \tag{13}$$

where $K_0(sr)$ and $K_1(sr)$ are the modified Bessel function of zero and first order, respectively, and of second kind.

The nodal values of the pseudo-forces $\boldsymbol{f}_{v_0}^{\,j}$ and $\boldsymbol{f}_{u_0}^{\,j}$ are computed from the known initial velocity $\boldsymbol{v}_0^j$ and initial displacement $\boldsymbol{u}_0^j$, respectively, at each j cell as:

$$\boldsymbol{f}_{v_0}{}_0^j(X) = \frac{1}{c^2\,\Delta t}\,\boldsymbol{v}_0^j(X) \quad \text{and} \quad \boldsymbol{f}_{u_0}{}_k^j(X) = c^2\,\nabla^2\boldsymbol{u}_0^j(X), \qquad X \in \Omega_j \cup \Gamma_j \tag{14}$$

Eq. (9) can be rewritten in matricial form as follows:

$$\boldsymbol{u}^n = \sum_{k=0}^{n} \boldsymbol{G}^{n-k}\boldsymbol{p}^k - \sum_{k=0}^{n} \boldsymbol{H}^{n-k}\boldsymbol{u}^k + \boldsymbol{M}^0\boldsymbol{f}_{v_0}^{\,0} - \sum_{k=0}^{n} \boldsymbol{M}^{n-k}\boldsymbol{f}_{u_0}^{\,k} \tag{15}$$

where $\boldsymbol{G}$, $\boldsymbol{H}$ and $\boldsymbol{M}$ are the final boundary element influence matrices and indices n and k correspond to the discrete times $t_n = n\Delta t$ and $t_k = k\Delta t$, respectively. For details concerning the CQM-BEM and ICPF-CQM-BEM references [3, 11] are indicated.

4 Time derivative of the integral equation

In order to obtain the time derivative of the function $u(\xi,t)$, Eq. (7) is differentiated. Applying Eq. (4) to each integral term leads to:

$$\frac{\partial u(\xi,t)}{\partial t} = \int_{\Gamma_p}\int_0^{t^+} \frac{\partial u^*(r,t-\tau)}{\partial t}\,p(X,\tau)\,d\tau d\Gamma_p - \int_{\Gamma_u}\int_0^{t^+} \frac{\partial p^*(r,t-\tau)}{\partial t}\,u(X,\tau)\,d\tau d\Gamma_u$$
$$+\int_{\Omega}\int_0^{t(\Delta t)} \frac{\partial u^*(r,t-\tau)}{\partial t}\,f_{v_0}(X,t)\,d\tau d\Omega - \int_{\Omega}\int_0^{t^+} \frac{\partial u^*(r,t-\tau)}{\partial t}\,f_{u_0}(X,t)\,d\tau d\Omega \tag{16}$$

Following the CQM procedure, the time derivative discretized version of the Eq. (9) is:

$$\left.\frac{\partial u(\xi,t)}{\partial t}\right|_{t=tn} = \sum_{j=1}^{J}\sum_{k=0}^{n} \bar{\boldsymbol{g}}_{n-k}^{\,j}(\xi,\Delta t)\,\boldsymbol{p}_k^j(X) - \sum_{j=1}^{J}\sum_{k=0}^{n} \bar{\boldsymbol{h}}_{n-k}^{\,j}(\xi,\Delta t)\,\boldsymbol{u}_k^j(X)$$
$$+\sum_{j=1}^{Ce} \bar{\boldsymbol{m}}_n^{\,j}(\xi,\Delta t)\,\boldsymbol{f}_{v_0}{}_0^j(X) - \sum_{j=1}^{Ce}\sum_{k=0}^{n} \bar{\boldsymbol{m}}_{n-k}^{\,j}(\xi,\Delta t)\,\boldsymbol{f}_{u_0}{}_k^j(X) \tag{17}$$

The CQM weights in Eq. (17) are computed by the following expressions:

$$\bar{\boldsymbol{g}}_n^{\,j}(\xi,\Delta t) = \frac{\rho^{-n}}{L}\sum_{l=0}^{L-1}\int_{\Gamma_j} s_l\,\hat{u}^*(r,s_l)\,\Phi^j(X)\,d\Gamma\,e^{-i n l 2\pi/L} \tag{18}$$

$$\bar{\boldsymbol{h}}_n^{\,j}(\xi,\Delta t) = \frac{\rho^{-n}}{L}\sum_{l=0}^{L-1}\int_{\Gamma_j} s_l\,\hat{p}^*(r,s_l)\,\Phi^j(X)\,d\Gamma\,e^{-i n l 2\pi/L} \tag{19}$$

$$\bar{\boldsymbol{m}}_n^j(\xi,\Delta t) = \frac{\rho^{-n}}{L}\sum_{l=0}^{L-1}\int_{\Omega_j} s_l \hat{u}^*(r,s_l)\, \boldsymbol{\Phi}_{ce}^j(X)\, d\Omega\, e^{-i n l 2\pi/L} \tag{20}$$

where the parameter s_l is the same as indicated before.

5 Space derivatives of the integral equation

The space derivative in the direction m of the function $u(\xi,t)$ can be obtained differentiating Eq. (9):

$$\begin{aligned}\frac{\partial u(\xi,t_n)}{\partial m} &= \sum_{j=1}^{J}\sum_{k=0}^{n} \bar{\bar{\boldsymbol{g}}}_{n-k}^j(\xi,\Delta t)\, \boldsymbol{p}_k^j(X) - \sum_{j=1}^{J}\sum_{k=0}^{n} \bar{\bar{\boldsymbol{h}}}_{n-k}^j(\xi,\Delta t)\, \boldsymbol{u}_k^j(X) \\ &+ \sum_{j=1}^{Ce} \bar{\bar{\boldsymbol{m}}}_n^j(\xi,\Delta t)\, \boldsymbol{f}_{v_0 0}^j(X) - \sum_{j=1}^{Ce}\sum_{k=0}^{n} \bar{\bar{\boldsymbol{m}}}_{n-k}^j(\xi,\Delta t)\, \boldsymbol{f}_{u_0 k}^j(X)\end{aligned} \tag{21}$$

and the CQM weights of above expression can be calculated as:

$$\bar{\bar{\boldsymbol{g}}}_n^j(\xi,\Delta t) = \frac{\rho^{-n}}{L}\sum_{l=0}^{L-1}\int_{\Gamma_j} \frac{\partial \hat{u}^*(r,s_l)}{\partial m}\, \boldsymbol{\Phi}^j(X)\, d\Gamma\, e^{-i n l 2\pi/L} \tag{22}$$

$$\bar{\bar{\boldsymbol{h}}}_n^j(\xi,\Delta t) = \frac{\rho^{-n}}{L}\sum_{l=0}^{L-1}\int_{\Gamma_j} \frac{\partial \hat{p}^*(r,s_l)}{\partial m}\, \boldsymbol{\Phi}^j(X)\, d\Gamma\, e^{-i n l 2\pi/L} \tag{23}$$

$$\bar{\bar{\boldsymbol{m}}}_n^j(\xi,\Delta t) = \frac{\rho^{-n}}{L}\sum_{l=0}^{L-1}\int_{\Omega_j} \frac{\partial \hat{u}^*(r,s_l)}{\partial m}\, \boldsymbol{\Phi}_{ce}^j(X)\, d\Omega\, e^{-i n l 2\pi/L} \tag{24}$$

The directional derivative of the fundamental solution of Eq. (13) is:

$$\frac{\partial \hat{u}^*(r,s)}{\partial m} = \frac{\partial \hat{u}^*(r,s)}{\partial r}\frac{\partial r}{\partial m} = -2sK_1(sr)\frac{\partial r}{\partial m} \tag{25}$$

and for the normal derivative of the fundamental solution one has:

$$\frac{\partial \hat{p}^*(r,s)}{\partial m} = 2s^2\left(\left(K_0(sr) + \frac{1}{sr}K_1(sr)\right)\frac{\partial r}{\partial m}\frac{\partial r}{\partial n} - K_1(sr)\frac{\partial}{\partial m}\left(\frac{\partial r}{\partial n}\right)\right) \tag{26}$$

where the directional derivatives are:

$$\frac{\partial r}{\partial m} = \frac{(\xi - X)\cdot m}{r}$$

6 Examples

Two numerical examples corresponding to a rod with the boundary conditions $\partial u(0, y,t)/\partial n = 0$ and $u(a, y,t) = 0$ are analyzed in order to validate the proposed method. The ICPF-CQM-BEM as presented in [11] was used to solve the scalar wave problem and Eqs. (17) and (21) were used to calculate time and space derivatives for interior points.

Derivatives of time and space in x direction were calculated at the central point $O = (a/2,0)$. These results were compared with respect to the results obtained with an explicit Finite Difference Method (FDM) approach for a very fine mesh (reference solution).

The following parameter was adopted for the examples: the wave velocity $c = 1$, and the dimensionless parameter $\beta = c\Delta t/l$ was used to estimate the time-step length (l is the smallest boundary element length) in the BEM formulation.

6.1 Rod under initial displacement condition prescribed over a subdomain

This example consists of a rod under initial displacement, as follows:

$$u_0(X) = \frac{P}{E}\left(\frac{a}{4} - x\right) \quad \left(0 \le x \le \frac{a}{4}, -\frac{a}{4} \le y \le \frac{a}{4}\right),$$

where $\frac{P}{E} = 1$, P is the force and E is the Young modulus. Fig. 1 depicts the boundary element mesh $ABCD$ and the internal point O selected. In this analysis 100 linear boundary elements were used and the subdomain $AMND$ was divided in 1024 linear triangular cells.

In this example line forces appear where $grad(u_0(X))$ is discontinuous and the volume integral in Eqs. (12), (20) and (24) become a line integral on AD and a line integral on MN as indicated in Fig. 1.

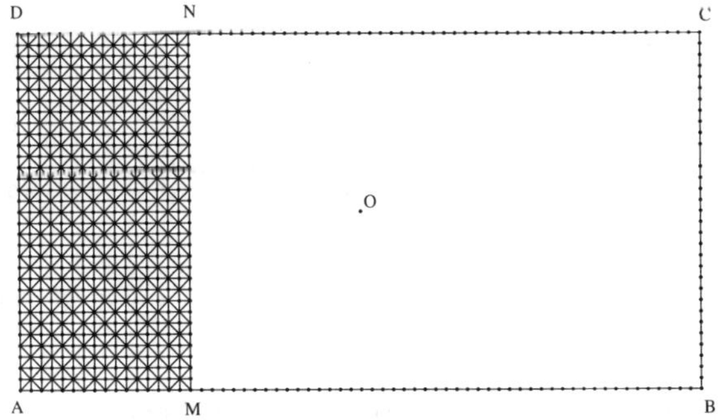

Figure 1: Rod under initial conditions prescribed over a subdomain: boundary discretization and cells.

The calculated derivatives $\partial u/\partial x$ and time $\partial u/\partial t$ at interior point O are shown in Fig. 2 and 3, respectively.

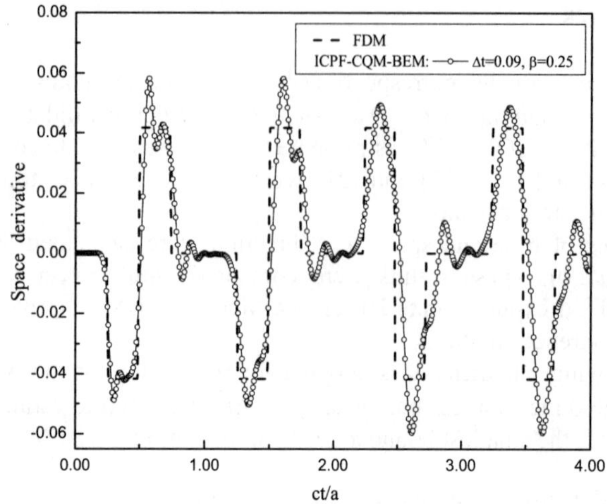

Figure 2: Calculated space derivative $\partial u/\partial x$ at interior point O for the rod of example 6.1.

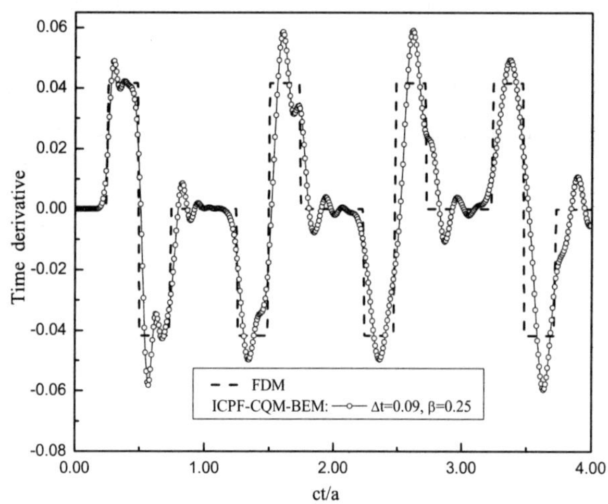

Figure 3: Calculated time derivative at interior point O for the rod of example 6.1.

6.2 Rod under initial velocity condition prescribed over a sub-domain

This example consists of a one-dimensional rod under initial velocity as follows: $v_0(X) = \frac{Pc}{E}$ $(0 \le x \le \frac{a}{4}, -\frac{a}{4} \le y \le \frac{a}{4})$. The same mesh used in the previous example was adopted (see Fig. 1). The calculated derivatives $\partial u/\partial x$ and time $\partial u/\partial t$ at interior point O are shown in Fig. 4 and 5, respectively.

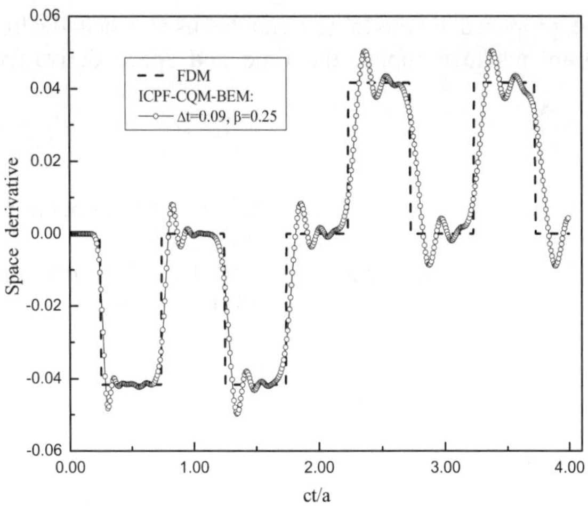

Figure 4: Calculated space derivative $\partial u/\partial x$ at interior point O for the rod of example 6.2.

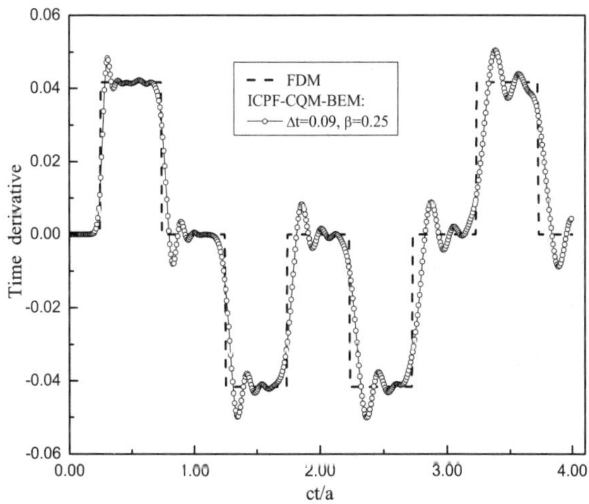

Figure 5: Calculated time derivative at interior point O for the rod of example 6.2.

7 Conclusions

In the present work, a method to compute the time and space derivatives of the displacement function of the scalar wave propagation problem was presented. The method uses the boundary element method based on the convolution quadrature method. The CQM-BEM takes into account non-homogeneous initial conditions by means of the general procedure known as initial condition pseudo-force (ICPF). The obtained results employing the ICPF-CQM-BEM for the studied examples agree with the reference solution obtained by the FDM and shown that the proposed methodology can be used when dealing with wave propagation problems to compute the time and space derivatives at interior points.

References

[1] Lubich, C., Convolution quadrature and discretized operational calculus I. *Numerische Mathematik*, **52**, pp. 129-145, 1988.

[2] Lubich, C., Convolution quadrature and discretized operational calculus II. *Numerische Mathematik*, **52**, pp. 413-425, 1988.

[3] Abreu, A. I., Carrer, J. A. M. & Mansur, W. J., Scalar wave propagation in 2d: a BEM formulation based on the operational quadrature method. *Engineering Analysis with Boundary Elements*, **27**, pp. 101-105, 2003.

[4] Schanz, M., *Wave propagation in viscoelastic and poroelastic continua: a boundary element approach*. Springer, Berlin, New York, 2001.

[5] Gaul, L. & Schanz, M., A comparative study of three boundary element approaches to calculate the transient response of viscoelastic solids with unbounded domains. *Computer Methods in Applied Mechanics and Engineering*, **179**, pp. 111-123, 1999.

[6] Antes, H., Schanz, M. & Alvermann, S., Dynamic analyses of plane frames by integral equations for bars and Timoshenko beams. *Journal of Sound and Vibration*, **276**, pp. 807–836, 2004.

[7] Moser, W., Antes, H. & Beer, G., A Duhamel integral based approach to one-dimensional wave propagation analysis in layered media. *Computational Mechanics*, **35**, pp. 115–126, 2005.

[8] Moser, W., Antes, H. & Beer, G., Soil-structure interaction and wave propagation problems in 2D by a Duhamel integral based approach and the convolution quadrature method. *Computational Mechanics*, **36** (6), pp. 431-443, 2005.

[9] Dobromil, P. & Schanz, M., Comparison of mixed and isoparametric boundary elements in time domain poroelasticity. *Engineering Analysis with Boundary Elements*, **30**, pp. 254-269, 2006.

[10] Schanz, M., Rüberg, T. & Struckmeier, V., Quasi-static poroelastic boundary element formulation based on the convolution quadrature method. *Compt Mech*, **37**, pp. 70-77, 2005.

[11] Abreu, A. I., Mansur, W. J. & Carrer, J. A. M., Initial conditions contribution in a BEM formulation based on the operational quadrature method. *International Journal for Numerical Methods in Engineering*, **67** (3), pp. 417-434, 2006.

[12] Mansur, W. J., Abreu, A. I. & Carrer, J. A. M., Initial conditions contribution in frequency-domain BEM analysis. *Computer Modeling in Engineering and Sciences*, **6** (1), pp. 31-42, 2004.

[13] Mansur, W. J., Soares Jr., D. & Ferro, M. A. C., Initial conditions in frequency domain analysis: the FEM applied to the scalar wave equation. *Journal of Sound and Vibration*, **270**, pp. 767-780, 2004.

[14] Morse, P. M. & Feshbach, H., *Methods of theoretical physics*, McGraw-Hill, New York, 1953.

A method for obtaining a sparse matrix from the volume integral equation for elastic wave propagation

T. Touhei
Department of Civil Engineering, Tokyo University of Science, Japan

Abstract

The advantage of the volume integral equation is that it is possible to clarify the relationship between fluctuations of the wave field and radiation of scattered waves. This paper proposes a method to obtain a sparse matrix for the volume integral equation for elastic wave propagation. The formulation employed here is based on the wavenumber domain solution together with usage of Haar scaling functions. The unitarity of the Fourier transform in terms of the Haar scaling function reveals that the integral equation is transformed into a linear algebraic equation with a sparse matrix. Numerical calculations are carried out to verify the proposed formulation.
Keywords: elastic waves, volume integral equation, sparse matrix, unitary transform, Haar scaling function.

1 Introduction

Since 1980s, the boundary element technique has been recognized as an efficient tool for the analysis of wave propagation (for example, Brebbia and Walker [3]). On the other hand, the volume integral equation methods have not been used very often except for some cases (for example, Kitahara et al [5]). The advantage of the volume integral equation such as the Lippmann–Schwinger equation (Colton and Kress [4]) is in that it clarifies the relationship between the fluctuation of the medium and the radiation of scattered waves. Standard technique for discretizing the equation, however, leads to a large and dense matrix for the volume integral equation, that makes sometimes numerical analysis impossible even by recent high performance computers.

WIT Transactions on Modelling and Simulation, Vol 44,
www.witpress.com, ISSN 1743-355X (on-line)
doi:10.2495/BE070241

In this paper, a method for obtaining a sparse matrix from the volume integral equation is presented. The method is based on the usage of the Fourier transform and the Haar scaling function (Williams and Amaratunga [7]). The volume integral equation in the wavenumber domain is discretized by means of the Haar scaling function. The unitarity of the Fourier transform shows that the Fourier inverse transform of the Haar scaling functions form the orthonormal basis and as a result, a sparse matrix is found to be derived from the volume integral equation in the case that the spectral structure of the fluctuation of the wave field is narrow band. Several numerical calculations are carried out to verify the accuracy of the present method.

2 Theoretical formulation

2.1 Basic equations

An elastic full space of three dimension is considered in this paper. The elastic wave field is assumed to have a fluctuation represented by the Lamé constants such that

$$\begin{aligned} \lambda(x) &= \lambda_0 + \tilde{\lambda}_1(x) \\ \mu(x) &= \mu_0 + \tilde{\mu}_1(x), \quad (x \in \mathbb{R}^3) \end{aligned} \tag{1}$$

where λ_0 and μ_0 are the background Lamé constants and $\tilde{\lambda}$ and $\tilde{\mu}$ are their fluctuations. Note that $x \in \mathbb{R}^3$ denotes the spatial point. The governing equation for the elastic wave propagation for the medium can be derived according to the literature (for example, Aki and Richards [1]). The governing equation is expressed by

$$(\lambda_0 + \mu_0)\partial_i\partial_j u_j + \mu_0\partial_k\partial_k u_i + \rho\omega^2 u_i = N_{ij}u_j \tag{2}$$

where u_i is the displacement field whose subscript denotes the component of the Cartesian coordinate, ρ is the mass density, ω is the circular frequency, ∂ is the partial differential operator whose subscript denotes the parameter for the differentiation and N_{ij} is the operator describing the fluctuation of the elastic medium. The summation convention is applied to the subscript index for the component of the coordinate. The explicit form of N_{ij} is expressed by

$$\begin{aligned} N_{ij} = &-(\tilde{\lambda}(x) + \tilde{\mu}(x))\partial_i\partial_j - \delta_{ij}\tilde{\mu}(x)\partial_k\partial_k \\ &- \partial_i\tilde{\lambda}(x)\partial_j - \delta_{ij}\partial_k\tilde{\mu}(x)\partial_k - \partial_j\tilde{\mu}(x)\partial_i \end{aligned} \tag{3}$$

where δ_{ij} is the Kronecker delta. The volume integral equation for the elastic wave field is directly derived from Eq. (2), which is

$$u_i(x) = f_i(x) - \int_{\mathbb{R}^3} g_{ij}(x,y)N_{jk}(y)u_k(y)dy \tag{4}$$

where f_i is the plane incident wave and $g_{ij}(x,y)$ is the Green's function expressed by

$$g_{ij}(x,y) = \frac{1}{4\pi\mu_0}\delta_{ij}\Big(Z_1^{(T)}(r) + Z_2^{(T)}(r) - (c_T/c_L)^2 Z_2^{(L)}(r)\Big) + \frac{1}{4\pi\mu_0}\partial_i r\,\partial_j r\,\Big(Z_3^{(T)}(r) - (c_T/c_L)^2 Z_3(r)\Big) \tag{5}$$

In Eq. (5), $r = |x-y|$ and $Z_j^{(p)}(r), j = 1,2,3, p = T, L$ is the function defined by

$$\begin{aligned} Z_1^{(p)}(r) &= \frac{1}{r}\exp(ik_p r) \\ Z_2^{(p)}(r) &= -\frac{1}{r}\Big(\frac{1}{k_p^2 r^2} + \frac{i}{k_p r}\Big)\exp(ik_p r) \\ Z_3^{(p)}(r) &= -Z_1^{(p)}(r) - 3Z_2^{(p)}(r) \end{aligned} \tag{6}$$

where k_T and k_L are the wavenumber of the S and P waves, respectively.

2.2 Fourier transform of Integral equation

The formulation presented here employs the Fourier integral transform for the volume integral equation. The Fourier and its inverse integral transforms (Reed and Simon [6]) are respectively expressed as

$$\begin{aligned} \hat{f}(\xi) &= \frac{1}{\sqrt{2\pi}^3}\int_{\mathbb{R}^3} f(x)\exp(-i\xi\cdot x)dx \\ \check{f}(x) &= \frac{1}{\sqrt{2\pi}^3}\int_{\mathbb{R}^3} f(\xi)\exp(i\xi\cdot x)d\xi \end{aligned} \tag{7}$$

where $\hat{f}$ denotes the Fourier transform of f and $\check{f}$ denotes the Fourier inverse transform of f. Note that $\xi \in \mathbb{R}^3$ is used for the point in the wavenumber space. The Fourier transform of the volume integral equation in terms of scattered wave $v_i(x) = u_i(x) - f_i(x)$ becomes as

$$\hat{v}_i(\xi) = -\hat{h}_{ij}(\xi)\hat{q}_j(\xi - \xi_p) - \hat{h}_{ij}(\xi)\hat{w}_j(\xi) \tag{8}$$

$$w_j(x) = N_{jk}(x)v_k(x) \tag{9}$$

where $\hat{h}_{ij}(\xi)$ is related to the Fourier transform of the Green's function that is expressed by

$$\hat{h}_j(\xi) = \frac{\delta_{ij}}{\mu_0(|\xi|^2 - k_T^2 + i\epsilon)} - \frac{\xi_i\xi_j}{2\mu_0(1-\nu)(|\xi|^2 - k_T^2 + i\epsilon)(|\xi|^2 - k_L^2 + i\epsilon)} \tag{10}$$

Note that ϵ in Eq. (10) is the infinitesimal positive number and ν is the Poisson ratio. In addition, ξ_p in Eq.(8) is the wavenumber vector of the plane incident wave and $\hat{q}$ is the function related to the fluctuation of the medium and plane incident wave.

2.3 Discretization of the volume integral equation

The Fourier transformed volume integral equation shown in Eq. (8) can be discretized by means of the Haar scaling functions. Let $\{\phi_\alpha(\xi)\}_{\alpha=1}^{N}$ is the set of the Haar scaling functions embedded in the wavenumber space, where α is the integer to identify the element of the set of the Haar scaling functions. In the following formulation, the Greek character used for the subscript is for identifying the Haar scaling functions. For each α, ϕ_α has the resolution and integer shift vector such that

$$\phi_\alpha(\xi) = 2^{m_\alpha/2} \prod_{j=1}^{3} \phi(2^{m_\alpha}\xi_j - \Gamma_{j(\alpha)}) \tag{11}$$

where m_α is the resolution, $\Gamma_{j(\alpha)}, j = 1, 2, 3$ is the component of the integer shift vector and ϕ is the Haar box function such that

$$\phi(x) = \begin{cases} 1 \ \ (0 \leq x < 1) \\ 0 \ \ \text{otherwise} \end{cases} \qquad (x \in \mathbb{R}) \tag{12}$$

The set of the Haar scaling functions is set up so that the support of each element is disjoint each other. Namely,

$$\text{supp } \phi_\alpha(\xi) \cap \text{supp } \phi_\beta(\xi) = \emptyset \ \ (\alpha \neq \beta) \tag{13}$$

In addition, the set of the Haar scaling function fills a region in the wavenumber space C densely,

$$C \setminus \bigcup_{\alpha=1}^{N} \text{supp } \phi_\alpha = \emptyset \tag{14}$$

The region C is set up such that

$$C = \{\xi = (\xi_1, \xi_2, \xi_3) \mid -L \leq \xi_j \leq L, \ j = 1, 2, 3\} \tag{15}$$

where L is the positive number, which is taken large enough for the numerical calculation. Due to the above properties, the set of the Haar scaling functions forms the orthogonal basis in the wavenumber domain. Namely,

$$\langle \phi_\alpha(\xi), \phi_\beta(\xi) \rangle = \delta_{\alpha\beta} \tag{16}$$

where $\langle \cdot, \cdot \rangle$ is the scalar product of the functions defined by

$$\langle f(\xi), g(\xi) \rangle = \int_{\mathbb{R}^3} f^*(\xi) g(\xi) d\xi \tag{17}$$

Note that the Fourier inverse transform of the Haar scaling functions also form the orthonormal basis according to the unitarity of the Fourier transform. Therefore, the following equation can be established:

$$\langle \check{\phi}_\alpha(x), \check{\phi}_\beta(x) \rangle = \delta_{\alpha\beta} \tag{18}$$

where $\check{\phi}_\alpha$ is the Fourier inverse transform of the Haar scaling function:

$$\check{\phi}_\alpha(x) = \frac{1}{\sqrt{2^{m_\alpha+1}\pi}^3} \times \prod_{k=1}^{3} \frac{\sin\theta_{k(\alpha)}}{\theta_{k(\alpha)}} \exp\Big(i\theta_{k(\alpha)}(2\Gamma_{k(\alpha)} + 1)\Big) \tag{19}$$

where

$$\theta_{k(\alpha)} = \frac{x_k}{2^{m_\alpha+1}} \tag{20}$$

At this stage, we are in a situation in that we can discretize the integral equation (8). We expand each variable of Eq. (8) such that

$$\begin{aligned} \hat{v}_i(\xi) &= \sum_\alpha \hat{V}_{i(\alpha)}\phi_\alpha(\xi) \\ \hat{h}_{ij}(\xi) &= \sum_\alpha \hat{H}_{ij(\alpha)}\phi_\alpha(\xi) \\ \hat{q}_j(\xi) &= \sum_\alpha \hat{Q}_{j(\alpha)}\phi_\alpha(\xi) \\ \hat{w}_j(\xi) &= \sum_\alpha \hat{W}_{j(\alpha)}\phi_\alpha(\xi) \end{aligned} \tag{21}$$

Then, Eq. (8) can be modified into

$$\hat{V}_{i(\alpha)} = -c_\alpha \hat{H}_{ij(\alpha)}\hat{Q}_{j(\alpha)} - c_\alpha \hat{H}_{ij(\alpha)}\hat{W}_{j(\alpha)} \tag{22}$$

For Eq. (22), the following equation:

$$\phi_\alpha(\xi)\phi_\beta(\xi) = c_\alpha \delta_{\alpha\beta} \tag{23}$$

where

$$c_\alpha = \left(2^{m_\alpha/2}\right)^3 \tag{24}$$

is used. To discretize Eq. (9), the Fourier inverse transform of $\hat{v}_i$ and $\hat{w}_j$ which has the following forms

$$\begin{aligned} v_i(x) &= \sum_\alpha \hat{V}_{i(\alpha)}\check{\phi}_\alpha(x) \\ w_j(x) &= \sum_\alpha \hat{W}_{j(\alpha)}\check{\phi}_\alpha(x) \end{aligned} \tag{25}$$

are used. Substituting Eq. (25) into Eq. (9) as well as using the orthogonality relation shown in Eq. (18) leads to the following equation:

$$\hat{W}_{i(\alpha)} = S_{ij(\alpha\beta)} \hat{V}_{j(\beta)} \tag{26}$$

where $S_{ij(\alpha\beta)}$ is due to the following operation:

$$S_{ij(\alpha\beta)} = \langle \check{\phi}_\alpha, N_{ij} \check{\phi}_\beta \rangle \tag{27}$$

As a result, the following linear algebraic equation in terms of $\hat{V}_{i(\alpha)}$ is derived:

$$\hat{V}_{i(\alpha)} = F_{i(\alpha)} - A_{ik(\alpha\beta)} \hat{V}_{k(\beta)} \tag{28}$$

where

$$\begin{aligned} F_{i(\alpha)} &= -c_\alpha \hat{H}_{ij(\alpha)} \hat{Q}_{j(\alpha)} \\ A_{ik(\alpha\beta)} &= c_\alpha \hat{H}_{ij(\alpha)} S_{jk(\alpha\beta)} \end{aligned} \tag{29}$$

The scattered wave field can be derived by solving Eq. (28), since we have

$$v_i(x) = \sum_\alpha \hat{V}_{i(\alpha)} \check{\phi}(x) \tag{30}$$

At the end of the formulation, note that the operation of the product shown in Eq. (27) can be carried out without difficulty. To carry out the operation, the fluctuation of the lamé constants $\tilde{\lambda}$ and $\tilde{\mu}$ are expanded such that

$$\begin{aligned} \tilde{\lambda}(x) &= \sum_{\beta \in B} \Lambda_\beta \check{\phi} \\ \partial_j \tilde{\lambda} &= \sum_{\beta \in B} \Lambda_\beta^{(j)} \check{\phi}(x) \\ \tilde{\mu}(x) &= \sum_{\beta \in B} M_\beta \check{\phi} \\ \partial_j \tilde{\mu} &= \sum_{\beta \in B} M_\beta^{(j)} \check{\phi}(x) \end{aligned} \tag{31}$$

where B is the set of index for expressing the fluctuation of the Lamé constants.

Substituting Eq, (31) into Eq. (27) clarifies that the operation for Eq. (27) can be constituted by the following integral formulas:

$$S_{ij(\alpha\gamma)} = \sum_{\beta \in B} \sum_p \gamma_p \int_{\mathbb{R}^3} \check{\phi}_\alpha^*(x) \check{\phi}_\beta(x) \partial^p \check{\phi}_\gamma(x) dx \tag{32}$$

where $p = (p_1, p_2, p_3)$ and $\partial^p = \partial_1^{p_1} \partial_2^{p_2} \partial_3^{p_3}$. In addition, γ_p in Eq. (32) is uniquely determined by N_{ij} and coefficients Λ_β, $\Lambda_\beta^{(j)}$, M_β and $M_\beta^{(j)}$ in Eq. (31). The closed form of the result of the integral of Eq. (32) is possible, which leads to the fact that the matrix $S_{ij(\alpha\gamma)}$ is sparse in the case that the range of set of B is narrow. Namely, in the case that the spectral structure of the fluctuation of the Lamé constants is narrow band, a sparse matrix is obtained.

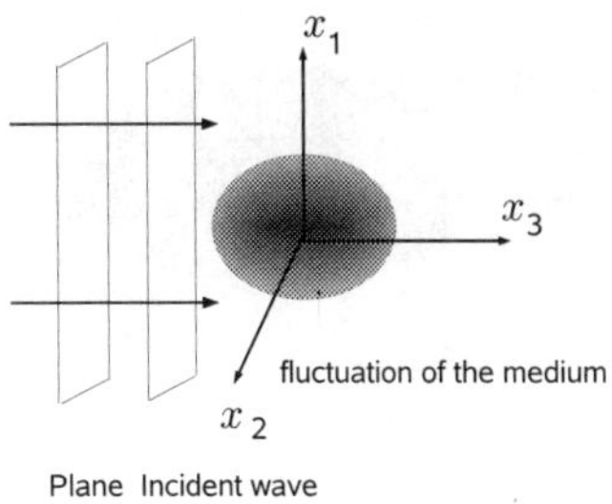

Figure 1: Analyzed model.

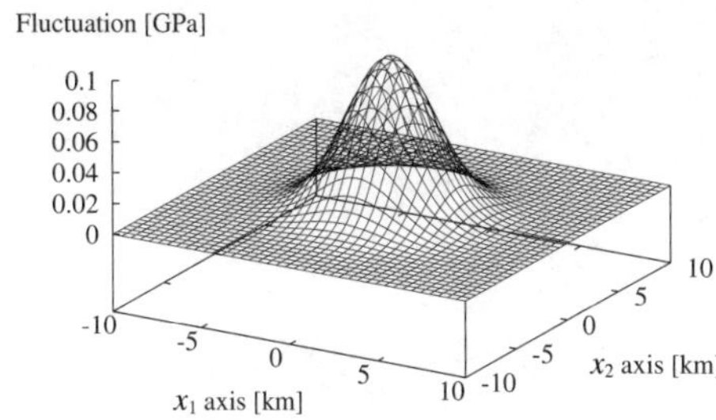

Figure 2: Fluctuation of the Lamé constants.

3 Numerical examples

3.1 Analyzed model

The concept of the analyzed model is shown in Figure 1, in that the plane incident wave is propagating to the fluctuated area along x_3 axis. The background Lamé constants are set at λ_0 =2 [GPa], μ_0=1 [GPa] and the mass density of the wave field is $\rho = 1$ [g/cm^3]. In addition, the frequency of the wave field is 1 [Hz]. Therefore, the P wavenumber of the background wave field is $k_L = 3.14$ [km^{-1}] and the S wavenumber of that is $k_T = 6.28$ [km^{-1}]. The fluctuation of the Lamé constants are set at

$$\tilde{\lambda}(x) = 0.1\exp(-0.1|x|^2)$$
$$\tilde{\mu}(x) = 0.1\exp(-0.1|x|^2) \text{ [Gpa]} \tag{33}$$

The fluctuation of Lamé constant at $x_3 = 0$ plane is shown in Figure 2. As shown in Figure 2, the fluctuation gradually decreases towards the far field range. The Fourier transform of the fluctuation of the Lamé constants for Eq. (33) becomes

$$\hat{\tilde{\lambda}}(\xi) = \hat{\tilde{\mu}}(\xi) = \frac{1}{\sqrt{2\eta}^3}\exp(-|\xi|^2/(4\eta)), \quad (\eta = 0.1) \tag{34}$$

As can be seen in the following, the spectral structure of the fluctuation of the Lamé constants is narrow enough to generate a sparse matrix.

3.2 Haar scaling functions used for the analysis

As shown in the formulation, the volume integral equation is discretized by the Haar scaling functions in the wavenumber space. Figures 3 to 6 are the location of the Haar scaling functions in the wavenumber space, where $s_j, j = 1, 2, 3$ indicates the dimensionless wavenumber defined by $s_j = \xi/k_T$. The wavenumber space is spanned by the Haar basis for the region $-10 \leq s_j \leq 10 (j = 1, 2, 3)$.

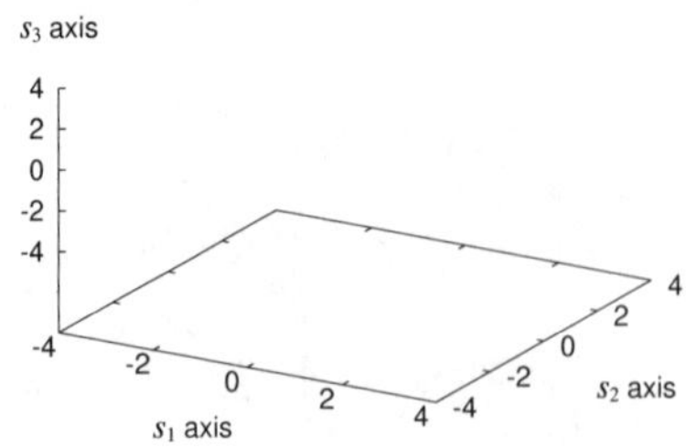

Figure 3: Haar scaling functions in the wavenumber space ($m = 1 \sim 3$).

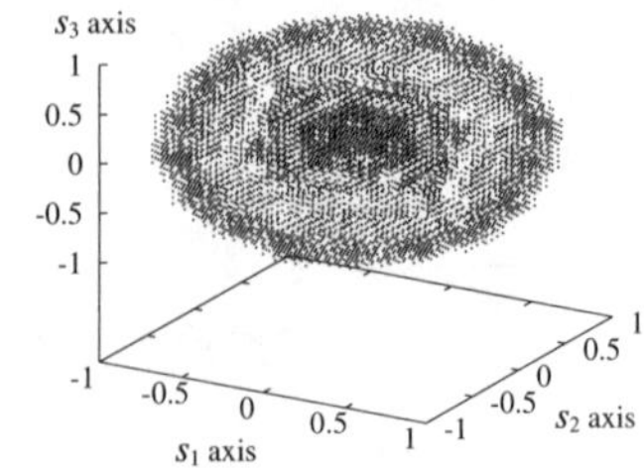

Figure 4: Haar scaling functions in the wavenumber space ($m = 4$).

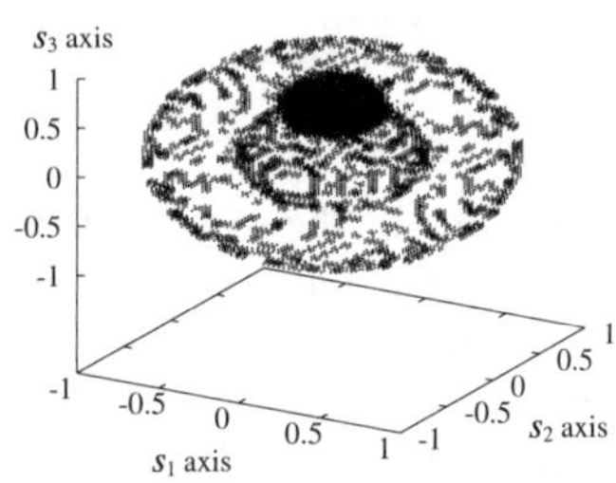

Figure 5: Haar scaling functions in the wavenumber space ($m = 5$).

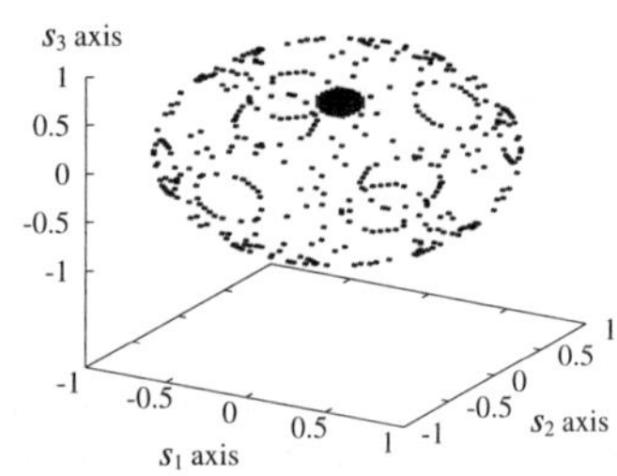

Figure 6: Haar scaling functions in the wavenumber space ($m = 6$).

The resolution m ranges from 1 to 6, that is for the dimensionless wavenumber. In these figures, the center of the supports of the Haar scaling functions are plotted. The size of the supports become smaller as the resolution increases. As a result, the resolution of the Haar basis is described by the density of the points. The higher resolution of the Haar basis is used for the region close to singular point of the Green's function in the wavenumber domain. To simplify the view of the location of the Haar basis, the Haar basis is shown in the region $-4 \leq s_j \leq 4 (j = 1, 2, 3)$ in figure 3. Note that the resolution of the Haar scaling functions that located farthest outside the region is $m = 1$. The number of the Haar scaling functions used for the analysis here are 97824.

3.3 Sparse matrix obtained from the present procedure

Figure 7 shows the structure of the matrix obtained form the Haar scaling functions shown in Figures 3-6. In figure 7, the non-zero elements are plotted. Those non-zero elements are judged from Eq. (32). The size of the matrix is about $290,000 \times 290,000$ with the sparse ratio 2.8%. To solve the linear algebraic

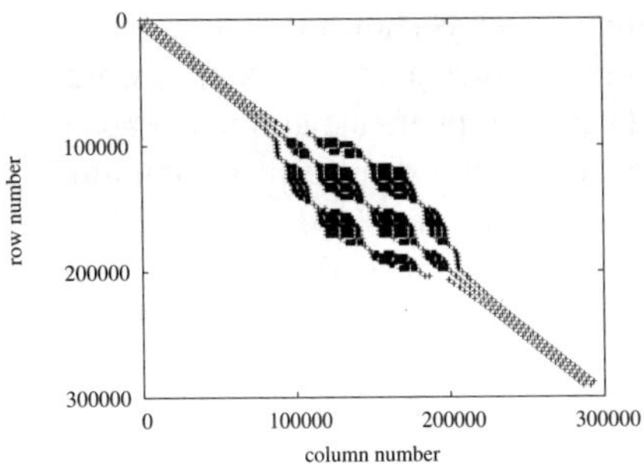

Figure 7: Structure of the matrix.

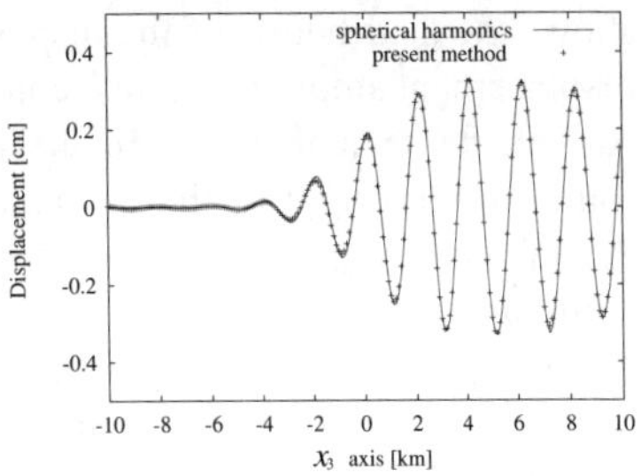

Figure 8: Comparison of displacement.

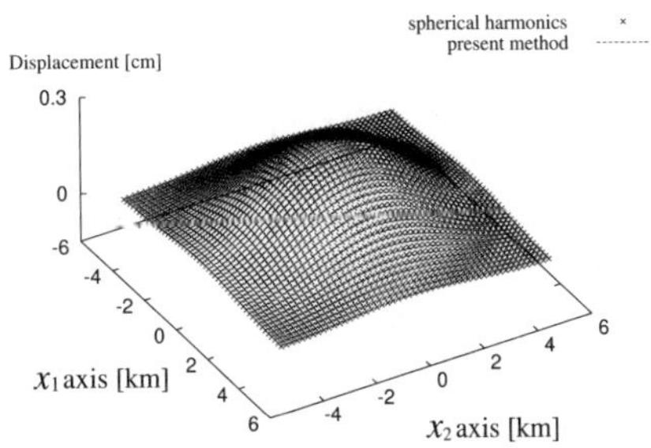

Figure 9: Comparison of displacement at $x_3 = 2$ km.

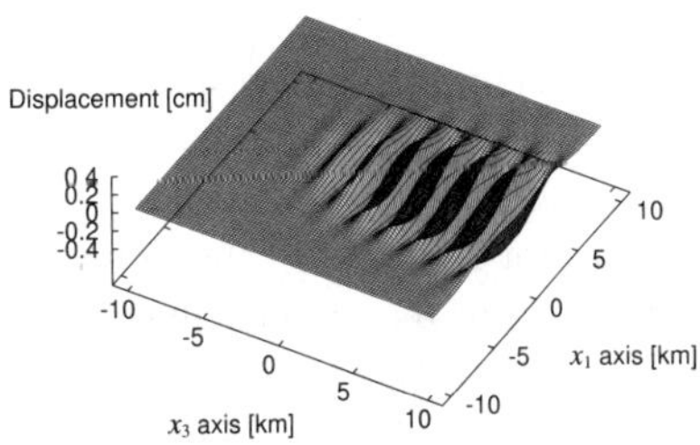

Figure 10: Displacements in $x_1 - x_3$ plane.

equation, the iterative scheme of Bi-CGSTAB method (Barrett et al [2]) is employed. The elapsed time for solving the matrix was about 5 minutes by IBM pSeries 690 provided by Tokyo University of Science, for the case of 32-cpu parallel processing. The number of iteration for the convergence of the solution was three. The condition for the convergence of the linear algebraic equation $\boldsymbol{Ax} = \boldsymbol{b}$ is $|\boldsymbol{Ay} - \boldsymbol{b}| < 0.001|\boldsymbol{b}|$, where $\boldsymbol{A}$ is the coefficient matrix and $\boldsymbol{y}$ is the approximate solution.

3.4 Properties of the scattered wave field

Now, assume that the plane incident wave propagating along x_3 axis is P wave. The comparison of the displacement of the scattered wave field between the present method and the spherical harmonics expansion is shown in figure 8. In the figure, the displacement component of x_3 direction are compared along x_3 axis. Good agreements can be found in both two methods in figure 8, which validates the accuracy of the present method. The amplitude of the backward scattering is found to be very small in figure 8, while that of the forward scattering is relatively large and decreases slowly towards the far field region. The slow decrease of the displacement amplitude is due to the slow decrees of the fluctuation of the

Lamé constants towards the far field region, which is shown in figure 2. Figure 9 shows the distribution of the displacement amplitude at $x_3 = 2$ km, in which the displacement amplitude is also compared with the spherical harmonics expansion. As can be seen in figure 9, the peak value of the displacement amplitude is found in the origin of the x_3 plane. The displacement slowly decreases toward the far field region. The direction of the displacement is outstanding in the x_3 component, which indicates that the scattered waves are mainly constituted by P wave. Figure 10 shows the distribution of the displacement of x_3 component in $x_1 - x_3$ plane. It is found from figure 10 that the scattered wave does not spread widely in the forward region, indicating that the scattered waves here has rather strong directionality.

4 Conclusion

A method for obtaining a sparse matrix was presented in this paper for the volume integral equation. The Fourier transform was employed to the volume integral equation. The usage of the Haar scaling functions in the wavenumber domain as well as the unitarity of the Fourier transform revealed that a sparse matrix was derived from the volume integral equation. A iterative scheme for solving the linear algebraic equation was found to be successfully applied to the sparse matrix. Numerical results ensured the validity and accuracy of the present method.

References

[1] Aki, K. and Richards, P.G. (1980): *Quantitaive Seismology. Theory and Methods*, W.H. Freeman and Comnpany.

[2] Barrett, M., Berry, M., Chan,T.F., Demmel, J., Donato, J. M., Dongarra, J., Eijkhout, V., Pozo,R., Romine, C. and Van der Vorst, H. (1994). *Templates for the solution of Linear Systems: Building Blocks for Iterative Methods*, SIAM.

[3] Brebbia, C. A. and Walker, S. (1980). *Boundary element techniques in engineering*. London, Butterworth and Co. Ltd.

[4] Colton, D. and Kress, R. (1998). *Inverse acoustic and electromagnetic scattering theory*, Berlin, Springer.

[5] Kitahara, M., Niwa, Y., Hirose, S. and Yamazaki, M. (1984). Coupling of numerical Green's matrix and boundary integral equations for the elastodynamic analysis of inhomogeneous bodies on an elastic half-space, *Applied Mathematical Modelling*, **8**, 397-407.

[6] Reed, M. and Simon, B. (1975). *Method of Modern Mathematical Physics*, Vol. II, Fourier Analysis and Self-adjointness, Academic Press.

[7] Williams, J.R. and Amaratunga, K.: Introduction to wavelets in engineering, *International Journal for Numerical Methods in Engineering*, Vol. 37, pp. 2365-2388, 1994.

Section 9
Damage mechanics and fracture

Two-parameter concept for anisotropic cracked structures

P. Brož
Czech Institution of Structural and Civil Engineers,
Prague, Czech Republic

Abstract

For two-dimensional anisotropic elasticity the M-integral was evolved, in terms of the boundary element method serving for numerical determination of the T stress which represents a notable parameter for fracture evaluation of cracked solid bodies on top of the stress intensity factor. T-stress issues for the crack face pressure instances are employed as the reference solutions to infer weight functions, e.g. to obtain T-stress results for thick-walled cylinders weakened by an internal edge crack subject to any complicated loading. Some examples are demonstrated.
Keywords: complex stress distribution, crack face, material anisotropy, remote load, self-regularization.

1 Introduction

Besides the stress intensity factor, the T-stress is the other parameter reflected in fracture evaluation. The path-independent *M-integral* approach to interpret the T-stress, in conformity with Shah et al. [1], is augmented to use plane, in general anisotropic cracked constituents. It is realized in the boundary element method. An example is demonstrated to indicate the accuracy of the quantification formulated and its suitability. The numerical projects derived are illustrative of the fact that material anisotropy has indeed a substantial influence on the T stress.

To assess the *M integral*, conceptions for the stresses and displacements at interior points are required. When being the interior points very close to the boundary, some inconveniences in the standard BEM representation arise. A technique to get over this near-singularity problem was issued as early as

WIT Transactions on Modelling and Simulation, Vol 44,
www.witpress.com, ISSN 1743-355X (on-line)
doi:10.2495/BE070251

several years ago, namely for isotropic analysis. In so doing, a chart was worked out to get the continuous form of the Somigliana's identity by using a simple solution commensurable to a rigid body movement. The self-regularization scheme can be performed to obtain interior point field quantities in an anisotropic body. It makes the application of comparatively coarse mesh layouts possible, for the boundary even if the interior point is located extremely near it.

The implementation is based on the quadratic isoparametric element. Standard and traction singular quarter point elements were applied contiguous to the crack tip for fracture mechanics study. To check up path independence of the T-stress solutions obtained applying the *M integral*, no less than two circular round the crack tip were selected for the interpretation of the integral.

For numerical determination of the T stress, the *M integral* represents an effective methodology. This paper launches the *M integral* for two-dimensional anisotropic elasticity being realized in context with the BEM.

Collapse of thick-walled pressurized cylinders is frequently owing to the presence of internal or external cracks in practice. A long internal single radial crack may be treated as an edge crack in two dimensions. It has been shown that an internally pressurized cracked cylinder can be considered as a "low-constrained" geometry. The fracture toughness measured from "high-constrained" test specimens may be conservative when applied to this constituent. On that account, precision T stress solutions for thick-walled cylinders with an internal radial crack are desirable to reliably predict the failure loading.

This paper presents T-stress solutions for a cylinder afore-said, namely using the BEM and the contour integral approach, with the loading being an example of crack face pressures which are realized by polynomial stress distributions.

Notwithstanding that the superposition method can be applied to estimate the T-stress using available T-stress solutions for simple loading conditions, it is not possible to cover the complete range of loading conditions in engineering. The weight function method was initiated to be one of the most powerful concepts to obtain the stress intensity factors for more intricate problems in Li et al. [2]. The T-stress weight functions are derived from T-stress solutions for two reference loading conditions, which correspond to the cases when the cracked cylinder is subject to a constant and to a linear variation of the applied stress on the crack faces.

2 BEM interpretation

It is known that the solution for the displacements $u_i(p)$ and stress $\sigma_{ij}(p)$ at an interior point, p, of a domain can be derived from the Somigliana´s identities, namely in the form

$$u_j(p) = \int_S t_i(Q) U_{ji}(p,Q) dS(Q) - \int_S u_i(Q) T_{ji}(p,Q) dS(Q) \qquad (1)$$

and

$$\sigma_{ij}(p)=\int_S u_k(Q)S_{kij}(p,Q)dS(Q)-\int_S t_k(Q)D_{kij}(p,Q)dS(Q), \tag{2}$$

where U_{ji}, T_{ji}, D_{kij} and S_{kij} are the fundamental solutions, $t_i(p)$ the traction vector and S the boundary of the domain.

A procedure to overcome the near-singularity problem was proposed by Richardson and Cruse for isotropic variant. The developed an outline to get the continuous form of the Somigliana's displacement identity by using a solution corresponding to a rigid body motion to the identity. Like this, it holds

$$u_j(p)-u_j(P)=\int_S t_i(Q)U_{ji}(p,Q)dS(Q)-\int_S [u_i(Q)-u_i(P)]\,T_{ji}(p,Q)dS(Q) \tag{3}$$

A weakly singular form of the stress identity is obtained by a simple technique which is equivalent to subtracting and adding back a simple solution corresponding to a state of constant stress in the body that equals the boundary stress at point P close to the interior point p, on condition that the stress at the boundary point P is continuous. The self-regular stress identity can be written down

$$\sigma_{ij}(p)=\sigma_{ij}(P)+\int_S \left[u_k(Q)-u_k^L(P,Q)\right]S_{kij}(p,Q)dS(Q)$$
$$-\int_S \left[t_k(Q)-t_k^L(P,Q)\right]D_{kij}(p,Q)dS(Q) \tag{4}$$

In Eq. (4), t_k^L and u_k^L are the linear state tractions and displacements; they are related to a constant stress state in the body corresponding to the stress at the boundary point P, as follows

$$u_k^L(P,Q)=u_k(P)+u_{km}(P)\{x_m(Q)-x_m(P)\} \tag{5}$$

and

$$t_k^L(P,Q)=\sigma_{mk}(P)\,n_m \tag{6}$$

As the linear state traction density (t_k - t_k^L) and the displacement density (u_k - u_k^L) are $O(r)$ and $O(r^2)$, respectively, the integrals in Eq. (6) work out regular or weakly singular.

3 T-stress assessment

The path-independent mutual *M integral* to determine T stress was generalized in Shah et al. [1] to the anisotropic eventuality in two dimensions. In so doing, in the analytical adaptation of the anisotropic case, Lekhnitskii's guidelines are applied. In fig. 1 a cracked body Ω limited by the boundary Γ_o is indicated. The path-independent *J integral* may be expressed in the form

$$J=\int_{\Gamma_0}\left(W_{n_1}-t_i u_{j,1}\right)d\Gamma \tag{7}$$

where W is the strain energy density:

$$W = \int_0^{\varepsilon_{ij}} \sigma_{ij} d\varepsilon_{ij} = \frac{1}{2}\sigma_{ij}\varepsilon_{ij} \tag{8}$$

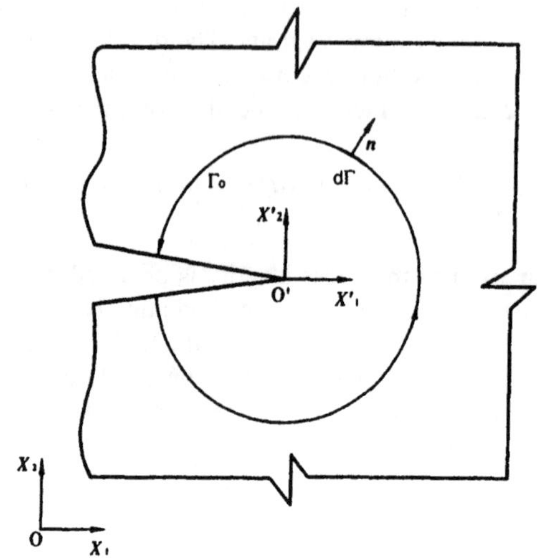

Figure 1: Contour Γ_o round the crack tip.

Contemplate two independent equilibrium states (σ_{ij}^A, ε_{ij}^A, u_i^A) and (σ_{ij}^B, ε_{ij}^B, u_i^B). The first state (*A*) befits to the boundary value problem that is analyzed with the unknown T stress. The second state (*B*) corresponds to the solution of a semi-infinite crack loaded by a point (line) force *f* acting on the crack tip in the direction parallel to the crack plane. The first state *A* is relevant to the stress and displacement fields near a crack tip.

After getting the required field expressions for states *A* and *B*, the contour *M-integral* expression for determining the T stress can be derived. For the corresponding *J integrals* with regard to the local coordinates, x_i, for states *A* and *B* it holds

$$J^{(A)} = \int_{\Gamma_0} \left[\frac{1}{2}\sigma_{ij}^A \varepsilon_{ij}^A n_1 - \sigma_{ij}^A n_j u_{i,1}^A \right] d\Gamma \tag{9}$$

and

$$J^{(B)} = \int_{\Gamma_0} \left[\frac{1}{2}\sigma_{ij}^B \varepsilon_{ij}^B n_1 - \sigma_{ij}^B n_j u_{i,1}^B \right] d\Gamma \tag{10}$$

and further, consider $J^{(A+B)}$, which denotes the value of *J integral* when both *A* and *B* fields are superimposed. Consequently,

$$J^{(A+B)} = \int_{\Gamma_0} \left[\frac{1}{2} \left(\sigma_{ij}^A + \sigma_{ij}^B \right) \left(\varepsilon_{ij}^A + \varepsilon_{ij}^B \right) n_1 - \left(\sigma_{ij}^A + \sigma_{ij}^B \right) n_j \left(u_{i,1}^A + u_{i,1}^B \right) \right] d\Gamma \qquad (11)$$

In conformity with Sládek et al. [3], the *M integral* in local coordinates is expressed by the relation

$$M = J^{(A+B)} - J^{(A)} - J^{(B)} = \int_{\Gamma_0} \left[\frac{1}{2} \left(\sigma_{ij}^A \varepsilon_{ij}^B + \sigma_{ij}^B \varepsilon_{ij}^A \right) n_1 - \sigma_{ij}^A n_j u_{i,1}^B - \sigma_{ij}^B n_j u_{i,1}^A \right] d\Gamma \quad (12)$$

Since the loading states are applied to the same elastic body,

$$\sigma_{ij}^A \varepsilon_{ij}^B = \sigma_{ij}^B \varepsilon_{ij}^A \qquad (13)$$

That is why

$$M = \int_{\Gamma_0} \left[\left(\sigma_{ij}^A \varepsilon_{ij}^B n_1 - \sigma_{ij}^A n_j u_{i,1}^B - \sigma_{ij}^B n_j u_{i,1}^A \right) \right] d\Gamma \qquad (14)$$

As the *M integral* is expressed by virtue of the path independent *J integral*, it is also path independent. Thus the integration contour can be arbitrarily chosen, say, a circle with radius ε, which is then shrunk to zero.
Next, the *M integral* reads

$$M = \lim_{\varepsilon \to 0} \int_{\Gamma_\varepsilon} \left[\left(\sigma_{ij}^A \varepsilon_{ij}^B n_1 - \sigma_{ij}^A n_j u_{i,1}^B - \sigma_{ij}^B n_j u_{i,1}^A \right) \right] d\Gamma \qquad (15)$$

Since the *J integrals* are bounded, so is the *M integral*; it is then possible to infer that there are no contributions to the *M integral* from the singular stress terms of the asymptotic expansion. The asymptotic displacements and stresses may be separated into singular and non-singular parts in the following form:

$$\sigma_{ij}^A = \sigma_{ij}^s + \sigma_{ij}^T \qquad (16.a)$$

and

$$u_i^A = u_i^s + u_i^T \qquad (16.b)$$

In Eq. (16), the superscript *s* denotes the terms of the asymptotic expansion containing the stress intensity factor, and the terms with superscript *T* are proportional to the non-singular T stress. The circular contour integral from θ = -π to + π of the angular functions of the singular terms of the auxiliary field in Eq. (15) cancel out preserving only the non-vanishing contribution from the T stress. From Eq (15) it results

$$M = \lim_{\varepsilon \to 0} \int_{\Gamma_\varepsilon} \left[\left(\sigma_{ij}^T \varepsilon_{ij}^B \delta_{1k} - \sigma_{ij}^T u_{i,1}^B - \sigma_{ik}^B u_{i,1}^T \right) \right] n_k d\Gamma \qquad (17)$$

where

$$\sigma_{ij}^T = T \delta_{i1} \delta_{j1} \qquad (18)$$

The elastic strains corresponding to the uniform T stress σ_{ij}^{T} under plane stress conditions are given as

$$\varepsilon_{11}^{T} = a'_{11} \tag{19}$$

and

$$u_{i,1}^{T} = u_{1,1}^{T}\delta_{i1} = \varepsilon_{11}^{T}\delta_{i1} = a'_{11}T\delta_{i1} \tag{20}$$

Thus, the *M integral* is reduced and may be re-arranged into the form

$$T = M\frac{1}{a'_{11}f} \tag{21}$$

Eq. (21) yields the relationship between the *M integral*, which may be evaluated using Eq. (14) and the T stress. It can also be used to evaluate T stress in plane strain conditions provided that a_{11} is replaced by b_{11}.

It should be reminded that the terms in Eq.(14) are given in the local coordinate system about the crack tip. In the numerical implementation, the *M integral*, given by Eq. (14) is obtained in global coordinates and then transformed into the local coordinates. The transformation for the *J integral* was determined by Kishimoto et al., when similarly applied to *M integral*, it provides in local coordinates the form:

$$M_{k(GLOBAL)} = M_{1(GLOBAL)}\cos\omega + M_{2(GLOBAL)}\sin\omega \tag{22}$$

where

$$M_{k(GLOBAL)} = \int_{\Gamma_0}\left[\left(\sigma_{ij}^{A}\varepsilon_{ij}^{B}n_k - \sigma_{ij}^{A}n_j u_{i,k}^{B} - \sigma_{ij}^{B}n_j u_{i,k}^{A}\right)\right]d\Gamma\,. \tag{23}$$

4 Discussion

The implementation of the statements for the *M integral* and self-regularized Somigliana´s identities in two-dimensional anisotropic elasticity is based on the quadratic isoparametric elements. Usual and traction-singular quarter point elements were applied contiguous to the crack tip for fracture mechanics analysis. To check up path independence of the T-stress solutions gained using the *M integral*, at least two circular contours round the crack tip were selected for the assessment of the integral.

The radii of these circular contours were characteristically 0.4 – 0.6 times the simulated crack length. Each of these contours was divided into smaller circular arcs and the *M integral* was evaluated by Gaussian quadrature over each of them and summed. In general, the deviations between the numerical solutions obtained for the T stress from the different contours were less than 2 per cent.

An example considered embraces an orthotropic plate with a single edge crack that is inclined (SECP) – fig. 2. To examine the effect of the degree of orthotropy on the T-stress solutions for various relative crack lengths, the parametric study was performed. The analysis was carried out under plane stress conditions and the results of the T-stress are normalized regarding the applied stress σ_0.

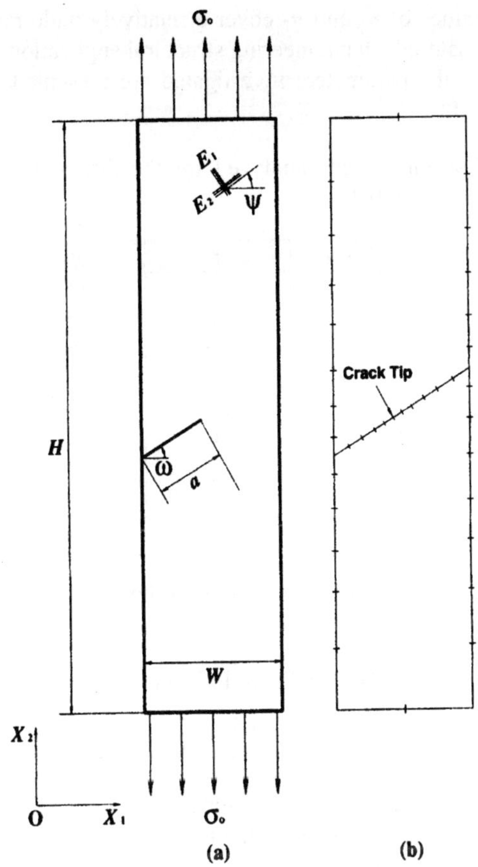

Figure 2: (a) A single-edge cracked plate subject to remote load; (b) BEM mesh taken.

For that purpose, the orthotropic material properties are defined by virtue of purely imaginary roots of the characteristic equation, denoted by $i\eta_1$ and $i\eta_2$, like this

$$\eta_1\eta_2 = \left(\frac{E_1}{E_2}\right)^{1/2} \tag{24.a}$$

$$\eta_1 + \eta_2 = \sqrt{2}\left[\left(\frac{E_1}{E_2}\right)^{1/2} + \frac{E_1}{2\mu_{12}} - \nu_{12}\right]^{1/2} \tag{24.b}$$

The quantities of these parameters investigated are: η_1 =1.5, 3, 4.5 and 6 and η_2 = 0.5 and 0.75; the Poisson´s ratio was taken ν_{12} = 0.3. Altogether, eight cases with the different combinations of values of η_1 and η_2 were studied, as listed in

Table 1. These values of η_1 and η_2 cover a relatively wide range of commonly used orthotropic materials in engineering structural applications.
Specific values of the parameters investigated are presented for the geometric eventuality of $\omega = 0°$.

Table 1: Orthotropic events analyzed for the different combinations of η_1 and η_2; $v_{12} = 0.3$.

Cases	η_1	η_2
1	1.5	0.5
2	3	0.5
3	4.5	0.5
4	6	0.5
5	1.5	0.75
6	3	0.75
7	4.5	0.75
8	6	0.75

Table 2: Characteristic properties for the orthotropic material Kevlar and the values of η_1 and η_2

	E_{11} [GPa]	E_{22} [GPa]	G_{12} [GPa]	v_{12}	η_1	η_2
Kevlar 49/epoxy	86	5.5	2.1	0.34	6.32	0.63

For the sake of brevity and primarily for the purpose of illustration here, only the results for a few specific values of the parameters investigated are presented; they are all for the geometric case of $\omega = 0°$. To be representative of a 'long' plate even for the biggest crack size analyzed, $H/W = 4$ was considered for SECP. In the case when $\psi = 0°$, variations of the T stress with relative crack size, a/W, ranging from $a/W = 0.1$ to 0.5, in SECP specimen are shown in fig. 3. The T-stress results for all the crack sizes treated reveal significant increase in the level of constraint at the crack tip with increasing values E_1/E_2, as supported by the decreasing magnitudes of their negative values.

5 Conclusion

Moreover the stress intensity factor, the T stress is extensionally acclaimed being a significant second parameter for fracture evaluation of cracked bodies. To stipulate T stress numerically, the M integral is an effective way, notably, when performing with the BEM. The paper presents the M integral for two-dimensional anisotropic elasticity and it is implemented being related to the BEM. An example demonstrates the rightness of the formulations and their suitability. It became apparent that T stress for a given cracked pattern may really be greatly influenced by anisotropy.

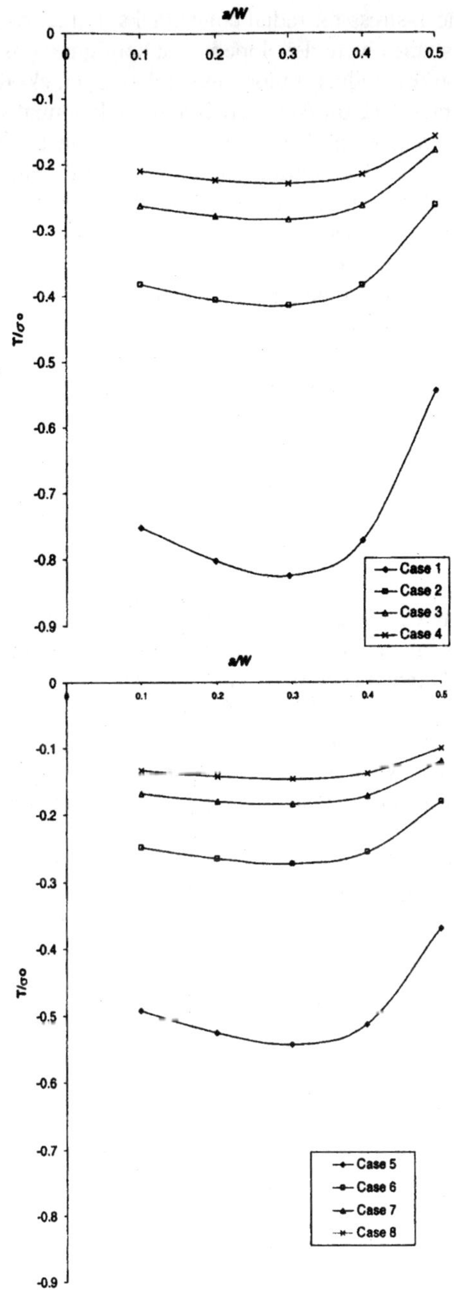

Figure 3: (Taken from [1]): (a) alteration of normalized T stress, T/σ_0, with relative crack length, a/W, ω = 0° and ψ = 0°; (b) alteration of normalized T stress, T/σ_0, with relative crack length, a/W, ω = 0° and ψ = 0°.

www.witpress.com, ISSN 1743-355X (on-line)

To determine the T-stress for radial edge cracks in thick-walled cylinders, the boundary element studies were developed. The configurations incorporated both the wide extension of radius ratios and relative crack depths. The loads contemplated are crack-face pressures embracing polynomial stress distributions. Later, the T-stress weight functions were derived from two reference T-stress solutions for crack pressures of uniform and linear distributions.

The inferred weight functions were verified for miscellaneous loading events. A satisfactory correspondence between the weight functions predictions and solutions obtained directly from the boundary element analysis was acquired. The methodology of weight functions is amenable for T-stress calculations even in conditions of more complicated loading.

Acknowledgement

The author gratefully acknowledges the financial support of the presented research by the Grant Agency of the Czech Republic (project No. 103/06/1382).

References

[1] Shah, P.D., Tan, C.L. and Wang, X., T-stress solutions for two dimensional crack problems in anisotropic elasticity using the boundary element method. *Fatigue Fract Engng Mater Struct,* **29**, pp. 343-356, 2006.

[2] Li, J., Tan Ch., Wang, X., Weight functions for T-stress for edge cracks in thick-walled cylinders. *Journal of Pressure Vessel Technology*. **127**, 2005.

[3] Sladek, J., Sladek, V. and Fedlinski, P., Contour integrals for mixed-mode crack analysis effect of nonsingular terms. *Theor Appl Fract Mech*, **27**, 115-127, 1997.

[4] Wang, X., Elastic T-stress for crack in test specimens subjected to non-uniform stress distributions. *Eng Fract Mech*, **69**, pp. 1339-1352, 2002.

Coupled FEM-BEM crack growth analysis

L. Zhang & R. A. Adey
Wessex Institute of Technology, Ashurst Lodge, Southampton, UK

Abstract

The increasing interest in large structure fracture analysis has heightened the need for efficient numerical computational tools suitable to predict crack propagation. While a sub modelling approach can be used in some cases it does not take into account the redistribution of the loads in the structure thus requiring a large part of the structure to be included in the crack growth model.

Generally large structures are modelled with finite element methods (FEM) because of the many varied types of structural element. Modelling crack growth with FEM results in a particularly complex remeshing process as the crack propagates. Hence, self-adaptive remeshing is one of the major features that must be incorporated in the construction of a computational tool to properly perform crack propagation analysis with the FE method.

The boundary element method (BEM) has attracted lots of attention in the field of fracture mechanics as it simplifies the meshing process and has the ability to accurately represent the singular stress fields near the crack front. One challenge is how the two methods can work together efficiently for a large structure. A new FEM-BEM method is therefore proposed to perform such crack growth analyses.

This paper describes the methodology of a coupled FEM-BEM crack growth analysis for a large scale structure. Both finite element software ABAQUS and boundary element software BEASY were used in the analysis. Several examples are presented at the end of the paper including crack growth in a gear tooth and in a stiffened panel respectively.

Keywords: finite element, boundary element, coupling method, fracture mechanics, crack growth, stress intensity factor.

WIT Transactions on Modelling and Simulation, Vol 44,
www.witpress.com, ISSN 1743-355X (on-line)
doi:10.2495/BE070261

1 Introduction

Fracture mechanics represents the applied mechanics framework necessary for the description of the behaviour of cracked components under applied loads. It can be used to predict how cracks will affect the durability and life of components and structure. The finite element method is employed in many engineering areas including fracture mechanics. Many methods have been developed to solve fracture problems during the last twenty years [1]. However, for crack growth problems, the continuous remeshing process has been a difficulty for most finite element methods [2]. Dual Boundary element, on the other hand, is more flexible as only the boundary needs to be discretized during the analysis [3–6]. The purpose of the current research is to minimize the extent of the remeshing process, yield as accurate results as possible and enable crack propagation to be simulated in large scale structures with different element types. An automatic FE crack propagation code was developed and based on this an FE-BE sub-modelling code has been developed to perform automatic crack propagation analysis. The FE method will be used to compare the results with coupled FE-BE method and the latter incorporates the capabilities of both BEASY10.0 and ABAQUS 6.5. This paper focuses on Mode I and II crack propagation. We will show some examples of edge crack propagation in 2D structures and edge crack propagation in 3D thin structures to illustrate the applications of the proposed method.

2 Methodology of the coupled FE-BE crack growth method

The finite element method is a robust method for elastic and nonlinear material problems. There are numerous pre-processing programmes capable of translating CAD models into finite element models such as PATRAN, GID, etc. The boundary element method can model cracks without remeshing the domain, which significantly simplifies the analysing process. Using coupled FE-BE automatic method allows us to employ advantages of both methods in fracture mechanics problems, especially for the models in which cracks only exist in local areas. In the proposed coupled FE-BE method, instead of applying direct coupling of the BE and FE solution matrices by either presenting the BE matrices as stiffness matrices or transforming the FE forces into tractions and linking them with the tractions in the BE matrices, two models are created, the original FE model and a local BE model representing the crack. The displacements or hybrid displacement-traction values calculated from the FE model are used as prescribed boundary conditions for the local BE model [3]. FE stress values, however, are not used since in finite element analysis stresses are obtained by differentiating displacements, in which process computational errors might be introduced. As the stresses and the displacements change during the crack propagation process, the boundary conditions on the local model must be updated in order to take into account the redistribution of stresses as the crack grows. In the proposed approach changes in strain energy is used as a criterion to determine whether the boundary conditions of the sub-model need updating. This will be explained in more detail in chapter 4. The following is a flowchart of the proposed coupled FE-BE method:

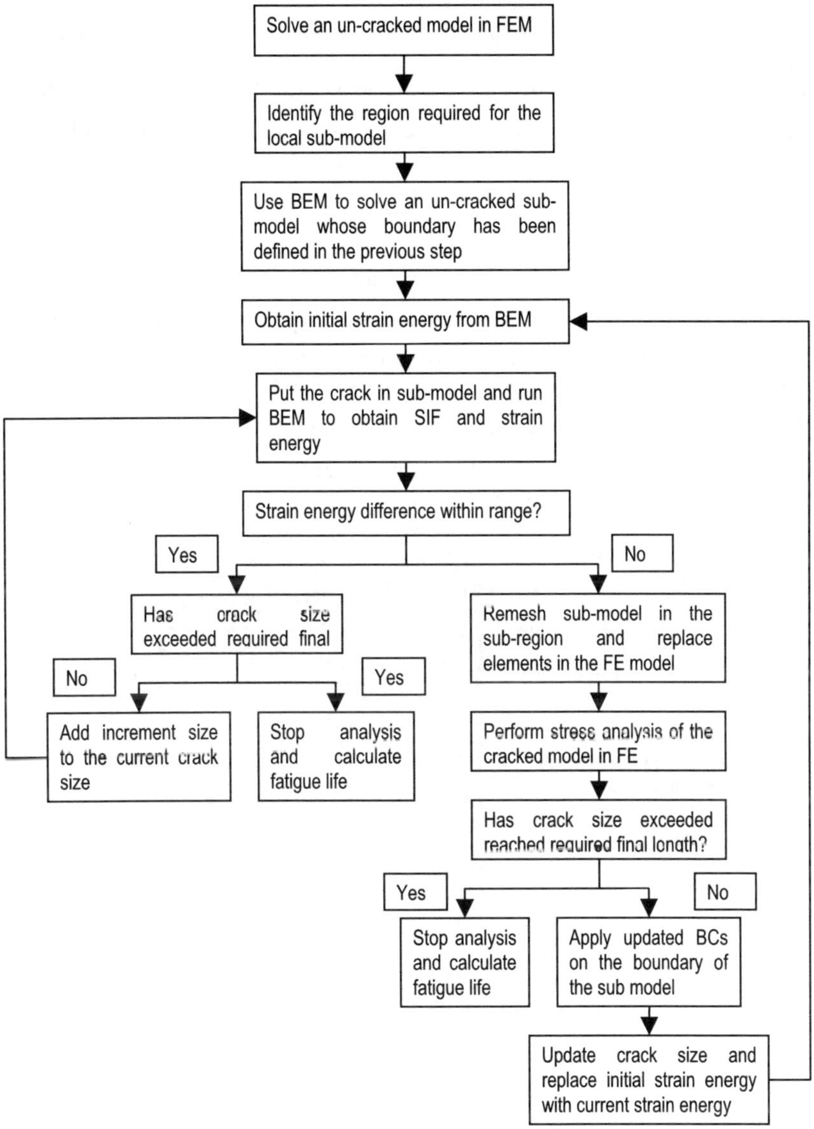

Figure 1: Flowchart of the directly coupled FE-BE crack growth method.

3 BEM sub-model generation and local remeshing

A region is first created to identify the sub-model boundary (Figure 2). The length of the edge can be expressed in the following equation:

$$L_S = L_{fCrack} + D_{gap} \tag{6}$$

where L_S represents the side length of the square, L_{fCrack} is the length of the final crack size and D_{gap} defines the region as measured form the crack that is to be included in the sub model. The second term on the right hand side of the equation can be approximated by iteratively using the following formula:

$$D = \frac{l_1 - ql_n}{1-q} \tag{7}$$

where D is the gap distance, l_1 is the crack increment size and l_n is the side length of the elements that intersect the circle and q is the aspect ratio of the gap mesh. In general, aspect ratios should be less than 2.5:1 in order to avoid large calculation errors. Take a flat-plate under tension for example. Assuming the initial crack size is 2mm and the estimated final crack size is 10mm, so L_{fCrack} equals 10mm. After iteration, if the minimum gap found is 5mm, then the final side length of the square is 10+5=15mm.

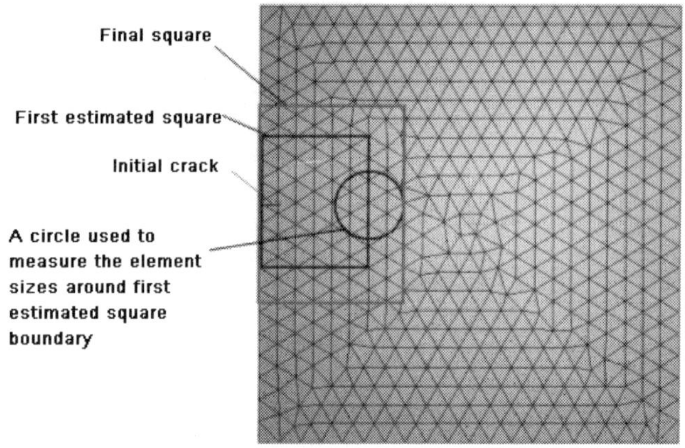

Figure 2: Determination of a square used to generate the sub-model boundary.

Once the sub-model boundary has been determined, the next step to insert the initial crack in the model and perform a fatigue crack growth analysis. The crack type chosen for the analysis is a straight line, 2-element crack. Before putting the crack inside the sub-model boundary, it is necessary to compare the element size on the initial crack and the size of the elements connecting the initial crack to the boundary. If the original element size is too big, it must be divided into smaller pieces and map them onto the original curve.

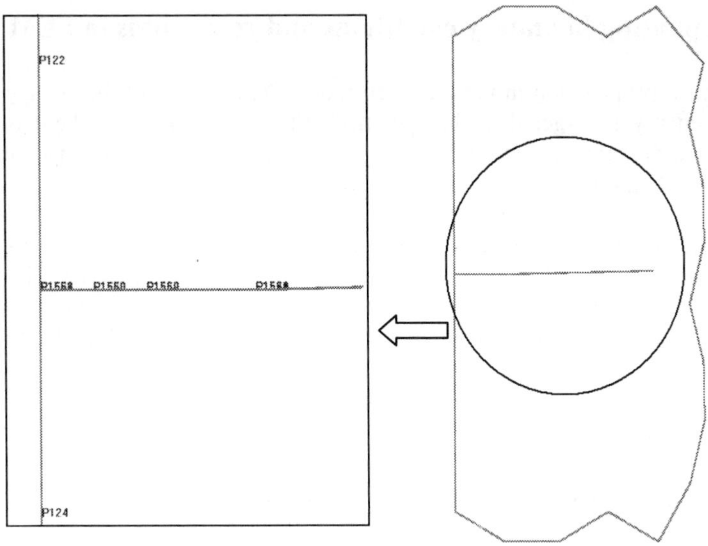

Figure 3: A focused view of a BEM sub-model abstracted from: a plate.

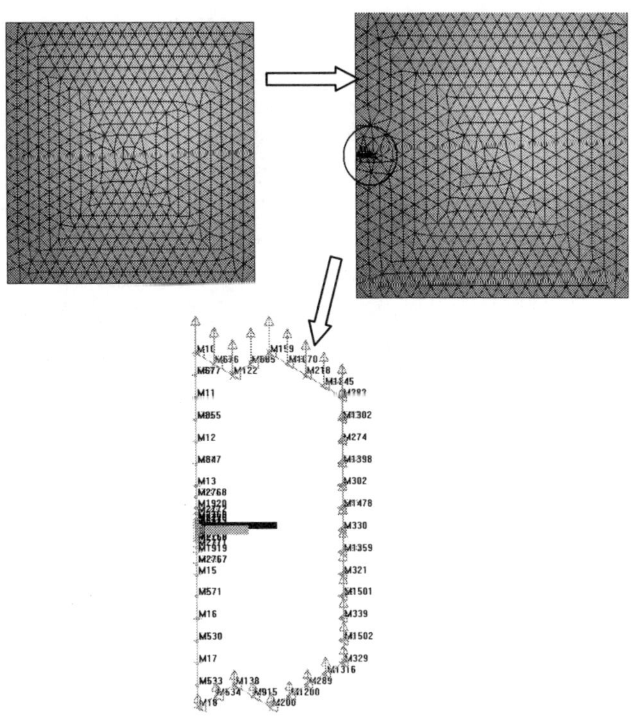

Figure 4: Boundary condition abstracting process.

www.witpress.com, ISSN 1743-355X (on-line)

4 Updating boundary conditions and re-analysis in FEM

The crack propagation in the local sub model continues until the change in the strain energy is larger than the specified criteria. At this point the sub-model including the crack is remeshed as a FE model so it can be added back into the original FE model. Because this new model will not be used to calculate any fracture mechanics data it is not necessary to use any special crack elements or use a refined grid near the crack. The mesh has simply to be sufficient to model the general stress distribution near the crack but not at the crack tip. The whole model is then re-analysed to provide a new solution which can be used to identify the new boundary conditions for the sub-model, which is illustrated in Figure 4.

5 Examples

5.1 Example-edge crack at the root of a gear tooth

This test will investigate crack propagation in a mixed mode crack growth situation. The test results will be compared with those obtained from the boundary element method.

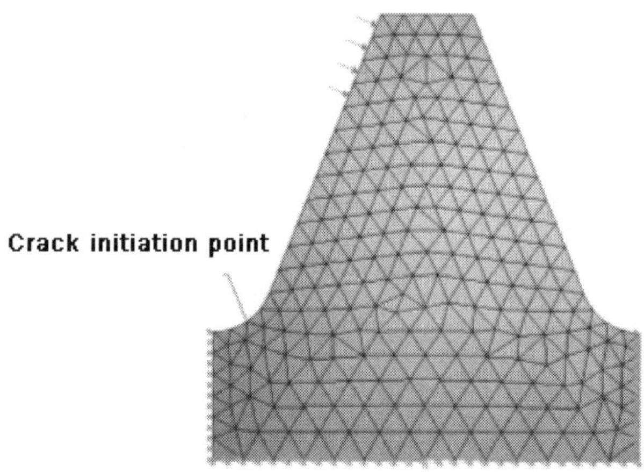

Figure 5: A gear tooth under contact pressure during engagement.

5.1.1 Input parameters

a) Geometry definitions:

Height of the tooth: 14 mm, Width of the tooth: 3mm (top); 14mm (bottom), Fillet radius at the root of the gear tooth: 2 mm.

b) Parameters in automatic coupled FE-BE method:

Geometry: as shown in Figure 5.

Boundary conditions: 1500MPa pressure applied on one side the tooth to simulate the engagement of the gear tooth, Crack initiation angle: 315 degrees, Initial crack length=0.1mm, Number of elements on initial crack: 2 for AutoFEBE, Number of steps: 30, Increment size: 0.1mm, Element type: 6-node quadratic plane strain triangle (CPE6) [7], Re-analysis criterion: Strain energy difference>=2%

5.1.2 Results

Figure 6 shows a comparison of the stress intensity factor vs crack size for the three different methods. As can be seen there is close agreements for the SIF and the crack path as shown in Figure 7. The FE mesh required in the FEBE approach is clearly much less refined than that required when the finite elements are used to approximate the stress field close to the crack.

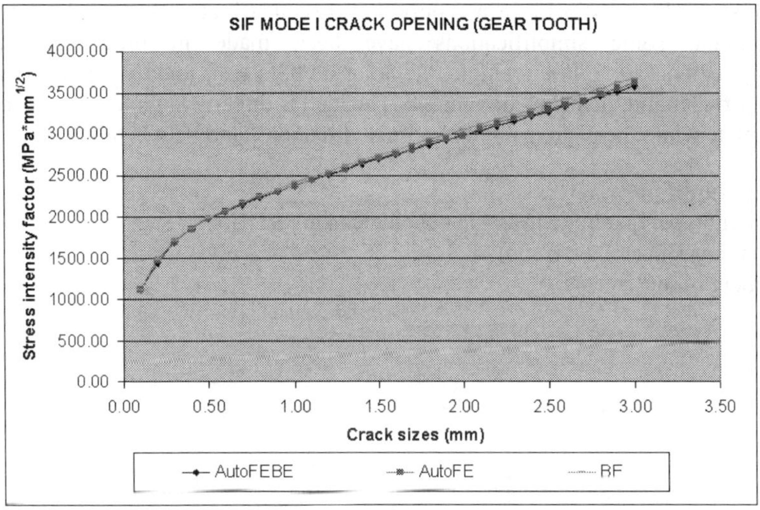

Figure 6: Stress intensity factor vs. crack size for mode I crack opening.

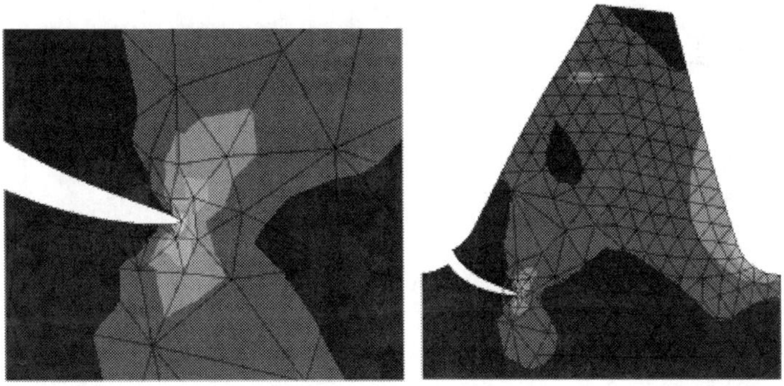

Figure 7: Deformed gear tooth in coupled FEBE method.

5.2 Example-stiffened panel with an imbedded crack in the centre

5.2.1 Test set-up

The two stringer specimens (2SP), Figure 8, with a sheet and outer flange thicknesses of 2.5 and 0.5 mm respectively, and a stringer height of 25 mm, were tested on a servo-hydraulic machine by Llopart et al [8]. The applied load ratio (R) was 0.1 with a maximum load of 60.3KN. The initial crack length was $2a_0 = 3$mm. An anti-bending device was used in the original tests to ensure only mode I crack opening exists.

5.2.2 Model simplification

As the model is symmetric with regard to the geometrical centre, it can be simplified as a plate with an edge crack in the middle. In addition, as the plate is very thin in comparison to its width and length, the whole structure is modelled with shell elements in FE. Since only the stress distribution around the crack is important, some simplifications have been made in order to reduce computational time. For example, the holes on both ends and the fillets between the stiffener and the panel are ignored. The load is directly applied on the edge of the plate while the plate is fixed in normal direction to prevent bending.

5.2.3 Input parameters

Element type: quadratic triangular shell elements STRI65 [7]
Initial crack size: $2a_0$=10mm
Number of increments: 5

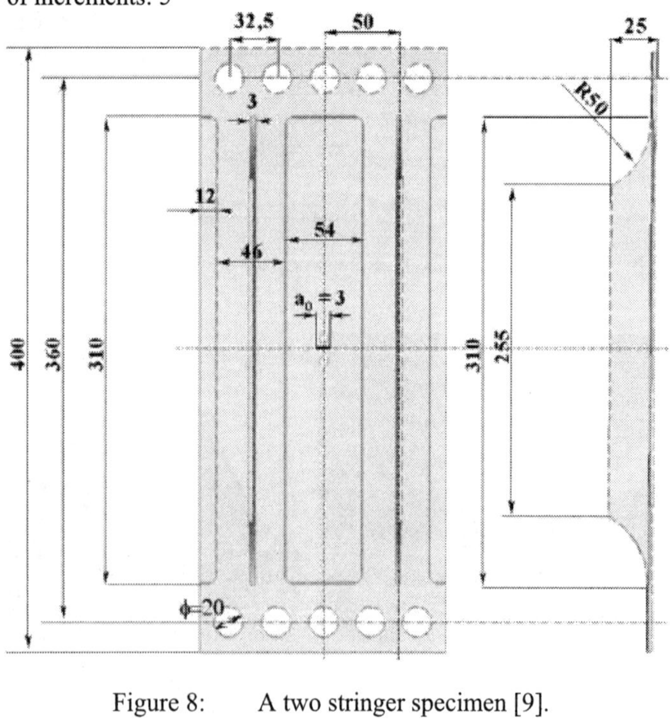

Figure 8: A two stringer specimen [9].

5.2.4 Results

Figure 9 shows the stress distribution and deformation of the stiffened panel after crack propagation. A comparison has been made between the numerical results from the proposed method and the test results from literature [8]. A good agreement is found between the two methods, which can be observed from Figure 10.

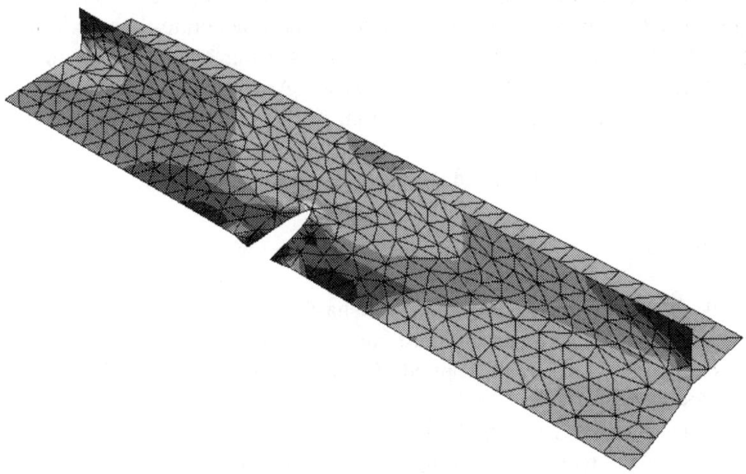

Figure 9: Stress distribution of the stiffened panel after crack propagation.

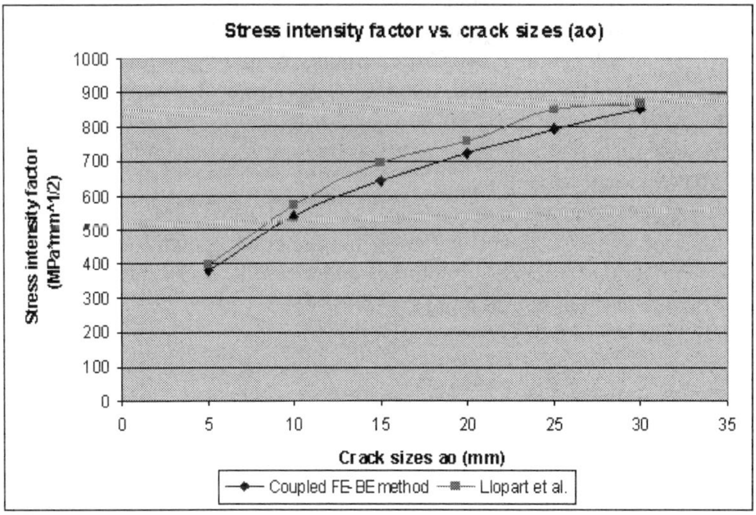

Figure 10: Stress intensity factors given by coupled FE-BE method and literature.

6 Summary

The advantage of the coupled BE-FE method is obvious. If a proper re-analysis criterion is selected, the coupled method can avoid continuous remeshing of the sub-model after each increment which was employed in automatic finite element crack propagation method. Also, no rosette-like singular elements are needed in the analysis and for the same level of mesh complexity the stress intensity factor calculated with boundary element method is more accurate than finite element method. With coupled method, fatigue analysis using BE in the vicinity of the crack of a non-linear structure becomes easier. As the crack is only located in a local area, the general non-linear behaviour of the structure can still be captured by finite element method therefore it is possible to take advantage of both the non-linear capabilities of FE and the accuracy of fatigue analysis in BE.

References

[1] P. O. Bouchard, F. Bay, and Y. Chastel, Numerical modelling of crack propagation: automatic remeshing and comparison of different criteria. Computer Methods in Applied Mechanics and Engineering, 2003. 192: p. 3887-3908.

[2] P. O. Bouchard, et al., Crack propagation modelling using an advanced remeshing technique. Computer Methods in Applied Mechanics and Engineering, 2000. 189: p. 732-742.

[3] A. A. Becker, The Boundary Element Method in Engineering. 1992: McGraw-Hill Book Company.

[4] A. Portela, Dual Boundary Element Analysis of Crack Growth. 1993, Southampton, UK: Computational Mechanics Publications.

[5] M. D. Snyder and T. A. Cruse, Boundary-integral equation analysis of cracked anisotropic plates. International Journal of Fracture, 1975. 11(2): p. 315-328.

[6] T. A. Cruse, Two-dimensional BIE fracture mechanics analysis. Applied Mathematical Modelling, 1978. 2: p. 287-293.

[7] ABAQUS Analysis User's Manual. 2005, ABAQUS Inc.

[8] Ll. Llopart, et al., Investigation of fatigue crack growth and crack turning on integral stiffened structures under mode I loading. Engineering Fracture Mechanics, 2006(73): p. 2139-2152.

Section 10
Electrical engineering and electromagnetics

Electromagnetic modeling of a lightning rod

D. Poljak[1], M. Birkic[2], D. Kosor[3], C. A. Brebbia[4] & V. Murko[5]
[1]*Department of Electronics, University of Split, Croatia*
[2]*Technical Department, Air Traffic Center Pula, Croatia*
[3]*Technical Department, Air Traffic Center Split, Croatia*
[4]*Wessex Institute of Technology, UK*
[5]*Iskra Zascite, Slovenia*

Abstract

This paper deals with a frequency domain analysis of a lightning rod using the antenna theory model. The lightning rod struck by lightning is represented by a straight thin wire antenna excited by an equivalent current source. The current induced along a lightning rod due to a direct lightning strike is determined by solving the homogeneous integro-differential equation of the Pocklington type. Once obtaining the current distribution along the rod provides the calculation of the charge induced along the rod and related irradiated electric field. The corresponding Pocklington equation and field integral relationships are handled via the Galerkin-Bubnov scheme of the Indirect Boundary Element Method (GB-IBEM).

1 Introduction

Lightning flash surges are common sources of electromagnetic interferences (EMI) induced on electrical and electronic systems and equipment, thus producing many undesired effects, or even malfunction of systems and equipment [1–3]. The purpose of a lightning protection system (LPS) in terms of lightning rod is to capture a direct lightning strike. The main parameter of LPS is its efficiency which measure is related to the probability of direct strike to LPS instead of strike to the object within the protected volume defined by the protection zone. The protection zone is considered to be substantially immune to lightning strike due to the air terminal, i.e. the lightning rod. The protection zone around the rod can be assessed by applying the full wave analysis [2]. In particular, the protection zone can be determined if the electric field in the vicinity of the rod is known [4, 5].

WIT Transactions on Modelling and Simulation, Vol 44, © 2007 WIT Press
www.witpress.com, ISSN 1743-355X (on-line)
doi:10.2495/BE070271

This paper first deals with the calculation of electric field around the rod using the antenna theory and boundary element analysis. The antenna model of the lightning rod used in this work is based on the homogeneous Pocklington integro-differential equation [2, 3] by which the current distribution along the rod is governed. The Pocklington equation is solved via the Galerkin-Bubnov variant of the indirect Boundary Element Method (GB-IBEM) [6].

Once the induced current along the rod is obtained one may calculate the related induced charge along the rod featuring the continuity equation and the irradiated electric field from the corresponding integral formulas using BEM formalism.

2 Equivalent antenna model of the lightning rod

The geometry of the problem is shown in Fig. 1. The lightning rod of length L and radius a, sits vertically on a perfectly conducting (PEC) ground. According to the image theory the equivalent representation in terms of straight wire antenna is presented in Fig. 2. The wire is assumed to be perfectly conducting and its dimensions satisfy the thin wire approximation (TWA) conditions [2, 3].

The calculation of the induced current along the lightning rod due to the lightning return stroke is carried out by using the end-fed thin wire model [1–3], Fig. 2.

2.1 Integral equation for a current distribution along the rod

The homogeneous Pocklington integro-differential equation for the current distribution along the lightning rod can be derived by expressing the electric field in terms of the magnetic vector potential and by satisfying certain continuity conditions for the tangential field components at the PEC cylinder surface [5].

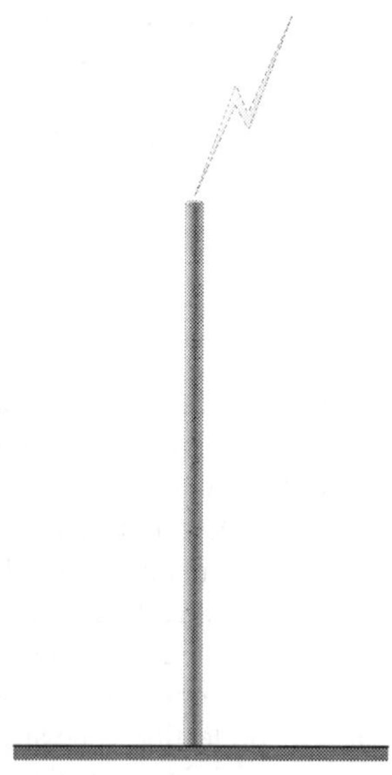

Figure 1: Lightning rod.

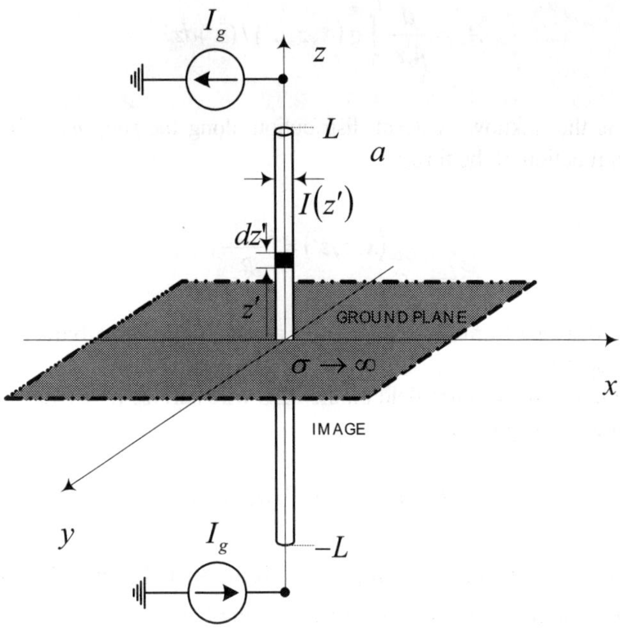

Figure 2: Equivalent antenna model of a lightning rod.

The electric field vector expressed in terms of magnetic vector potential [6], is given by:

$$\vec{E} = \frac{1}{j\omega\mu\varepsilon_0} \nabla\left(\nabla\vec{A}\right) - j\omega\vec{A} \tag{1}$$

where k is the phase constant of free space:

$$k^2 = \omega^2 \mu_0 \varepsilon_0 \tag{2}$$

while ε_0 and μ_0 denotes the permitivity and permeability of the free space.

Due to rotational symmetry of the problem the radiated electric field does not depend on azimuth variable Φ and the electric field components are given by:

$$E_\rho = \frac{1}{j\omega\mu\varepsilon_0} \frac{\partial^2 A_z}{\partial\rho\partial z} \tag{3}$$

$$E_z = \frac{1}{j\omega\mu\varepsilon_0} \frac{\partial^2 A_z}{\partial z^2} - j\omega A_z \tag{4}$$

The vector potential z-component is given by the particular integral [6]:

$$A_z = \frac{\mu}{4\pi} \int_{-L}^{L} g(x,z,z') I(z') dz' \tag{5}$$

where $I(z')$ is the unknown current distribution along the rod, $g(x,z,z')$ is the free space Green function of the form:

$$g(x,z,z') = \frac{e^{-jkR}}{R} \tag{6}$$

and R is the distance from the source point on the rod to the arbitrary observation point in free space.

The total tangential electric field on the PEC wire surface ($\rho=a$) vanishes, i.e. the interface condition is given by:

$$E_z^{exc}(a,z) + E_z^{sct}(a,z) = 0 \tag{7}$$

where E_z^{exc} is the excitation function and E_z^{sct} is the related scattered field.

Combining the relations (4) to (7) results in the Pocklington integro-differential equation for the unknown current:

$$E_z^{exc} = -\frac{1}{j4\pi\omega\varepsilon_0} \int_{-L}^{L} \left[\frac{\partial^2}{\partial z^2} + k^2 \right] g_a(z,z') I(z') dz' \tag{8}$$

where g_a is the integral equation kernel:

$$g_a(x,z,z') = \frac{e^{-jkR}}{R} \tag{9}$$

R_a is the distance from the source point in the wire axis to the arbitrary observation point at the wire surface:

$$\mathrm{R_a} = \sqrt{(\mathrm{z}-\mathrm{z'})^2 + a^2} \tag{10}$$

As the excitation function is not available in the form of electric field, i.e. the equivalent antenna is neither driven by a voltage source, nor excited by a plane wave, the left-hand side of the equation (8) vanishes:

$$-\frac{1}{j4\pi\omega\varepsilon_0} \int_{-L}^{L} \left[\frac{\partial^2}{\partial z^2} + k^2 \right] g_a(z,z') I(z') dz' = 0 \tag{11}$$

As it is well-known from the antenna theory the integro-differential equation kernel becomes quasisingular [6] due to the presence of the second-order differential operator. This problem can be overcome by applying the so-called weak formulation of the problem and Galerkin Bubnov indirect Boundary Element Method (GB-IBEM). Thus, utilizing the property of the kernel

$$\frac{\partial g_a(z,z')}{\partial z} = -\frac{\partial g_a(z,z')}{\partial z'} \tag{12}$$

the alternative form of the integro-differential equation is obtained:

$$-\frac{1}{j4\pi\omega\varepsilon_0}\left[\int_{-L}^{L}\frac{\partial I(z')}{\partial z'}\frac{\partial g_a(z,z')}{\partial z}dz'+k^2\int_{-L}^{L}I(z')g_a(z,z')dz'\right]=0 \tag{13}$$

Solving the Pocklington equation the antenna current is obtained. The mathematical details on the BEM solution of equation (13) are presented in Appendix A.

2.2 The boundary (end) conditions

The equivalent antenna is excited by an ideal current generator with one terminal connected to the antenna and the other one grounded in the remote point in the space. This current source can be included into the integral equation scheme through the forced boundary condition applied at the top of rod. In accordance to the image theory the imaged current source is also injected on the top of the image rod, as shown in Fig.2.

Therefore, the applied symmetric boundary condition can be written, as follows:

$$I(-L) = I(L) = I_g \tag{14}$$

where I_g denotes the equivalent current source.

The proposed model can be further upgraded by taking into account the influence of the lossy earth [6, 7], and by including the lightning channel attached to the structure [1].

2.3 The induced charge along the rod

A deeper insight into the lightning phenomena and behaviour of LPS is provided by the assessment of the charge distribution induced along the cylinder rod struck by lightning. The linear charge density along the rod can be readily computed from the continuity equation [7], as follows:

$$q(z) = -\frac{1}{j\omega}\frac{dI(z)}{dz} \tag{15}$$

where $I(z)$ is the current distribution along the rod.

2.4 The electric field integral formulas

The electric field irradiated by the equivalent antenna representing the lightning rod can be determined knowing the current distribution induced along the rod.

The radial (normal) field component can be determined inserting the expression for the magnetic vector potential (5) into equation (3):

$$E_\rho = \frac{1}{j4\pi\omega\varepsilon_0} \int_{-L}^{L} I(z') \frac{\partial^2 g(z,z',\rho)}{\partial\rho\partial z} dz' \tag{16}$$

Performing the integration by parts equation (16) becomes:

$$E_\rho = \frac{1}{j4\pi\omega\,\varepsilon_0} \int_{-L}^{L} \frac{\partial I(z')}{\partial z'} \frac{\partial g(z,z',\rho)}{\partial\rho} dz' \tag{17}$$

The z-component of the electric field is defined by equations (4) and (5), i.e.:

$$E_z = -\frac{1}{j4\pi\omega\varepsilon_0} \int_{-L}^{L} \left[\frac{\partial^2}{\partial z^2} + k^2 \right] g_0(z,z',\rho) I(z') dz' \tag{18}$$

and after integration by parts it follows:

$$E_z = -\frac{1}{j4\pi\omega\,\varepsilon_0} \left[\int_{-L}^{L} \frac{\partial I(z')}{\partial z'} \frac{\partial g(z,z',\rho)}{\partial z} dz' + k^2 \int_{-L}^{L} I(z') g(z,z',\rho) dz' \right] \tag{19}$$

The integrals in expressions (17) to (19) contain quasi-singular kernel due to the presence of differential operator [6]. This quasi-singularity can be efficiently treated by the boundary element/finite differences approach [6, 7]. The mathematical details on computation of field components are given in Appendix B.

3 Numerical results

First computational example is related to the single lightning rod of length L=10m and radius a=0.019m. Figures 3 and 4 show the induced current and charge along the rod excited by the unit current source at frequency f=1kHz. Figure 5 shows the tangential and the normal component of the irradiated field ρ=10m away from the rod. Figure 6 and 7 show the induced current and charge, respectively, along the rod with the same dimensions excited by the unit current source at the frequency f=1MHz, while Fig 8 shows the related tangential and normal field component.

The last set of Figs is related to the lightning rod protecting the radar antenna system of length L=7.5m, radius a=0.5m excited by the unit current source at frequency f=1MHz.

Figs 9 and 10 show the normal and axial field component, respectively.

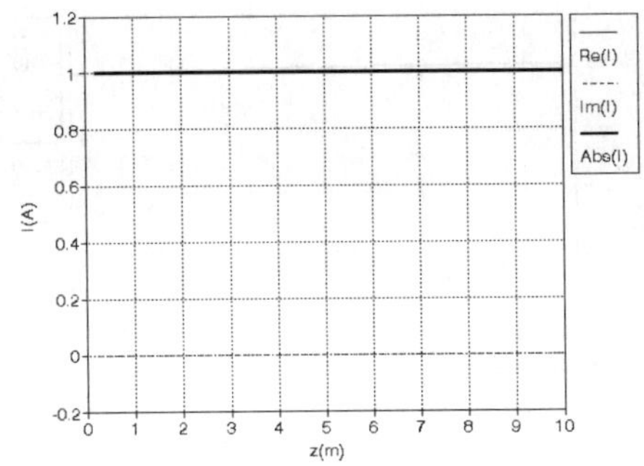

Figure 3: Real, imaginary and absolute value of the induced current (L=10m, a=0.019m, f=1kHz).

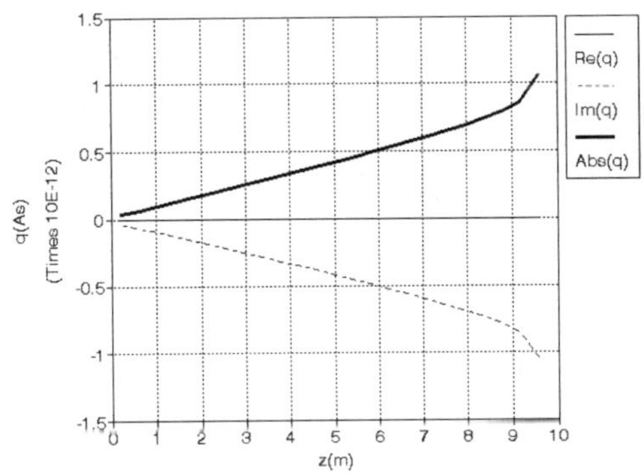

Figure 4: Real, imaginary and absolute value of the induced charge (L=10m, a=0.019m, f=1kHz).

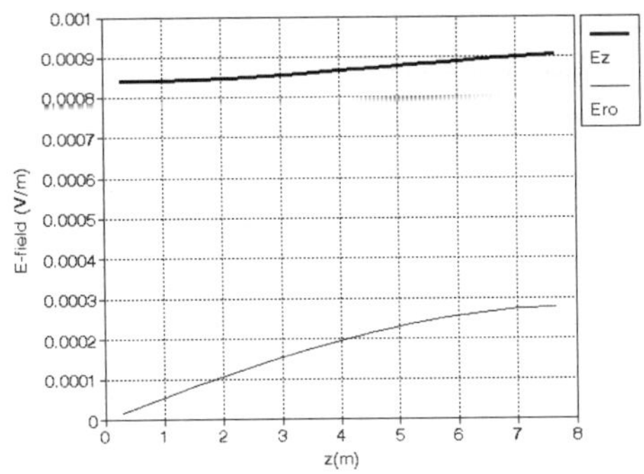

Figure 5: Absolute value of the tangential and normal field components (L=10m, a=0.019m, f=1kHz).

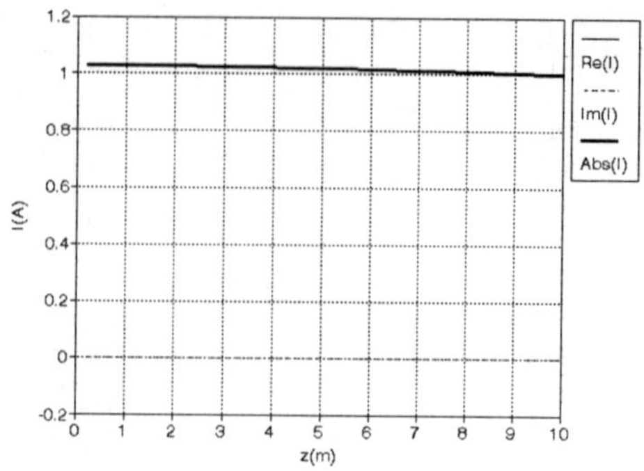

Figure 6: Real, imaginary and absolute value of the induced current (L=10m, a=0.019m, f=1MHz).

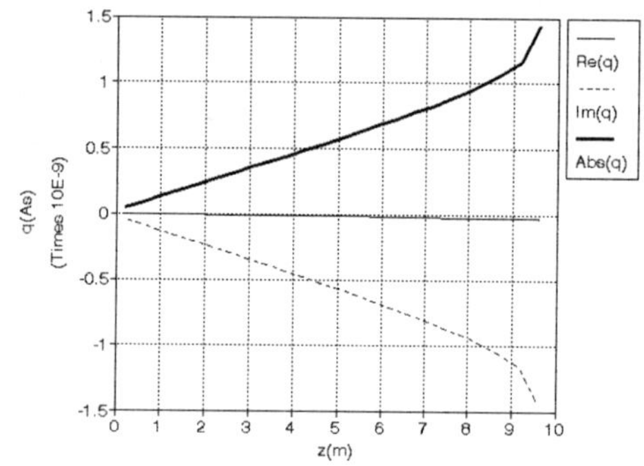

Figure 7: Real, imaginary and absolute value of the induced charge (L=10m, a=0.019m, f=1MHz).

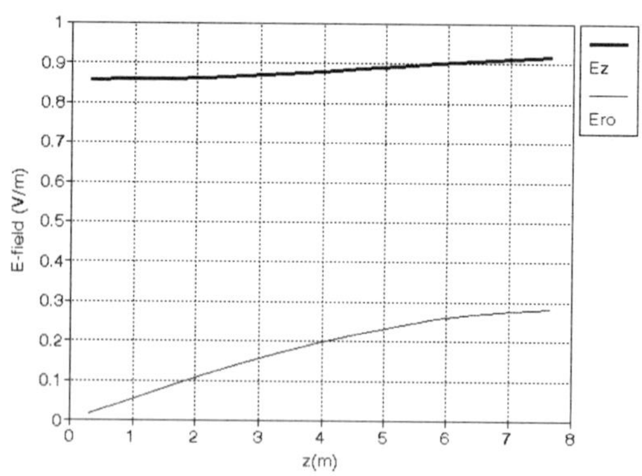

Figure 8: Absolute value of the tangential and normal field components (L=10m, a=0.019m, f=1MHz).

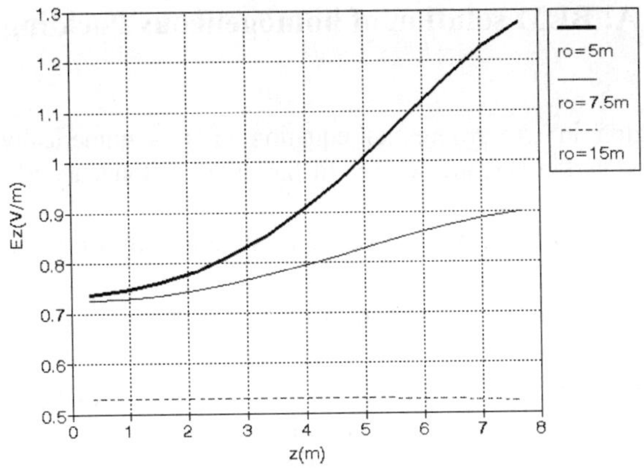

Figure 9: Absolute value of the tangential field component for different distances from the rod (L=7.5m, a=0.5m, f=1MHz).

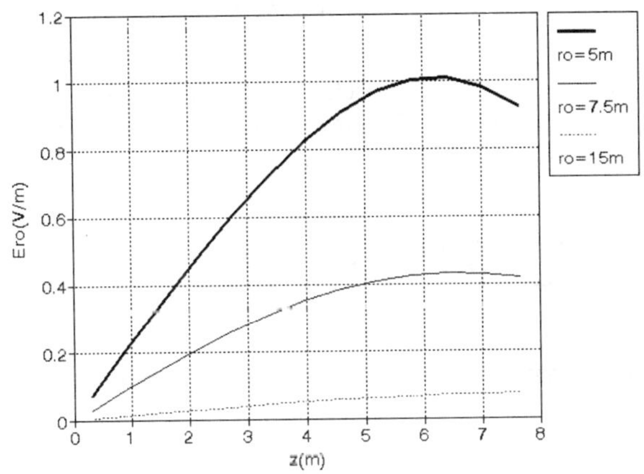

Figure 10: Absolute value of the normal field component for different distances from the rod (L=7.5m, a=0.5m, f=1MHz).

4 Concluding remarks

The frequency domain analysis of the single lightning rod representing the simple lightning protection system (LPS) is undertaken in this work. The full wave model is based on the wire antenna theory. The lightning induced current along the rod is obtained as the BEM solution of the corresponding homogeneous integro-differential equation of the Pocklington type. Having obtained the current distribution it is possible to compute the induced charge along the rod and the related irradiated electric field. Further extension of the present analysis will involve the treatment of complex lightning protection systems consisting of conductors in vertical and horizontal arrangement. A particular feature of the future work will be related to the direct time domain analysis of the problem.

Appendix A: BEM solution of homogeneous Pocklington equation

The Pocklington integro-differential equation (13) is numerically handled by means of the Galerkin-Bubnov scheme of the indirect Boundary Element Method [6].

The operator form of equation (13), can be symbolically written as:

$$K(I)=0 \tag{A1}$$

where K is a linear operator and I is the unknown current to be determined.

The unknown current is expressed by the sum of n linearly independent basis functions $\{f_i\}$ with unknown complex coefficients I_i, i.e.:

$$I \cong I_n = \sum_{i=1}^{n} I_i f_i \tag{A2}$$

Applying the weighted residual approach and choosing the test functions to be the same as basis functions (Galerkin-Bubnov procedure) the operator equation (A1) is transformed into a system of algebraic equations:

$$\sum_{i=1}^{n} I_i \int_0^L K(f_i) f_j dz = 0 \quad j=1,2,\ldots,n \tag{A3}$$

Performing certain mathematical manipulations the following matrix equation is obtained:

$$\sum_{i=1}^{M} [Z]_{ji} \{I\}_i = 0, \qquad j=1,2,\ldots,M \tag{A4}$$

where the vector {I} contains the unknown coefficients, and M is the total number of boundary elements.

The mutual impedance matrix $[Z]_{ji}$ representing the interaction of the i-th source boundary element with the j-th observation element is given by:

$$[Z]_{ji} = -\frac{1}{j4\pi\omega\varepsilon} \begin{pmatrix} \int_{z_j}^{z_{j+1}} \{D\}_j \int_{z_i}^{z_{i+1}} \{D\}_i^T g_a(z,z')dz'dz + \\ +k^2 \int_{z_j}^{z_{j+1}} \{f\}_j \int_{z_i}^{z_{i+1}} \{f\}_i^T g_a(z,z')dz'dz \end{pmatrix} \tag{A5}$$

where z_i, z_{i+1}, z_j and z_{j+1} are the coordinates of *i*-th and *j*-th wire segment, respectively.

A linear approximation over each boundary element has been used in this paper as this choice had already been shown to provide accurate and stable results [6, 7]. Matrices *{f}* and *{f'}* contain the linear shape functions:

$$f_i(z) = \frac{z_{i+1} - z}{\Delta z} \qquad f_{i+1}(z) = \frac{z - z_i}{\Delta z} \tag{A6}$$

while *{D}* and *{D'}* contain their derivatives:

$$\left.\frac{\partial I(z')}{\partial z}\right|_{z=z_i} = \frac{I_{i+1} - I_i}{\Delta z} \tag{A7}$$

Appendix B: BEM evaluation of the field integrals

The radial field component, defined by the integral:

$$E_\rho = \frac{1}{j4\pi\omega\varepsilon_0} \int_{-L}^{L} \frac{\partial I(z')}{\partial z'} \frac{\partial g(z,z',\rho)}{\partial \rho} dz' \tag{B1}$$

can be evaluated using the BEM formalism, i.e. the current along the segment can be written as:

$$I(z) = I_i f_i(z) + I_{i+1} f_{i+1}(z) \tag{B2}$$

and, therefore, it follows:

$$E_x = \frac{1}{j4\pi\omega\varepsilon_0} \sum_{i=1}^{M} \frac{I_{i+1} - I_i}{\Delta z} \int_{z_i}^{z_{i+1}} \frac{\partial g(x,z,z')}{\partial x} dz' \tag{B3}$$

Furthermore, to overcome the quasisingularity problem, the kernel is approximated via central finite difference formula:

$$\frac{\partial f(x,y)}{\partial x} = \frac{f(x+\Delta x, y) - f(x-\Delta x, y)}{2\Delta x} \tag{B4}$$

and the final expression for the radial electric field is then:

$$E_\rho = \frac{1}{j4\pi\omega\varepsilon_0} \sum_{i=1}^{M} \frac{I_{i+1} - I_i}{\Delta\rho\Delta z} \int_{z_i}^{z_{i+1}} \left[G(z,z',\rho+\Delta\rho) - G(z,z',\rho)\right] dz' \tag{B5}$$

where $\Delta\rho$ denotes the finite difference step.

The tangential field component is given by:

$$E_z = -\frac{1}{j4\pi\omega\varepsilon_0}\left[\int_{-L}^{L}\frac{\partial I(z')}{\partial z'}\frac{\partial G(z,z',\rho)}{\partial z}dz' + k^2\int_{-L}^{L}I(z')G(z,z',\rho)dz'\right] \tag{B6}$$

Using linear interpolation for current over the segment yields:

$$E_z = -\frac{1}{j4\pi\omega\varepsilon_0}\sum_{i=1}^{M}\left\{\begin{array}{l}\frac{I_{i+1}-I_i}{\Delta z}\int_{z_i}^{z_{i+1}}\frac{\partial G(z,z',\rho)}{\partial z}dz' + \\ k^2\int_{z_i}^{z_{i+1}}\left[I_i f_i(z) + I_{i+1}f_{i+1}(z)\right]G(z,z',\rho)dz'\end{array}\right\} \tag{B7}$$

and approximating the kernel with finite differences the final formula for the axial field component is:

$$E_z = -\frac{1}{j4\pi\omega\varepsilon_0}\sum_{i=1}^{M}\frac{I_{i+1}-I_i}{\Delta z^2}\left\{\begin{array}{l}\int_{z_i}^{z_{i+1}}\left[G(z+\Delta z/2,z',\rho) - G(z-\Delta z/2,z',\rho)\right]dz' \\ +k^2\int_{z_i}^{z_{i+1}}\left[I_i f_i(z) + I_{i+1}f_{i+1}(z)\right]G(z,z',\rho)dz'\end{array}\right\} \tag{B8}$$

References

[1] F.C. Yang, K.S.H. Lee, Natural Frequencies of a Rod with a Lightning Return Stroke, pp. 75-86, in Lightning Electromagnetics, R. Gardner, (Ed) CRC Press 1990.

[2] D. Poljak, B. Jajac, "Lightning Induced Current on a Metallic Rod-Frequency Domain Analysis " ICECOM '97, pp 87-90 Dubrovnik, Croatia, Oct. 1997.

[3] D. Poljak, B. Jajac, "On the Use of Monopole Antenna Model in Lightning Protection System Analysis " EMC'98 ROMA, pp 370-374, Roma, Italy, Sept. 1998.

[4] K. Aniserowitz, "Methods of Creation of Lightning Protection Zones Near Tall Telecommunication Structures According to Different National Standards ", TCSET' 2002, Lviv- Slavsko, Ukraine, Feb 18-23, 2002.

[5] B. Jajac, D. Poljak, N. Kovac "Boundary Element Modelling of the Metallic Rod Protection Zone", BEM XXV, Southampton, UK, Boston, and USA: WIT Press, 2003.

[6] D. Poljak, C.A. Brebbia, "Boundary Element methods for Electrical Engineers ", WIT Press, Southampton-Boston, 2005.

[7] D. Poljak, V. Doric, S. Antonijevic "Computer aided Design of Wire Structures: Frequency and Time Domain Analysis", WIT Press, Southampton-Boston, 2007.

The analysis of TM-mode and TE-mode optical responses of metallic nanostructures by new surface integral equations

J.-W. Liaw
Department of Mechanical Engineering, Chang Gung University, Kwei-Shan, Tao-Yuan, Taiwan

Abstract

Two sets of new surface integral equations, based on the Stratton-Chu formulation, are developed to analyze the interactions of an incident electromagnetic wave with a 2D metallic nanostructure for the transverse-magnetic (TM) mode and the transverse-electric (TE) mode, respectively. For the former, the surface integral equations are in terms of the surface components of the tangential magnetic field H_z, the normal displacement field D_n, and the tangential electric field E_t. As to the latter, the equations are in terms of the surface components of the tangential electric field E_z, the normal magnetic flux density B_n, and the tangential magnetic field H_t. The numerical results show that for TM mode a standing wave pattern is observed on the backside of a metallic nanoscatterer with a size of several hundreds nanometers, which is caused by two surface plasmon waves creeping along the boundary clockwise and counterclockwise. However there is only a shadow zone on the backside of the metallic scatterer for TE mode.
Keywords: surface plasmon resonance, surface plasmon wave, metallic nanostructure, transverse magnetic mode, transverse electric mode, surface integral equations.

1 Introduction

Recently a new topic-plasmonics attracts much attention in nanooptics [1–4]. It is concerned with the interaction of light with metallic nanostructures. Because of the free electrons in metals, the permittivity of metal (e.g. Au and Ag) is not

WIT Transactions on Modelling and Simulation, Vol 44,
www.witpress.com, ISSN 1743-355X (on-line)
doi:10.2495/BE070281

only frequency-dependent but also a complex number with a negative real part and a positive imaginary part in certain range of VU to NIR. When a nanometer-sized metallic scatterer is illuminated by an incident light at a specific frequency, this highly dispersive property makes them exhibit a collective motion of electrons in the metal to induce a strong scattering and absorption of the illuminating light. The phenomenon is the so-called surface plasmon resonance (SPR). The main difference of the SPR of a metallic nanoscatterer caused by an electromagnetic (EM) wave from the regular resonance of the acoustic or elastic waves problems is that the size of the metallic nanoscatterer is much smaller than the wavelength of the incident light. In addition, the frequency of SPR depends on the metal's size, shape, and the permittivity of the surrounding medium sensitively. For a 2D metallic nanostructure (the length in z-direction is infinitely long), the SPR can be induced only by an EM wave of *p*-polarization (the polarization of the electric field is in-plane), but not by that of *s*-polarization (the polarization of the electric field is anti-plane). The unique behaviors of a nanometer-sized metallic scatterer at SPR are the significantly large scattering and absorption cross sections at the far field, and the strong electric field at near field. However, when the size of the metallic nanoscatterer becomes larger, the property of the frequency-selected SPR gradually disappears, and a surface plasmon wave gradually forms on the surface of the metal for a TM-mode problem only. Since the behaviors of a nanometer-sized scatterer (e.g. several tens nanometers) are intensively investigated in lots of papers [1–4], we focus on the studies of a 2D metallic nanoscatterer with a size of several hundreds nanometers in this paper to demonstrate the difference of the TM-mode and TE-mode responses. In this paper, two sets of surface integral equations derived from the Stratton-Chu formulation [5] are used and implemented by BEM to simulate the interactions of an EM wave in TM-mode and in TE-mode with a 2D metallic nanoscatterer, respectively.

2 Theory

In this paper, the time harmonic responses of an EM plane wave propagating in a host (exterior) to interact with a single metallic scatterer (interior) are considered; the time harmonic factor is $exp(-i\omega t)$, where $i=\sqrt{-1}$. The permittivity of each material is denoted by ε_j, and the permeability by μ_j, where j=1 is for the host, and j=2 for the scatterer.

2.1 TM mode

For a 2D TM-mode (*p*-polarization) problem, the electric field is in plane and can be expressed as $\mathbf{E}=E_x\mathbf{e}_x+E_y\mathbf{e}_y$, and the magnetic field is $\mathbf{H}=H_z\mathbf{e}_z$. Here, $\mathbf{e}_x$, $\mathbf{e}_y$ and $\mathbf{e}_z$ are the unit vectors of x, y and z directions, respectively. The total fields in the exterior domain can be decomposed of two parts; one is the incident field, and the other the scattering field, i.e. $\mathbf{E}=\mathbf{E}^{(i)}+\mathbf{E}^{(s)}$ and $H_z=H_z^{(i)}+H_z^{(s)}$. Throughout the paper, the superscript '(*i*)' represents the incident field, and '(*s*)' the scattering field. A set of surface integral equations can be derived from the Stratton-Chu formulation [5] as

$$H_z(\mathbf{x}) = H_z^{(i)}(\mathbf{x}) - \int_S H_z(\mathbf{x}')\mathbf{n}' \cdot \nabla'[G_1 - G_2]\, dl' - i\omega \int_S E_t(\mathbf{x}')[\varepsilon_1 G_1 - \varepsilon_2 G_2]\, dl', \quad \mathbf{x} \in S_{12} \tag{1}$$

$$D_n(\mathbf{x}) = D_z^{(i)} - i\omega \int_{S_{12}} H_z[\varepsilon_1\mu_1 G_1 - \varepsilon_2\mu_2 G_2]\mathbf{n} \cdot d\mathbf{l}' - \int_{S_{12}} D_n \mathbf{n} \cdot \nabla'[G_1 - G_2]\, dl' + \int_{S_{12}} E_t \mathbf{n} \cdot \mathbf{e}_z \times \nabla'[\varepsilon_1 G_1 - \varepsilon_2 G_2] dl', \ \mathbf{x} \in S_{12} \tag{2}$$

$$E_t(\mathbf{x}) = E_t^{(i)} - i\omega \int_{S_{12}} H_z[\mu_1 G_1 - \mu_2 G_2]\mathbf{t} \cdot d\mathbf{l}' - \int_{S_{12}} D_n \mathbf{t} \cdot \nabla'\left[\frac{G_1}{\varepsilon_1} - \frac{G_2}{\varepsilon_2}\right] dl' + \int_{S_{12}} E_t \mathbf{t} \cdot \mathbf{e}_z \times \nabla'[G_1 - G_2]\, dl', \ \mathbf{x} \in S_{12} \tag{3}$$

where S_{12} is the interface of the host and the scatterer, the unit normal vector **n** of S_{12} is in the inner direction, and the unit tangential vector is defined as $\mathbf{t} = \mathbf{n} \times \mathbf{e}_z$. These surface integral equations [2] are in terms of the surface components of the tangential magnetic field H_z, the normal displacement field D_n, and the tangential electric field E_t. Here, the Green functions G_j of medium j, j=1, 2, are written as

$$G_j(\mathbf{x}, \mathbf{x}') = \frac{\mathrm{i}}{4} H_0^{(1)}(k_j r) \tag{4}$$

where r is the distance $r = |\mathbf{x} - \mathbf{x}'|$, and the wavenumbers are $k_j = \omega\sqrt{\varepsilon_j \mu_j}$.

2.2 TE mode

For a 2D TE-mode (*s*-polarization) problem, the magnetic field is in plane and can be expressed as $\mathbf{H}=H_x\mathbf{e}_x+H_y\mathbf{e}_y$, and the electric field is $\mathbf{E}=E_z\mathbf{e}_z$. The total fields in the exterior domain can be decomposed of two parts; one is the incident field, and the other the scattering field, i.e. $\mathbf{H} = \mathbf{H}^i + \mathbf{H}^s$ and $E_z = E_z^i + E_z^s$. The surface integral equations derived from the Stratton-Chu formulation can be written in terms of the surface components of the tangential electric field E_z, the normal magnetic flux density B_n, and the tangential magnetic field H_t as,

$$E_z(\mathbf{x}) = E_z^{(i)} - \int_{S_{12}} E_z(\mathbf{x}')\mathbf{n}' \cdot \nabla'[G_1 - G_2]\, dl' + i\omega \int_{S_{12}} H_t(\mathbf{x}')[\mu_1 G_1 - \mu_2 G_2]\, dl' \ , \ \mathbf{x} \in S_{12} \tag{5}$$

$$B_n(\mathbf{x}) = B_n^{(i)} + i\omega \int_{S_{12}} E_z \left[\varepsilon_1 \mu_1 G_1 - \varepsilon_2 \mu_2 G_2\right] \mathbf{n} \cdot d\mathbf{l}' - \int_{S_{12}} B_n \mathbf{n} \cdot \nabla' \left[G_1 - G_2\right] dl' + \int_{S_{12}} H_t \mathbf{n} \cdot \mathbf{e}_z \times \nabla' \left[\mu_1 G_1 - \mu_2 G_2\right] dl', \mathbf{x} \in S_{12} \tag{6}$$

$$H_t(\mathbf{x}) = H_t^{(i)} + i\omega \int_{S_{12}} E_z \left[\varepsilon_1 G_1 - \varepsilon_2 G_2\right] \mathbf{t} \cdot d\mathbf{l}' - \int_{S_{12}} B_n \mathbf{t} \cdot \nabla' \left[\frac{G_1}{\mu_1} - \frac{G_2}{\mu_2}\right] dl' + \int_{S_{12}} H_t \mathbf{t} \cdot \mathbf{e}_z \times \nabla' \left[G_1 - G_2\right] dl', \mathbf{x} \in S_{12} \tag{7}$$

3 Numerical results and discussion

Consider a 2D scatterer of an infinite-long circular cylinder which is embedded in air. If an incident EM plane wave is in TM-mode (the polarization of the electric field is in-plane) to impinge upon the scatterer, Eqns. (1) to (3) are used to calculate the surface components along the boundary, and then the corresponding integral representations [2] are used to obtain the field values in the exterior and interior domains. Take a typical case as an example; an incident wave of TM-mode propagates from the left to the right hand side at ω=4.558×10^{15} rad/s (3eV) to impinge upon a silver cylinder (r=200nm). The distributions of the amplitude of the total electric field and the total magnetic field in the near field of the scatterer are depicted in Fig. 1(a) and 1(b), where ε_{1r}=1 and ε_{2r}=(-4.422171,0.73006) [6]. For this case, the wavelength of the incident wave is λ=413.3nm. The numerical results show an interference pattern on the backside of the metallic scatterer. This is because when the illuminating light impinges upon the silver scatterer, a part of the photon's energy is converted into two surface plasmon waves creeping along the circumference of the silver cylinder counterclockwise and clockwise, respectively. The two opposite-directional surface plasmon waves will interfere with each other to induce a standing wave on the backside of the scatterer.

If the incident EM plane wave is in TE mode (the polarization of the electric field is anti-plane), Eqns. (5) to (7) are used for calculation. Compared to the TM-mode, there is only a shadow zone, rather than a standing wave, to be observed on the backside of the metallic scatterer for TE-mode as depicted in Fig. 2(a) and 2(b). The phenomenon of shadow zone is consistent with the result of a perfect-conductor scatterer. In addition, due to the shielding effect of metals on the EM field, the EM field cannot directly transmit through the metallic scatterer, except within a thin skin depth, for both TM and TE modes, as shown in Figs. 1 and 2. Since the surface plasmon wave exists only in TM mode [2], the scattering responses of an EM wave in the near field of a metallic nanoscatterer are totally different in the TM and TE modes.

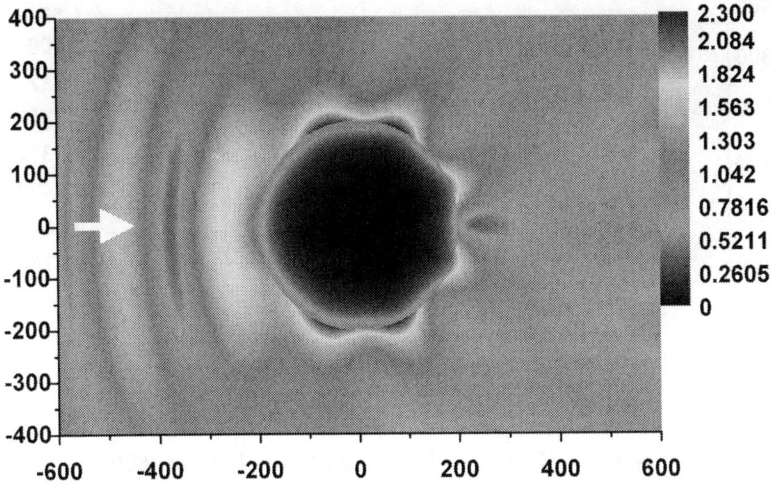

(a)

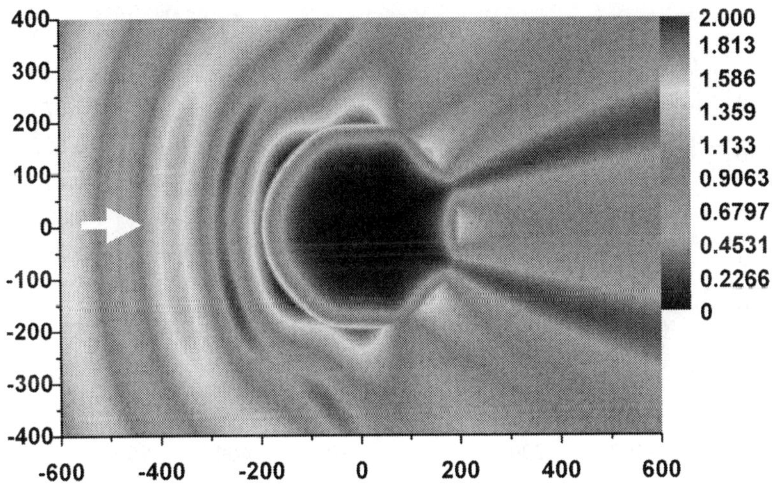

(b)

Figure 1: (a) The total electric field distribution in the vicinity of a silver cylinder with r=200nm irradiated by a TM-mode plane wave at $\omega=4.558\times10^{15}$ rad/s (3eV), where $\varepsilon_{1r}=1$, $\varepsilon_{2r}=(-4.422171,0.73006)$. (b) The total magnetic field distribution.

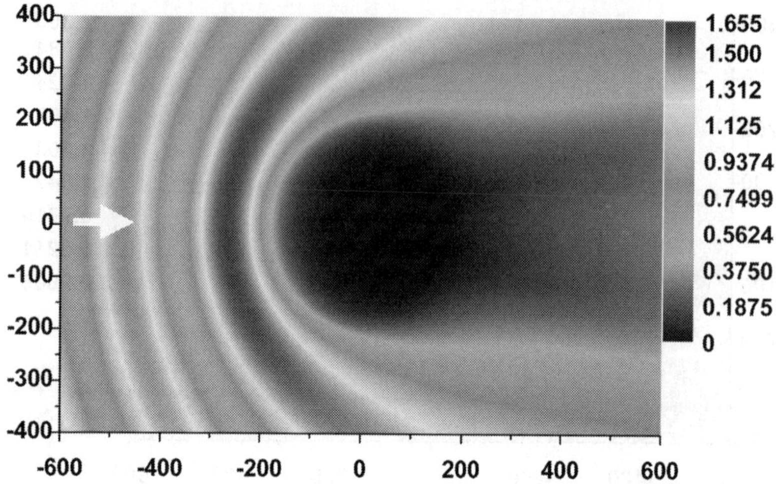

(a)

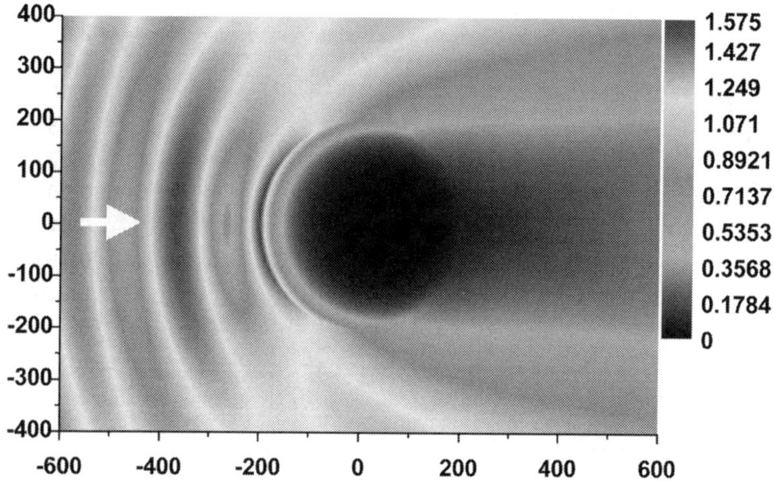

(b)

Figure 2: The total electric field distribution in the vicinity of a silver cylinder with r=200nm irradiated by a TE-mode plane wave at $\omega=4.558\times10^{15}$ rad/s (3eV), where $\varepsilon_{1r}=1$, $\varepsilon_{2r}=(-4.422171,0.73006)$. (b) The total magnetic field distribution.

References

[1] Kottmann, J.P., Martin, J.F., Smith, D. R., Schultz, S., Spectral response of plasmon resonant nanoparticles with a non-regular shape, Opt. Express **6**, pp. 213-219, 2000.

[2] Liaw, J.-W., Simulation of surface plasmon resonance of metallic nanoparticles by boundary-element method, J. Opt. Soc. Am., A **23(1)**, pp. 108-116, 2006.

[3] Liaw, J.-W., Analysis of the surface plasmon resonance of a single core-shelled nanocomposite by surface integral equations, Eng. Anal. with Boundary Elements, **30(9)**, pp. 734-745, 2006.

[4] Liaw, J.-W., New surface integral equations for the light scattering of multi metallic nanoscatterers, Eng. Anal. with Boundary Elements **31(4)**, pp. 299-310, 2007.

[5] Stratton, A., *Electromagnetic theory* (McGraw-Hill, New York, 1941), pp. 464-467.

[6] Palik, E. D., *Handbook of optical constants of solids* (Academic Press, New York, 1985)

Measures for the postprocessing of grounding electrodes transient response

D. Poljak[1], V. Dorić[1], V. Murko[2] & C. A. Brebbia[3]
[1]*Department of Electronics, University of Split, Croatia*
[2]*Iskra zascite, Slovenia*
[3]*Wessex Institute of Technology, UK*

Abstract

Various measures for quantifying the transient response of simple grounding systems are proposed in this paper. In addition to the standard transient impedance concept the suggested measures arising from the circuit theory are instantaneous power, average power and total energy stored in the near field of a grounding electrode. The frequency response of the grounding electrode is obtained by using the antenna model (AM) while the associated transient response is computed using the Inverse Fourier Transform. The integro-differential realtionships arising from the wire antenna theory are numerically handled via the Galerkin-Bubnov scheme of the Indirect Boundary Element Method (GB-IBEM). A number of illustrative numerical results are presented in the paper.

1 Introduction

Transient analysis of grounding systems, important for protection of personnel and equipment, is of widespread interest in electromagnetic compatibility (EMC) and high voltage (HV) engineering. Transient modeling of grounding systems can be carried out applying either the transmission line model (TLM) [1–3] or antenna (electromagnetic) model (AM) [4–6]. An important parameters arising from studies of transients in grounding systems is the transient impedance. Further to the transient impedance concept for postprocessing transient responses, widely adopted within EMC community, this work deals with some additional measures of a horizontal electrode transient response. These measures arise from the circuit theory and are, as follows: instantaneous power, average power and the total energy accumulated in the near field of the electrode.

WIT Transactions on Modelling and Simulation, Vol 44, © 2007 WIT Press
www.witpress.com, ISSN 1743-355X (on-line)
doi:10.2495/BE070291

The antenna model of the electrode used in in this work is based on the Pocklington integro-differential equation approach [4–6]. The current distribution along the horizontal grounding electrode is governed by the Pocklington integro-differential equation which is solved via the Galerkin-Bubnov scheme of the Indirect Boundary Element Method (GB-IBEM) [12, 13]. The effect of a dissipative half-space is taken into account via the corresponding reflection coefficient thus avoiding the solution of analytically demanding and numerically time consuming Sommerfeld integrals. The voltage at the feed point is obtaned by analytically integrating the normal electric field from the electrode surface to infinity. The input impedance of the electrode (transfer function of the system) is computed as a ratio of evaluated voltage and the feed point current.

The frequency response of the horizontal electrode to a particular current source excitation is obtained multiplying the input impedance spectrum with Fourier transform of the actual lightning current waveform. The transient response of the horizontal grounding electrode is assessed applying the inverse Fourier transform. Once determining the transient response of the electrode, the transient behaviour can be quantified using the standard concept of transient impedance and also by using the proposed circuit theory measures.

2 Equivalent antenna model of the grounding electrode

The geometry of interest, shown in Fig 1, is the horizontal grounding electrode of length L and radius a, buried in a lossy medium at depth d and excited by a current source. The wire is assumed to be perfectly conducting and its dimensions satisfy the thin wire approximation (TWA) conditions [12].

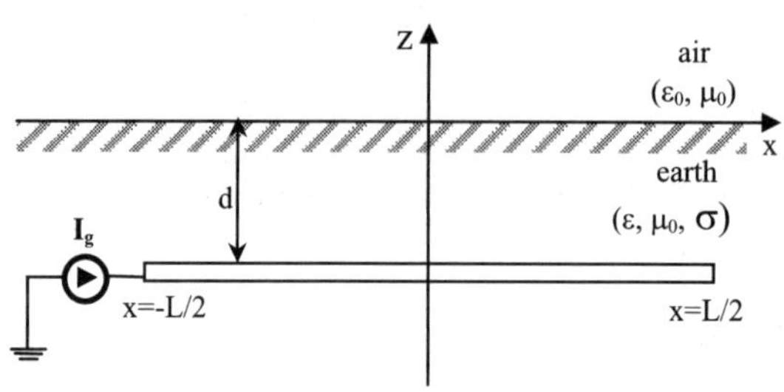

Figure 1: Horizontal grounding wire energized by a current generator I_g.

The Pocklington integro-differential equation for the current distribution along the horizontal grounding electrode can be derived by expressing the electric field in terms of the Hertz vector potential and by satisfying the given boundary conditions for the tangential field components at the electrode surface [4–6].

2.1 Thin wire integral equation for a horizontal electrode

The current induced along the horizontal grounding electrode is governed by the Pocklington integro-differential equation [6]:

$$E_x^{exc,H} = -\frac{1}{j4\pi\omega\varepsilon_{eff}} \int_{-L/2}^{L/2} \left[\frac{\partial^2}{\partial x^2} + k_1^2\right] [g_0^H(x,x') + \Gamma g_i^H(x,x')] I(x') dx' \tag{1}$$

where $I(x')$ is the unknown current distribution along the wire, $E_x^{exc,H}$ is the excitation function, $g_0^H(x,x',z)$ denotes the free space Green function of the form:

$$g_0^H(x,x',z) = \frac{e^{-jk_1 R_{1h}}}{R_{1h}} \tag{2}$$

while $g_i^H(x,x',z)$ arises from the image theory and is given by:

$$g_i^H(x,x',z) = \frac{e^{-k_2 R_{2h}}}{R_{2h}} \tag{3}$$

where R_{1h} and R_{2h} are the distances from the horizontal wire in the lossy ground and from its image in the air to the observation point in the lower medium, respectively. Furthermore, k_1 is the phase constant of a lossy ground:

$$k_1^2 = -\omega^2 \mu \varepsilon_{eff} \tag{4}$$

and ε_{eff} denotes the complex permitivity of the lossy ground:

$$\varepsilon_{eff} = \varepsilon_r \varepsilon_0 - j\frac{\sigma}{\omega} \tag{5}$$

where and ε_r and σ are relative permitivity and conductivity of the ground respectively, and ω denotes the operating frequency.

The presence of a lossy medium is taken into-account via the reflection coefficient while Γ is the corresponding reflection coefficient for the TM polarization [6]:

$$\Gamma = \frac{\frac{1}{\underline{n}}\cos\theta - \sqrt{\frac{1}{\underline{n}} - \sin^2\theta}}{\frac{1}{\underline{n}}\cos\theta + \sqrt{\frac{1}{\underline{n}} - \sin^2\theta}} \tag{6}$$

where θ and $\underline{n}$ are given by:

$$\theta = arctg\frac{|x - x'|}{2d}; \quad \underline{n} = \frac{\varepsilon_{eff}}{\varepsilon_0} \tag{7}$$

The rigorous AM approach leads to the repeated evaluation of Sommerefeld integrals representing rather time consuming task.

This work features the reflection coefficient (RC) approach [6, 11]. The main advantage of RC approach versus rigorous Sommerfeld integral approach is a simplicity of the formulation and significantly less computational cost. It is worth noting that the RC approach produces results roughly within 10% of these obtained via rigorous Sommerfeld integral approach [11, 14].

2.2 The current source excitation

Within the analysis of the grounding electrodes it is not possible to define the excitation function in the form of an electric field. The horizontal grounding electrode is energized by the injection of a corresponding current pulse represented by an ideal current source with one terminal connected to the grounding electrode and the other one grounded at infinity, as shown in Fig 1.

Consequently:

$$E_x^{exc} = 0 \tag{8}$$

and the Pocklington integro-differential equation (9) becomes homogeneous [4–6].

This current source is included into the integro-differential equation formulation through the following boundary condition:

$$I(-L/2) = I_g \tag{9}$$

where I_g denotes the impressed unit current generator.

3 Boundary element procedure

Solving the integral equation (1) via the GB-IBEM the equivalent current distribution along the horizontal grounding electrode is obtained. The numerical solution steps are outlined below.

Performing certain mathematical manipulations and boundary element discretization the solution for the unknown current $I^e(x)$ along the wire segment can then be written as:

$$I^e(x') = \{f\}^T \{I\} \tag{10}$$

Assembling the contributions from each element the integro-differential equation (9) is transferred into the following matrix equation:

$$\sum_{i=1}^{M} [Z]_{ji} \{I\}_i = 0, \quad \text{and} \quad j = 1,2,\ldots, M \tag{11}$$

where M is the total number of wire segments and $[Z]_{ji}$ is the mutual impedance matrix representing the interaction of the *i*-th source boundary element with the *j*-th observation boundary element:

$$[Z]_{ji} = -\frac{1}{j4\pi\omega\varepsilon_{eff}}\left\{ \int_{\Delta l_j}\int_{\Delta l_i} \{D\}_j \{D'\}_i^T g^H(x,x')dx'dx + k^2 \int_{\Delta l_j}\int_{\Delta l_i} \{f\}_j \{f\}_i^T g^H(x,x')dx'dx \right\} \tag{12}$$

Matrices $\{f\}$ and $\{f'\}$ contain the shape functions while $\{D\}$ and $\{D'\}$ contain their derivatives, M is the total number of finite elements, and Δl_i, Δl_j are the widths of *i*-th and *j*-th boundary elements.

The linear approximation over a boundary element is used in this work as it has been shown that this choice provides accurate and stable results for various wire configurations [12, 13].

The excitation function in the form of the current source I_g is taken into account as a forced boundary condition at the first node of the solution vector, i.e.:

$$I_1 = I_g \quad \text{and} \quad I_g = 1e^{j0} \tag{13}$$

providing the linear equation system to be solved properly.

4 The assessment of the transient response

Transient voltage at the feed point (x=-L/2) can be obtained from the convolution integral:

$$v(x,t)\Big|_{x=-L/2} = \int_0^t z_{in}(x,t)\Big|_{x=-L/2} \, i(x,t-\tau)\Big|_{x=-L/2} \, d\tau \tag{14}$$

The frequency response of the grounding system: is obtained by multiplying the frequency spectrum of the excitation function $I(f)$ with the frequency domain counterpart of the impulse response, i.e. the input impedance spectrum ($Z_{in}f$):

$$V(f) = I(f)Z_{in}(f) \tag{15}$$

This injected current waveform, i.e. the lightning channel current is given by:

$$i(t) = I_0 \cdot (e^{-\alpha t} - e^{-\beta t}), \quad t \geq 0 \tag{16}$$

where pulse rise time is shaped by constants α and β, while I_0 is the amplitude of the current waveform.

The Fourier transform of the excitation function (26) is given by [12]:

$$I(f) = I_0 \cdot \left(\frac{1}{\alpha + j2\pi f} - \frac{1}{\beta + j2\pi f} \right) \tag{17}$$

Instead of solving convolution integral, as the time domain waveform of the input impedance (the impulse response of the system) is not known, the transient voltage is computed by applying the Inverse Fourier Transform (IFT).

A time domain voltage counterpart, i.e. the IFT of the function *V*(*f*) is defined by the integral [16]:

$$v(t) = \int_{-\infty}^{\infty} V(f) e^{j2\pi ft} d\omega \tag{18}$$

as the frequency response *V*(*f*) is represented by a discrete set of values the integral (18) cannot be evaluated analytically and the Discrete Fourier transform, in this case the Inverse Fast Fourier Transform (IFFT) algorithm, is used, i.e.:

$$v(t) = IFFT\left[V(f)\right] \tag{19}$$

Implementation of this algorithm generates an error due to discretization and truncation of unlimited frequency spectrum. The discrete set of time domain voltage values is given by [16]:

$$v(n\Delta t) = F \cdot \sum_{k=0}^{N-1} V(k\Delta f) e^{jk\Delta fn\Delta t} \tag{20}$$

where F denotes the highest frequency taken into account, N is the total number of frequency samples, Δf is sampling interval and Δt is the time step.

Therefore, the grounding electrode problem is related to the assessment of the input impedance.

The input impedance is given by the ratio:

$$Z_{in} = \frac{V_g}{I_g} \tag{21}$$

where V_g and I_g are the values of the voltage and the current at the driving point.

The feed-point voltage due to the unit current source can be calculated by integrating the normal electric field component from the electrode surface to infinity, i.e.:

$$V_g = -\int_{\infty}^{a} \vec{E} d\vec{s} \tag{22}$$

Repeating this procedure in the wide frequency band gives the input impedance spectrum.

For the particular case of horizontal grounding electrode integral (22) becomes:

$$V_g = -\int_{\infty}^{a} E_z^H \left(x, z\right) dz \tag{23}$$

where E_z^H is the radial electric field component, normal to the electrode:

$$E_z^H(x,z) = \frac{1}{j4\pi\omega\varepsilon_{eff}} \int_{-L/2}^{L/2} I(x') \frac{\partial^2 G^H(x,x',z)}{\partial x \partial z} dx' \tag{24}$$

Inserting equation (24) into (23) and performing some mathematical manipulation [6] yields:

$$V_g = \frac{1}{j4\pi\omega\varepsilon_{eff}} [I(-L/2)G^H(x,-L/2,z) - \int_{-L/2}^{L/2} \frac{\partial I(x')}{\partial x'} G^H(x,x',z)dx'] \Big|_{z=a}^{z=\infty} \tag{25}$$

and the input impedance of the grounding wire is determined by the relation:

$$Z_{in} = \frac{1}{j4\pi\omega\varepsilon_{eff} I_g} [I(-L/2)G^H(x,-L/2,z) - \int_{-L/2}^{L/2} \frac{\partial I(x')}{\partial x'} G^H(x,x',z)dx'] \Big|_{z=a}^{z=\infty} \tag{26}$$

and the desired impedance spectrum can be computed.

5 Measures for postprocessing the transient response

The well-established measure for analyzing the transient behaviour of the horizontal grounding electrode is the transient impedance which is defined as a ratio of transient voltage and current at the input terminals [4]:

$$z(x,t)\Big|_{x=-L/2} = \frac{v(x,t)}{i(x,t)}\Big|_{x=-L/2} \tag{27}$$

where $i(t)$ is the current injected at an end of the horizontal electrode, Fig. 1.

Once obtaining the transient voltage flowing through the horizontal electrode it is possible to calculate certain parameters providing additional measures of this transient response. Such parameters can be found in the theory of electric circuits.

The convenient parameters quantifying the horizontal electrode transient response, suggested in this work, are: instantaneous power and the total energy accumulated in the near field of the electrode.

According to the theory of electric circuits the amount of delivered power strongly depends on the particular waveform. Thus, a time varying current delivers an average power to a grounding electrode and it is given by the product:

$$p(x,t)\Big|_{x=-L/2} = u(x,t) \cdot i(x,t)\Big|_{x=-L/2} \tag{28}$$

The corresponding average power P_{av} is determined by the integral relation:

$$P_{av} = \frac{1}{T_0}\int_0^{T_0} p(x,t)\big|_{x=-L/2}\,dt = \frac{1}{T_0}\int_0^{T_0} u(x,t)\cdot i(x,t)\big|_{x=-L/2}\,dt \tag{29}$$

In accordance to the circuit theory, the energy in the near field of the grounding electrode can be determined by temporally integrating the instantaneous power:

$$W_{tot}(x,t)\big|_{x=-L/2} = \int_0^t p(x,t)\big|_{x=-L/2}\,dt \tag{30}$$

Total stored energy can be obtained by specifying $t=T_0$.

6 Computational example

Numerical results shown in Figures 2 to 9 are related to the grounding electrode of radius a=5mm buried at depth d=5m in a lossy medium (σ=0.1mS/m, ε_r=10). Figures 2 to 9 show transient voltage, transient impedance, instantaneous power and energy accumulated in the grounding electrode near field within the considered time interval of 10μs for the various set of parameters. Analyzing the proposed energy measures it is obvious that the transient response of the electrode is particularly dependent on its length.

7 Concluding remarks

Some measures for postprocessing of the horizontal grounding electrode transient response are presented in this work. The transient response of the electrode is obtained using the frequency domain antenna model of the grounding electrode and the Inverse Fourier Transform. The integral relationships arising from the wire antenna model are numerically treated by using the GB-IBEM.

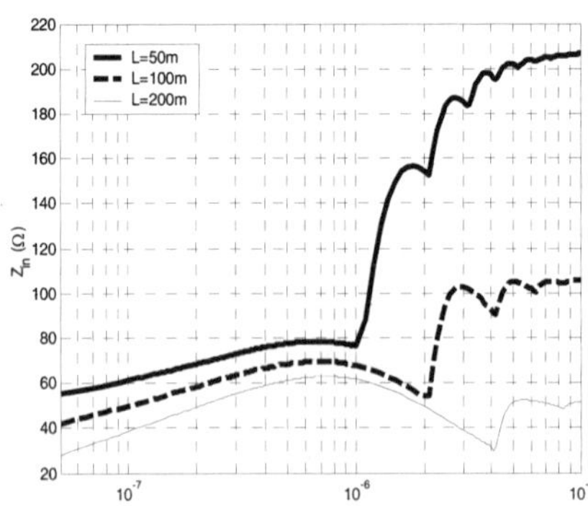

Figure 2: Transient impedance (d=5m, σ=0.0001S/m).

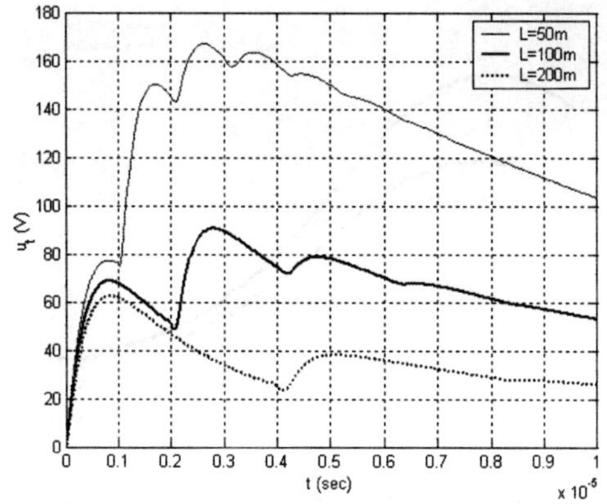

Figure 3: Transient voltage for various wire lengths (d=5m, σ=0.0001S/m).

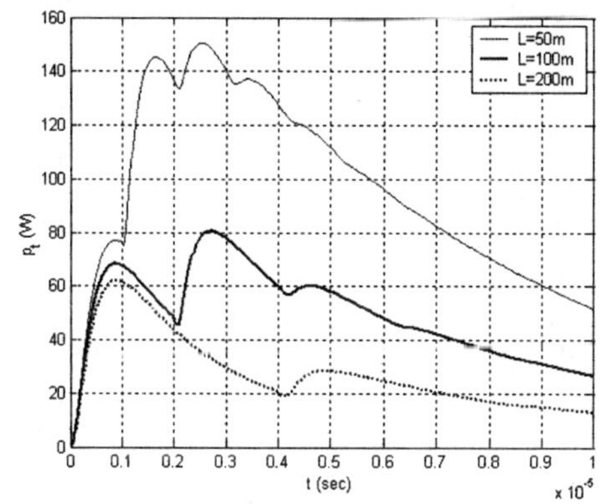

Figure 4: Instantaneous power for various wire lengths (d=5m, σ=0.0001S/m)

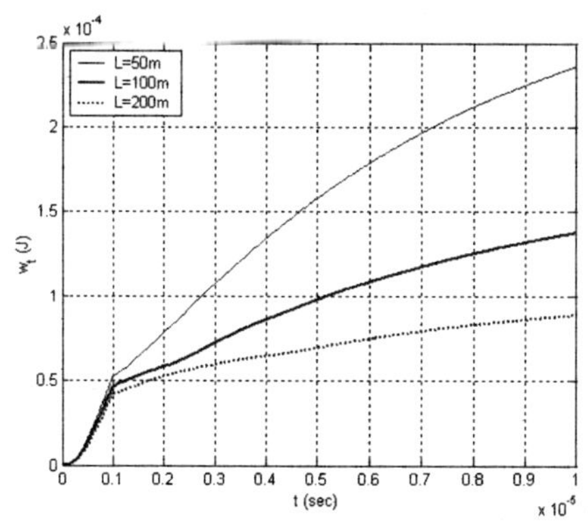

Figure 5: Total energy for various wire lengths (d=5m, σ=0.0001S/m).

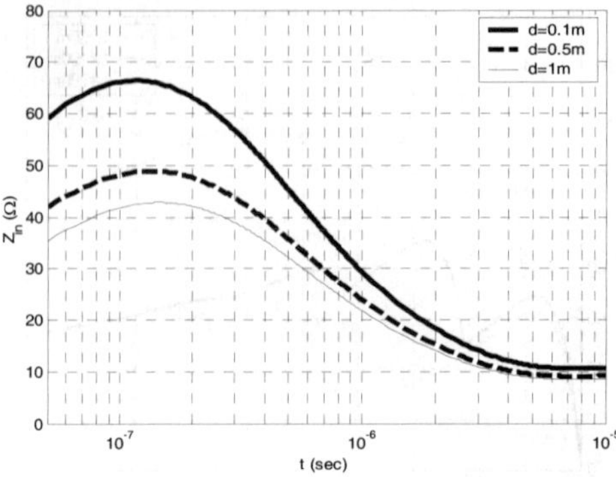

Figure 6: Transient impedance (L=100m, σ=0.001S/m).

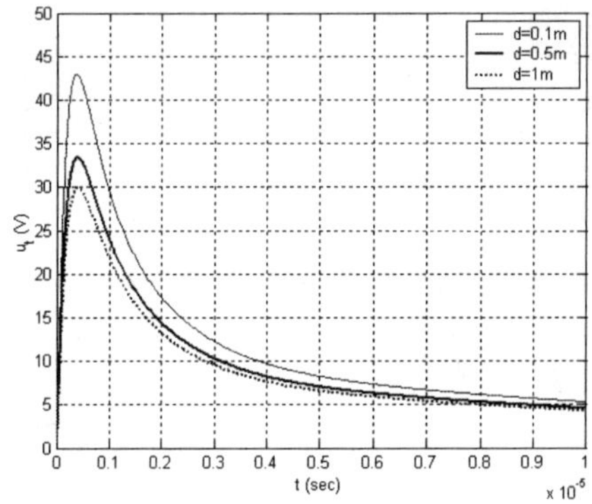

Figure 7: Transient voltage for various burial depths (L=100m, σ=0.001S/m).

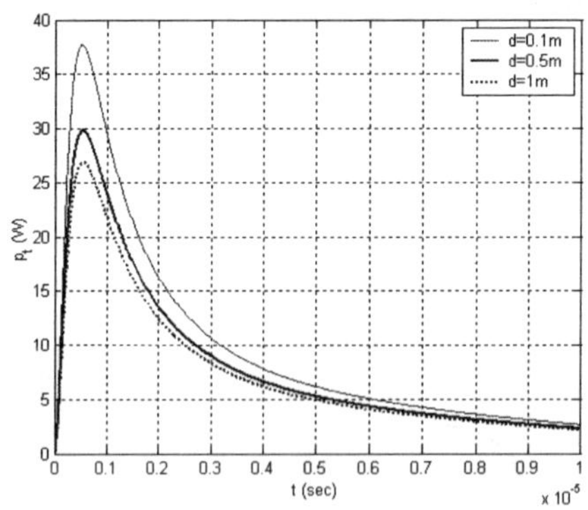

Figure 8: Instantaneous power for various burial depths (L=100m, σ=0.001S/m).

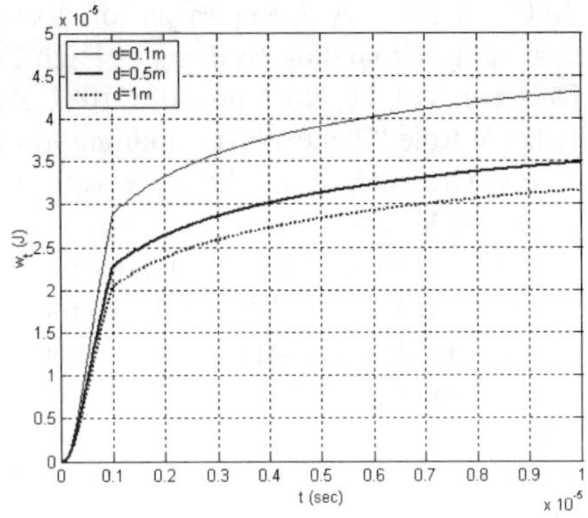

Figure 9: Total energy for various burial depths (L=100m, σ=0.001S/m).

Further to the standard transient impedance concept some additional measures for the transient response are presented in this paper. Once obtaining the transient response of the electrode it is possible to calculate the measures for quantifying the transient response in terms of the average power, instantaneous power and total energy stored in the electrode near field.

Further extension of the present analysis will involve the treatment of complex grounding systems including interconnected conductors.

References

[1] Y. Liu, M. Zitnik, R. Thottappillil, "An Improved Transmission Line Model of Grounding System", IEEE Trans. EMC, Vol.43, No.3, pp. 348-355, 2001.

[2] G. Ala, M. L. Di Silvestre, "A Simulation Model for Electromagnetic Transients in Lightning Protection Systems", IEEE Trans. EMC, Vol.44, No.4, pp.539-534, 2003.

[3] M.I. Lorentzou, N.D. Hatziargyriou, C.Papadias, "Time Domain Analyisis of Grounding Electrodes Impulse Response", IEEE Trans. Power Delivery, No 2., pp. 517-524, Apr. 2003.

[4] L. Grcev , F. Dawalibi, "An Electromagnetic Model for Transients in Grounding Systems", IEEE Trans. Power Delivery, No 4., pp. 1773-1781, Oct. 1990.

[5] L. D. Grcev, F.E. Menter, "Transient Electro-magnetic Fields Near Large Earthing Systems", IEEE Trans. Magnetics, Vol. 32, pp. 1525-1528, May 1996.

[6] D. Poljak, V. Roje, "The Integral equation method for ground wire impedance", Constanda, C., Saranen, J., Seikkala, S. (Ed), Integral methods in science and engineering, Vol. I, Longman, UK., 139-143, 1997.

[7] G. E. Bridges, "Transient Plane Wave Coupling to Bare and Insulated Cables Buried in a Lossy Half-Space", IEEE Trans. EMC, Vol. 37, No 1., pp. 62-70, Feb. 1995.

[8] R.G. Olsen, M.C. Willis, "A Comparison of Exact and Quasi-static Methods for Evaluating Grounding Systems at High Frequencies", IEEE Trans. Power Delivery, Vol. 11, No 2, pp. 1071-1081, April 1996.

[9] D. Poljak, I.Gizdic, V.Roje "Plane Wave Coupling to Finite Length Cables Buried in a Lossy Ground", Eng. Analysis with Boundary Elements, Vol.26, No.1, pp. 803-806, Jan . 2002.

[10] D. Poljak, R. Lucić, V. Roje, "Transient Analysis of Electromagnetic Field Coupling To Buried Cables" ICEAA 01 International Conference on Electromagnetics in Advanced Applications, Torino: 335-338, 2001.

[11] E.K. Miller , A.J. Poggio, G.J. Burke, E.S. Selden, "Analysis of Wire Antennas in the Presence of a Conducting Half-Space, Part II: The Horizontal Antenna in Free Space," Canadian Journal of Physics, 50, pp 2614-2627, 1972.

[12] D.Poljak, "Electromagnetic Modelling of Wire Antenna Structures", WIT Press, Southampton-Boston, 2002.

[13] D.Poljak, C.A. Brebbia, "Boundary Element methods for Electrical Engineers", WIT Press, Southampton-Boston, 2005.

[14] D.Poljak, V.Doric, "Time Domain Modeling of Electromagnetic Field Coupling to Finite Length Wires Embedded in a Dielectric Half-Space", IEEE Trans. EMC, 2005.

[15] R. Velazquez, D. Muhkedo, "Analytical Modeling of Grounding Electrodes Transient Behaviour", IEEE Trans. Power Appar. Systems, Vol. PAS-103, 1314-1322, June 1984.

[16] Ziemer, R.E., W.H. Tranter, Principles of Communications, Houghton Mifflin Company, Boston, 1995.

Section 11
Inverse problems

Singular superposition elastostatics BEM/GA algorithm for cavity detection

D. Ojeda[1,4], B. Gámez[1,4], E. Divo[2], A. Kassab[3] & M. Cerrolaza[4]
[1] *Department of Mechanical Design and Automation, University of Carabobo, Venezuela*
[2] *Department of Engineering Technology, University of Central Florida, USA*
[3] *Department of Mechanical, Materials, and Aerospace Engineering, University of Central Florida, USA*
[4] *National Institute of Bioengineering, Central University of Venezuela, Venezuela*

Abstract

A method for the efficient solution of the inverse geometric problem for cavity detection using a point load superposition technique in the elastostatics boundary element method (BEM) is presented in this paper. The superposition of point load clusters to simulate the presence of cavities offers tremendous advantages in reducing the computational time for the elastostatics field solution as no boundary re-discretization is necessary throughout the inverse problem solution process. The inverse solution is achieved in two steps: (1) fixing the location and strengths of the point loads, (2) locating the cavity(ies) geometry(ies). For a current estimated point load distribution, a first objective function measures the difference between BEM-computed and measured deformations at the measuring points. A Genetic Algorithm (GA) is employed to automatically alter the locations and strengths of the point sources to minimize the objective function. The GA is parallelized and dynamically balanced. Upon convergence, a second objective function is defined and minimized to locate the cavity(ies) location(s) indicated as traction-free surface(s). Results of cavity detection simulations using numerical experiments and simulated random measurement errors validate the approach in regular and irregular geometrical configurations with single and multiple cavities.
Keywords: boundary element method (BEM), cavity detection, genetic algorithm, elastostatics.

WIT Transactions on Modelling and Simulation, Vol 44,
www.witpress.com, ISSN 1743-355X (on-line)
doi:10.2495/BE070301

1 Introduction

The purpose of solving most inverse problem is to find the unknown: (a) in the governing equation for field variable, (b) physical properties, (c) boundary conditions, (d) initial condition(s), or (e) the system geometry using over-specified conditions. Typically, the over-specified conditions are provided by measuring a field variable at the exposed boundary, as in the case of the inverse geometric problem. However, in some inverse problems, the over-specified condition can be provided by internal measurements of field variable via embedded sensors (Divo *et al* [1], Ulrich *et al* [2], Kassab *et al* [3]). In this paper, such measurements, along with accompanying noise, are simulated numerically. The purpose of the inverse geometric problem, that concerns this study, is to determine the hidden portion of the system geometry by using over-specified boundary conditions on the exposed portion. This problem has gained importance in thermal and solid mechanics applications for nondestructive detection of subsurface cavities (Ulrich *et al* [2], Kassab *et al* [3, 4], Divo *et al* [5]). In thermal applications, the method requires over-specified boundary conditions at the surface, i.e., both temperature and flux must be given, Divo *et al* [1]. In elastostatics applications, the over-specified conditions are provided in terms of surface displacements and tractions. Generally, surface tractions are known boundary conditions, while the surface displacements are experimentally determined by measurements, (see Ulrich *et al* [2], Kassab *et al* [3]).

A variety of numerical methods have been used to solve the inverse geometric problem. This inverse problem has applications in the identification of surfaces flaws and cavities and in shape optimization problems (Divo *et al* [1, 5, 7–9], Ulrich *et al* [2], Kassab *et al* [3, 4], Bialecki *et al* [6]). The computational burden is intensive due to the inherent nature of the solution of inverse problem which requires numerous forward problems to be solved, regardless of whether a numerical or analytical approach is taken to solve the associated direct problem. Moreover, in the inverse geometric problem, a complete regeneration of the mesh is also necessary as the geometry evolves. Boundary Element Methods (BEM) lends themselves naturally to the numerical solution of the inverse problem (see Divo *et al* [1], Ulrich *et al* [2], Kassab *et al* [3], Cerrolaza *et al* [10], Müller-Karger *et al* [11], Annicchiarico *et al* [12–14], and Martinez and Cerrolaza [15]) and this is because the solution algorithms developed by researchers typically involve minimization of residuals, which measure the non-satisfaction of over-specified boundary conditions. Additionally, in the iterative solution of this problem the geometry is continuously updated. This places a premium on a numerical method, which does not require domain discretization (Brebbia and Dominguez [16]).

A method for the efficient solution of the inverse optimization problem of cavity detection using a point load superposition technique in elastostatics boundary element methods is presented in this paper. The superposition of point load clusters in the domain is posed as an alternative to satisfy the Cauchy conditions on the surface. The point loads must be located inside the eventual cavity or outside the domain in order to correctly satisfy the governing equation. Using

Genetics Algorithms, the point load distribution, strength, and location are altered to seek satisfaction of the over specified boundary displacements. Numerical results of direct 2D problems using the BEM are used as an alternative to validate the approach. Results of cavity detection problems simulated using numerical experiments and added random measurement errors validate the approach in regular and irregular geometrical configurations with single and multiple cavities.

2 Direct problem and BEM in elasticity

The solution of the forward elastostatics problem is expressed in terms of displacements which, for an isotropic, homogeneous, and linearly elastic medium imposed with an internal volumetric force b_i, is governed by Navier's equation as:

$$\mu\frac{\partial^2 u_i}{\partial x_i \partial x_j} + \left(\frac{\mu}{1-2\nu}\right)\frac{\partial^2 u_j}{\partial x_i \partial x_j} + b_i = 0 \tag{1}$$

Here, $\underset{\sim}{u}$ is the displacement vector, ν is Poisson's ratio, and μ is shear modulus. Introducing the fundamental solution to Navier's equation, a BEM formulation can be derived from the Somigliana identity providing an integral relation between the displacement vector u_i^p in a point collocation "p" and displacement vectors u_i and traction vectors t_i at the boundary Γ as well as the body forces b_i:

$$c_{ij}^p u_i^p + \oint_\Gamma H_{ij} u_i d\Gamma = \oint_\Gamma G_{ij} t_i d\Gamma + \int_\Omega G_{ij} b_i d\Omega \tag{2}$$

Here, Ω is the problem domain, G_{ij} and H_{ij} are the displacement and traction fundamental solutions (see Brebbia *et al* [16]). Establishing that internal force b_i is formed only by points loads, so that:

$$b_i = \sum_{l=1}^{NL} Q_i^l \delta(x_i, x_i^l) \tag{3}$$

where NL is the number of points loads, Q_i^l is the intensity of each load and $\delta(x_i, x_i^l)$ is the Dirac's delta function located in the impact point of each load x_i^l. Using the properties of the Dirac's delta function, the last integral equation term Eq. (2) is reduced to:

$$c_{ij}^p u_i^p + \oint_\Gamma H_{ij} u_i d\Gamma = \oint_\Gamma G_{ij} t_i d\Gamma + \sum_{l=1}^{NL} Q_i^l G_{ij}^l \tag{4}$$

Employing standard boundary element procedures, the above equation is written in discrete form as:

$$[H]\{u\} = [G]\{t\} + \{q\} \tag{5}$$

where the matrices $[H]$ and $[G]$, with dimensions $N \times N$, contain the influence coefficient that relate displacement and traction vectors $\{u\}$ and $\{t\}$ on the

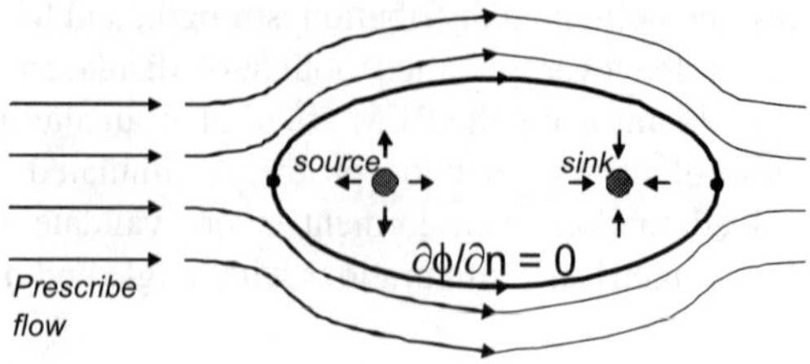

Figure 1: Simulation of elliptical surface with an incoming parallel flow by singularities superposition.

boundary. The size of N is $N = d \times NE \times NN$, where, d is the number of space dimensions (2 or 3), NE is the number of elements, and NN is the number of nodes per element. It is worth noting that all effects generated by points loads are located in the vector $\{q\}$, therefore, when point loads that are utilized in the inverse geometric problem solution are relocated in the evolving solution, there is no need for boundary remeshing.

Introducing the boundary condition $\overline{u_i}$ and $\overline{t_i}$ in Eq. (5), an algebraic system with the following form is obtained: $[A]\{x\} = \{b\} + \{q\}$. The vector $\{x\}$ contains the unknowns values of $\{u\}$ and $\{t\}$. This system of equations is solved using a standard method. In this paper, we use isoparametric-discontinuous-quadratic elements: the geometry and vectors $\{u\}$ and $\{t\}$ values are approximated using quadratic shape function locating the displacement and traction nodes within the element boundaries.

3 Cavity simulation with point loads

The approach proposed in this paper for the modeling of internal cavity(ies) is inspired from potential theory. For example, the superposition of a source and a sink with the same strength located a distance L in a prescribed parallel flow results in iso-flow lines and simulate the presence of a solid surface through the iso-lines containing stagnation points. This null-flow line can be interpreted as the presence of a solid surface or the artificial contour of an elliptical cavity, see Figure 1. The notion of utilizing point sources and sinks to model cavities has been successfully utilized by Divo *et al* [1] in solving the inverse geometric problem by thermal methods. This theory can be applied in elastostatics field where the interpretation is understood as superposition of point loads and the flow is considered to be that of elastic energy. With proper adjustment of the location, number, and intensities of these loads, one can generate a surface (or surfaces) that are traction-free and therefore interpreted as a cavity surface(s), this is illustrated in Figure 2, where a cluster of such point loads is shown.

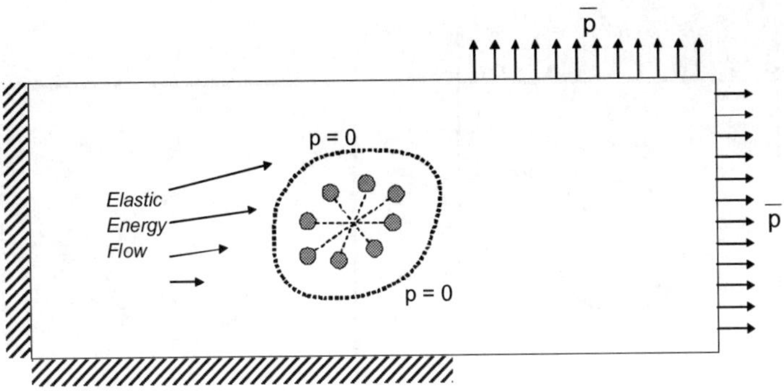

Figure 2: Geometric arrangement of points loads with a non-uniform elastic energy flow.

4 Inverse problem and objective function

The numerical inverse process for cavity detection is achieved in 2 steps: (1) fixing the location and strengths of the fictitious point loads, (2) locating the cavity(ies) geometry(ies). For a current estimated point load distribution, a first objective function measures the difference between BEM-computed and measured deformations at the measuring points. Since the governing equation for the elasticity problem is the Navier Equation without body forces, the fictitious point loads must be located outside the problem domain, that is outside the exposed bounding surface or within the subsurface cavities. As such, the first iteration process searches for locations and strengths of the fictitious point loads until a match is found between the tractions and deformations computed by the BEM and those measured on the boundary as additional information or over-specified conditions. This is achieved by the minimization of an objective function, $S1$, that quantifies the difference between the deformations u_i obtained by BEM (Eq. (5)) and measured deformations $\widehat{u}_i$ providing the additional information obtained through experimental measurements on the exposed boundaries (see Divo *et al* [1], Ulrich *et al* [2], Kassab *et al* [3, 4]). Upon convergence, a second objective function is defined and minimized to locate the cavity(ies) location(s) indicated as traction-free surface(s). A Genetic Algorithm (GA) is employed to solve both minimization problems and it is parallelized and dynamically balanced.

5 Method of optimization: GA

The GA used for this optimization process models the objective function as a haploid with a binary vector to model a single chromosome as described in Divo *et al* [1]. The length of the vector is dictated by the number of design variables and the required precision of each design variable. Each design variable has to be bounded with a minimum and a maximum value, and in the process the precision

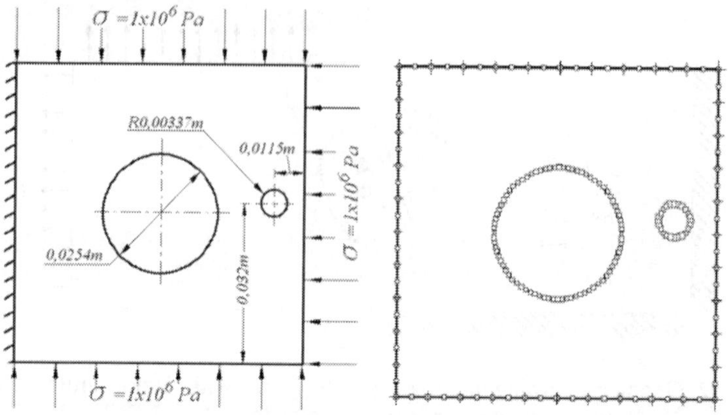

(a) Boundary conditions. (b) Discretization: 80 elements.

Figure 3: A square plate with centered circular hole of diameter $0.0254m$.

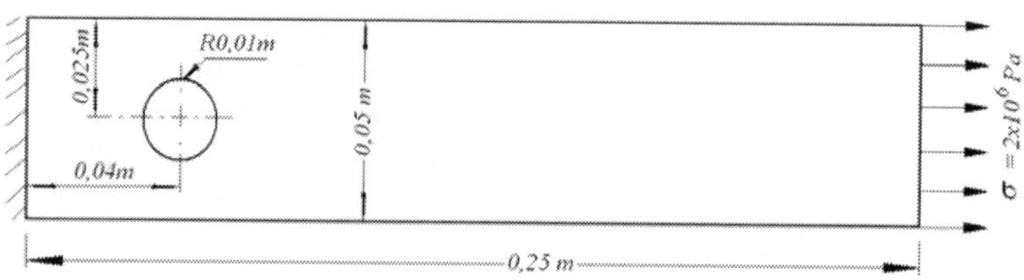

Figure 4: An elongated rectangular bar with a circular cavity under tension.

of the variable is determined. This procedure allows an easy mapping from real numbers to binary strings and vice versa. This coding process represented by a binary string is one of the distinguishing features of GA and differentiates them from other evolutionary approaches. The haploid GA place all design variables into one binary string, called a chromosome or offspring. The GA optimization process begins by setting a random set of possible solutions. Each individual is defined by parameters combinations, which in this case are $(Q_i^l, x_c, y_c, r_x, r_y, \theta)$ and is represented as a bit string or a chromosome. Since GA are used to maximize and not minimize, an aptitude function, Z, is formulated as the inverse of the objective function. This aptitude function Z is evaluated for every individual in the current population defining the fitness or their probability of survival. A series of parameters are initially set in the GA code, and these determine and affect the performance of the genetic optimization process.

Figure 5: Discretization of the rectangular bar under tension using 120 elements.

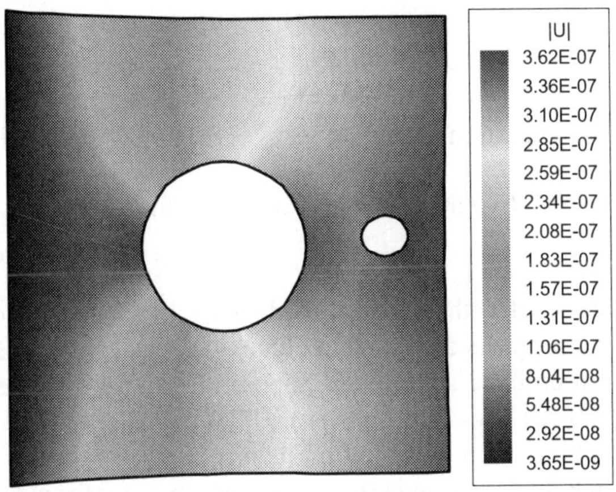

(a) Example 1.

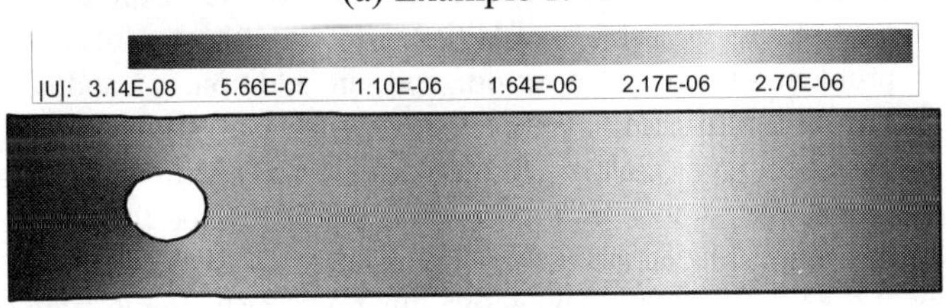

(c) Example 2.

Figure 6: Contour of BEM-computed displacements, in $[m]$.

6 Numerical examples

Results obtained from solving forward problems are used to generate the displacement results at the external surface to simulate experimental measurements. The latter are ladened with random error to simulate noise. Both forward and inverse problems use the BEM as the field solver. The BEM model uses discontinuous quadratic elements with adaptive quadratures. The first example, displayed in Figure 3, considers a $0.0635 \times 0.0635 m^2$ square plate with a $0.0254 m$ diameter centered hole and a $0.00674 m$ diameter cavity. The plate is

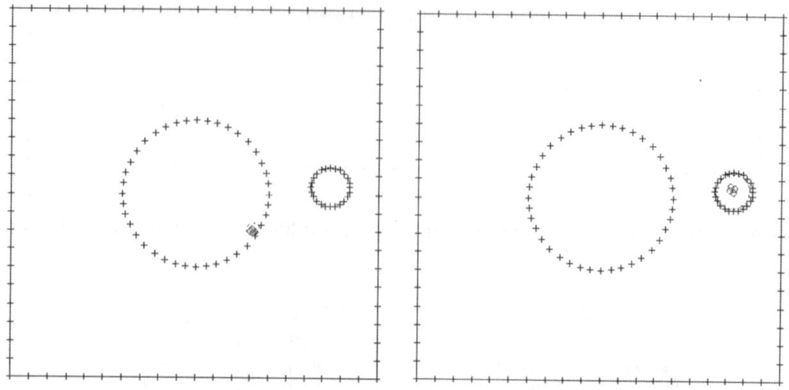

(a) 2nd generation (b) 2600th generation

Figure 7: Evolution of cavity detection for example.

clamped in one side and the others are imposed with a $10^6 Pa$ uniform compression loads. The second example is displayed in Figure 4: a rectangular $0.25 \times 0.05m^2$ clamped bar with a uniform $2 \times 10^6 Pa$ imposed traction load. The problem is discretized with 120 elements. The bar has a $0.02m$ diameter cavity located $0.4m$ from the left-end of the bar, see Figure 5. The displacements for the two forward problems shown in Figures 6(a) and 6(b) are used as inputs for the inverse problem.

The computed surface displacements at the traction-free exposed surfaces in both of the problems are used as additional information to solve the inverse geometric problem of cavity detection, and, in addition, a random error of $\pm 1x10^{-8}m$ is added to these surface displacements to mimic measurement error. The location of the cluster of singularities along with the strengths and cluster evolution, are shown in Figures 7(a) and 7(b) for the first example. Similarly, these are displayed in Figures 8(a) and 8(b) for the second example. The evolutions of the first and second objective functions for Example 1 are shown in Figures 9(a) and 9(b). In these two examples, the approach proposed in this paper is demonstrated to be successful in locating subsurface cavities using an inverse elastostatics BEM-point load superposition method.

7 Conclusions

A method for the efficient solution of the inverse optimization problem of cavity detection using a point load superposition technique in elastostatics boundary element methods is development in this paper. Two examples demonstrate the ability of the method to successfully locate single cavities in terms of their locations and size whilst using inputs ladened with simulated random error. The GA has been integrated as optimization tool using BEM. The technique is readily applicable to the closely related problem of shape optimization, in which the condition at the cavity side may be arbitrarily specified as a design target.

(a) 2nd generation

(b) 3200th generation

Figure 8: Evolution of cavity detection for example 2.

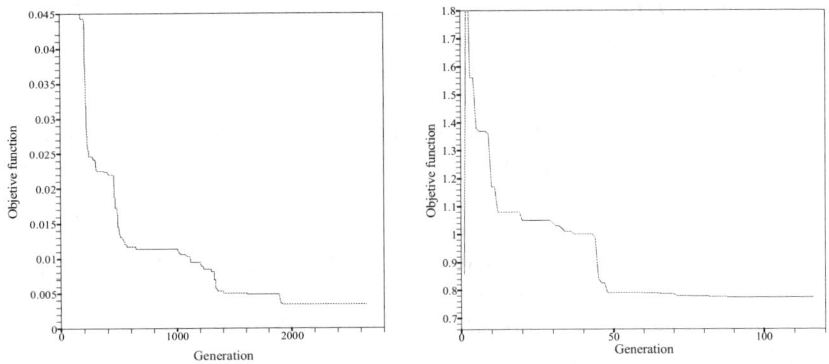

(a) 1st Objective function evolution. (b) 2nd Objective function evolution.

Figure 9: Objective function evolution for example 1.

Acknowledgements

The work undertaken in this project was carried out under the institutional and financial support provided by the University of Central Florida (USA), the University of Carabobo (Venezuela), and FONACIT (Venezuela).

References

[1] Divo E.A., Kassab A.J., Rodríguez F., An efficient singular superposition technique for cavity detection and shape optimization. *Numerical Heat Transfer*, Part B, 46: 1–30, 2004, Copyright Taylor & Francis Inc.

[2] Ulrich T.W., Moslehy F.A. and Kassab A.J., A BEM based pattern search solution for a class of inverse elastostatic problems. *Int. J. Solids Structures*, Vol. 33, No. 15, pp. 2123-2131, 1996, Copyright 1996.

[3] Kassab A.J., Moslehy F.A., Daryapurkar A.B., Nondestructive detection of cavities by an inverse elastostatics boundary element method. *J. Engineering Analysis with Boundary Elements* 13 (1994) 45–55.

[4] Kassab A.J., Moslehy F.A., Ulrich T.W., Inverse boundary element solution for locating subsurface cavities in thermal and elastostatic problems. In *Proc. IABEM-95, Computational Mechanics '95*, Hawaii, July 30-August 3 (ed. Atluri, Yagawa and Cruse), pp. 3024–3029, Springer, Berlin.

[5] Divo E.A., Kassab A.J., Rodríguez F., Characterization of space dependent thermal conductivity with a BEM-Based genetic algorithm. Institute for Computational Engineering, University of Central Florida, Orlando, Florida, 32816-2450, *Numerical Heat Transfer*, Part A: Applications, Vol. 37, No. 8, (2000), pp. 845–877.

[6] Bialecki R. Divo E. Kassab A., Reconstruction of time-dependent boundary heat flux by a BEM-based inverse algorithm. *J. Engineering Analysis with Boundary Elements* 30 (2006) 767–773.

[7] Divo E.A., Kassab A.J., A meshless method for conjugate heat transfer problems. *J. Eng. Analysis with Boundary Elements* 29 (2005) 136–149.

[8] Divo E.A., Kassab A.J., Ingber M.S., Shape optimization of acoustic scattering bodies. *J. Engineering Analysis with Boundary Elements* 27 (2003), 695–703.

[9] Divo E.A., Kassab A.J., Kapat J.S., Chyu Ming-King, Retrieval of multidimensional heat transfer coefficient distributions using an inverse BEM-based regularized algorithm: numerical and experimental results. *J. Engineering Analysis with Boundary Elements* 29 (2005), 150–160.

[10] Cerrolaza M., Annicchiarico W. and Martinez M., Optimization of 2D boundary element models using β-splines and genetic algorithms, *Engineering Anal. with Bound. Elem.*, 24(5): (2000), 427–440.

[11] Müller-Karger C., González C., Aliabadi M.H. and Cerrolaza M., Three dimensional BEM and FEM stress analysis of the human tibia under pathological conditions, *J. of Comp. Mod. In Eng. and Sciences*, 2(1): (2001), 1–13.

[12] Annicchiarico W. And Cerrolaza M., An Evolutionary Approach for the Shape Optimization of General Boundary Elements Models. *Electronic Journal of Boundary Elements*, Vol.2, (2002)

[13] Annicchiarico W. and Cerrolaza M., A 3D boundary element optimization approach based on genetic algorithms and surface modeling, *Eng. Anal. With Bound. Elem.*, Vol. 28(11), (2004), pp. 1351–1361

[14] Annicchiarico W., Martínez G., Cerrolaza M., Boundary elements and β-spline modeling for medical applications, *J. of App. Math. Mod.*, (2005), (in press)

[15] Martínez G. and Cerrolaza M., A bone adaptation integrated approach using BEM, *J. Eng. Anal. With Bound. Elem.* 30, (2006), 107–115

[16] Brebbia C.A., Dominguez J., Boundary element, An introductory course. *Computational Mechanics Publ.*, pp. 134–250, Boston, co-published with McGraw-Hill, New York, 1989.

Numerical solution of an inverse problem in magnetic resonance imaging using a regularized higher-order boundary element method

L. Marin[1], H. Power[1], R. W. Bowtell[2], C. Cobos Sanchez[2], A. A. Becker[1], P. Glover[2] & I. A. Jones[1]
[1] *School of Mechanical, Materials and Manufacturing Engineering, The University of Nottingham, Nottingham, UK*
[2] *Sir Peter Mansfield Magnetic Resonance Centre, School of Physics and Astronomy, The University of Nottingham, Nottingham, UK*

Abstract

We investigate the reconstruction of a divergence-free surface current distribution from knowledge of the magnetic flux density in a prescribed region of interest in the framework of static electromagnetism. This inverse problem is motivated by the design of gradient coils used in magnetic resonance imaging (MRI) and is formulated using its corresponding integral representation according to potential theory. A novel higher-order boundary element method (BEM) which satisfies the continuity equation for the current density, i.e. divergence-free BEM, is also presented. Since the discretised BEM system is ill-posed and hence the associated least-squares solution may be inaccurate and/or physically meaningless, the Tikhonov regularization method is employed in order to retrieve accurate and physically correct solutions.
Keywords: inverse problem, regularization, divergence-free BEM, magnetic resonance imaging (MRI).

1 Introduction

Magnetic resonance imaging (MRI) is a non-invasive technique for imaging the human body, which has revolutionised the field of diagnostic medicine. MRI relies on the generation of highly controlled magnetic fields that are essential to the process of image production. In particular, an extremely homogeneous, strong, static

WIT Transactions on Modelling and Simulation, Vol 44, © 2007 WIT Press
www.witpress.com, ISSN 1743-355X (on-line)
doi:10.2495/BE070311

field is required to polarize the sample and provide a uniform frequency of precession, while pure field gradients are needed to encode the spatial origin of signals. The field gradients are generated by carefully arranged wire distributions generally placed on cylindrical surfaces surrounding the imaging subject, known as gradient coils [1–3].

2 Mathematical formulation

In a non-magnetic material, as is the case of biological tissue, the magnetic flux density $\mathbf{B} = (\mathrm{B_x}, \mathrm{B_y}, \mathrm{B_z})^{\mathrm{T}}$ satisfies the following system of partial differential equations [4]:

$$\nabla \times \mathbf{B}(\mathbf{x}) = \mu_0 \mathbf{J}(\mathbf{x}), \qquad \nabla \cdot \mathbf{B}(\mathbf{x}) = 0, \qquad \mathbf{x} = (\mathrm{x}, \mathrm{y}, \mathrm{z})^{\mathrm{T}} \in \mathbb{R}^3. \quad (1)$$

Here $\mu_0 = 4\pi \times 10^{-7}\,\mathrm{N/A^2}$ is the permeability of the free-space and $\mathbf{J} = (\mathrm{J_x}, \mathrm{J_y}, \mathrm{J_z})^{\mathrm{T}}$ is the current density which is defined as a surface current density $\mathbf{J}^{\mathrm{coil}} = (\mathrm{J_x^{coil}}, \mathrm{J_y^{coil}}, \mathrm{J_z^{coil}})^{\mathrm{T}}$, i.e.

$$\mathbf{J}(\mathbf{x}) = \mathbf{J}^{\mathrm{coil}}(\mathbf{x}')\,\delta(\mathbf{x}', \mathbf{x}), \qquad \mathbf{x} \in \mathbb{R}^3, \quad \mathbf{x}' \in \Gamma_{\mathrm{coil}}, \quad (2)$$

where $\Gamma_{\mathrm{coil}} \subset \mathbb{R}^3$ is the coil surface and $\delta(\mathbf{x}', \mathbf{x})$ is the Kronecker delta function, such that

$$\nabla \cdot \mathbf{J}^{\mathrm{coil}}(\mathbf{x}) = 0, \qquad \mathbf{J}^{\mathrm{coil}}(\mathbf{x}) \cdot \boldsymbol{\nu}(\mathbf{x}) = 0, \qquad \mathbf{x} \in \Gamma_{\mathrm{coil}}, \quad (3)$$

with $\boldsymbol{\nu}$ the outward unit vector normal to the coil surface Γ_{coil}.

If the vector potential $\mathbf{A} = (\mathrm{A_x}, \mathrm{A_y}, \mathrm{A_z})^{\mathrm{T}}$ is introduced as:

$$\mathbf{B}(\mathbf{x}) = \nabla \times \mathbf{A}(\mathbf{x}), \qquad \mathbf{x} \in \mathbb{R}^3, \quad (4)$$

then the system of partial differential equations (1) reduces to the following Poisson equation for the vector potential $\mathbf{A}$:

$$\nabla^2 \mathbf{A}(\mathbf{x}) = \mu_0 \mathbf{J}(\mathbf{x}), \qquad \mathbf{x} \in \mathbb{R}^3. \quad (5)$$

In the *direct problem* formulation, the current density $\mathbf{J}^{\mathrm{coil}}$ is known on the coil surface Γ_{coil} and satisfies condition (3), whilst the vector potential $\mathbf{A}$ is determined from the Poisson equation (5) by employing its integral representation, namely

$$\mathbf{A}(\mathbf{x}) = \frac{\mu_0}{4\pi} \int_{\mathbb{R}^3} \frac{\mathbf{J}(\mathbf{x}')}{|\mathbf{x} - \mathbf{x}'|} \mathrm{d}\mathbf{x}' = \frac{\mu_0}{4\pi} \int_{\Gamma_{\mathrm{coil}}} \frac{\mathbf{J}^{\mathrm{coil}}(\mathbf{x}')}{|\mathbf{x} - \mathbf{x}'|} \mathrm{d}\Gamma(\mathbf{x}'), \qquad \mathbf{x} \in \mathbb{R}^3. \quad (6)$$

On using eqns. (4) and (6), the magnetic flux density may be recast as

$$\mathbf{B}(\mathbf{x}) = \frac{\mu_0}{4\pi} \int_{\Gamma_{\mathrm{coil}}} \frac{-(\mathbf{x} - \mathbf{x}') \times \mathbf{J}^{\mathrm{coil}}(\mathbf{x}')}{|\mathbf{x} - \mathbf{x}'|^3} \mathrm{d}\Gamma(\mathbf{x}'), \qquad \mathbf{x} \in \mathbb{R}^3. \quad (7)$$

Motivated by the design of gradient coils used in MRI, we investigate the reconstruction of the divergence-free surface current distribution $\mathbf{J}^{\mathrm{coil}}$ from knowledge of one component of the magnetic flux density $\mathbf{B}$ in a prescribed region of interest $\Omega \subset \mathbb{R}^3$, i.e. we focus on the following *inverse problem*:

$$\begin{aligned} &\text{Given } \widetilde{\mathrm{B}}_{\mathrm{z}}(\mathbf{x}),\ \mathbf{x}\in\Omega,\ \text{ find } \mathbf{J}^{\mathrm{coil}}(\mathbf{x}),\ \mathbf{x}\in\Gamma_{\mathrm{coil}},\ \text{ such that:}\\ &\mathrm{B}_{\mathrm{z}}(\mathbf{x}) = \widetilde{\mathrm{B}}_{\mathrm{z}}(\mathbf{x}),\ \mathbf{x}\in\Omega,\\ &\nabla\cdot\mathbf{J}^{\mathrm{coil}}(\mathbf{x}) = 0,\ \mathbf{J}^{\mathrm{coil}}(\mathbf{x})\cdot\boldsymbol{\nu}(\mathbf{x}) = 0,\ \mathbf{x}\in\Gamma_{\mathrm{coil}}. \end{aligned} \tag{8}$$

3 Divergence-free BEM

Assume that the coil surface Γ_{coil} is approximated as $\Gamma_{\mathrm{coil}} \approx \bigcup_{\mathrm{n}=1}^{\mathrm{N}} \Gamma_{\mathrm{n}}$, where Γ_{n}, $1 \le \mathrm{n} \le \mathrm{N}$, are triangular boundary elements (not necessarily flat). In the sequel, we use the following notation:

- $\Gamma_{\mathrm{n}} := \triangle\mathbf{x}^{\mathrm{n}1}\mathbf{x}^{\mathrm{n}2}\mathbf{x}^{\mathrm{n}3}$, $1 \le \mathrm{n} \le \mathrm{N}$, triangular boundary elements;
- $\mathbf{x}^{\mathrm{nj}}$, $1 \le \mathrm{j} \le \mathrm{N_e}$, local nodes corresponding to the triangular boundary element Γ_{n}, e.g. $\mathrm{N_e} = 3$, $\mathrm{N_e} = 6$ and $\mathrm{N_e} = 10$ in the case of linear, quadratic and cubic triangular boundary elements, respectively;
- $\mathbf{x}^{\mathrm{nj}}$, $1 \le \mathrm{j} \le 3$, vertices of the triangular boundary element Γ_{n};
- Γ_{nj}, $1 \le \mathrm{j} \le 3$, the edge of the triangular boundary element Γ_{n} opposite to the vertex $\mathbf{x}^{\mathrm{nj}}$, $1 \le \mathrm{j} \le 3$;
- N the number of triangular boundary elements;
- M the number of global nodes on the coil surface Γ_{coil};
- $\mathrm{N_e}$ the number of local nodes corresponding to each triangular boundary element Γ_{n}.

3.1 Geometry of the BEM

The parametrization of the triangular boundary elements is given by

$$\begin{aligned} &(\xi,\eta) \in \left\{ (\xi,\eta)\,|\,\xi \ge 0, \eta \ge 0, \xi+\eta \le 1 \right\} \longmapsto \mathbf{x}(\xi,\eta) \in \Gamma_{\mathrm{n}}\\ &\mathbf{x}(\xi,\eta) = \sum_{\mathrm{j}=1}^{\mathrm{N_e}} \mathrm{N_j}(\xi,\eta)\,\mathbf{x}^{\mathrm{nj}}\,, \end{aligned} \tag{9}$$

where $\mathrm{N_j}(\xi,\eta)$, $1 \le \mathrm{j} \le \mathrm{N_e}$, are given geometrical shape functions [5]. Consequently, the derivatives in the ξ- and η-directions may be recast as:

$$\begin{cases} \boldsymbol{\tau}^{\mathrm{n}\xi}(\xi,\eta) := \boldsymbol{\tau}^{\mathrm{n}\xi}\left(\mathbf{x}(\xi,\eta)\right) = \dfrac{\partial\mathbf{x}(\xi,\eta)}{\partial\xi} = \displaystyle\sum_{\mathrm{j}=1}^{\mathrm{N_e}} \dfrac{\partial\mathrm{N_j}(\xi,\eta)}{\partial\xi}\,\mathbf{x}^{\mathrm{nj}}\\ \boldsymbol{\tau}^{\mathrm{n}\eta}(\xi,\eta) := \boldsymbol{\tau}^{\mathrm{n}\eta}\left(\mathbf{x}(\xi,\eta)\right) = \dfrac{\partial\mathbf{x}(\xi,\eta)}{\partial\eta} = \displaystyle\sum_{\mathrm{j}=1}^{\mathrm{N_e}} \dfrac{\partial\mathrm{N_j}(\xi,\eta)}{\partial\eta}\,\mathbf{x}^{\mathrm{nj}}. \end{cases} \tag{10}$$

Then the surface metric (Jacobian) J^{n} and the outward unit vector normal $\boldsymbol{\nu}^{\mathrm{n}}$ to the triangular boundary element Γ_{n} are given by:

$$\mathrm{J}^{\mathrm{n}}(\xi,\eta) := \mathrm{J}^{\mathrm{n}}\left(\mathbf{x}(\xi,\eta)\right) = |\boldsymbol{\tau}^{\mathrm{n}\xi}(\xi,\eta) \times \boldsymbol{\tau}^{\mathrm{n}\eta}(\xi,\eta)| \tag{11}$$

and

$$\boldsymbol{\nu}^{\mathrm{n}}(\xi,\eta) := \boldsymbol{\nu}^{\mathrm{n}}\left(\mathbf{x}(\xi,\eta)\right) = \frac{1}{\mathrm{J}^{\mathrm{n}}(\xi,\eta)}\boldsymbol{\tau}^{\mathrm{n}\xi}(\xi,\eta) \times \boldsymbol{\tau}^{\mathrm{n}\eta}(\xi,\eta) \tag{12}$$

respectively.

3.2 Basis functions

On every triangular boundary element Γ_{n}, we define the following vectors:

$$\begin{cases} \mathbf{v}^{\mathrm{n}1}(\xi,\eta) := \mathbf{v}^{\mathrm{n}1}\left(\mathbf{x}(\xi,\eta)\right) = -\dfrac{1}{\mathrm{J}^{\mathrm{n}}(\xi,\eta)}\,\boldsymbol{\tau}^{\mathrm{n}\eta}(\xi,\eta) \\ \mathbf{v}^{\mathrm{n}2}(\xi,\eta) := \mathbf{v}^{\mathrm{n}2}\left(\mathbf{x}(\xi,\eta)\right) = \dfrac{1}{\mathrm{J}^{\mathrm{n}}(\xi,\eta)}\,\boldsymbol{\tau}^{\mathrm{n}\xi}(\xi,\eta) \\ \mathbf{v}^{\mathrm{n}3}(\xi,\eta) := \mathbf{v}^{\mathrm{n}3}\left(\mathbf{x}(\xi,\eta)\right) = \dfrac{1}{\mathrm{J}^{\mathrm{n}}(\xi,\eta)}\,\left[-\boldsymbol{\tau}^{\mathrm{n}\xi}(\xi,\eta) + \boldsymbol{\tau}^{\mathrm{n}\eta}(\xi,\eta)\right]. \end{cases} \tag{13}$$

From definition (13), it follows that the vectors $\mathbf{v}^{\mathrm{ni}}(\xi,\eta)$ satisfy the identity:

$$\sum_{\mathrm{i}=1}^{3} \mathbf{v}^{\mathrm{ni}}(\xi,\eta) = 0 \text{ for } \mathbf{x} = \mathbf{x}(\xi,\eta) \in \Gamma_{\mathrm{n}}. \tag{14}$$

Next, we define the incidence function i as follows:

$$\begin{gathered} i(\cdot,\cdot) : \{1,2,\ldots,\mathrm{M}\} \times \{1,2,\ldots,\mathrm{N}\} \longrightarrow \{0,1,2,3\} \\ (\mathrm{m},\mathrm{n}) \longmapsto i(\mathrm{m},\mathrm{n}) = \begin{cases} 0 & \text{if } \mathbf{x}^{\mathrm{m}} \neq \mathbf{x}^{\mathrm{nj}},\ \forall \mathrm{j} \in \{1,2,3\} \\ \mathrm{j} & \text{if } \exists \mathrm{j} \in \{1,2,3\} : \mathbf{x}^{\mathrm{m}} = \mathbf{x}^{\mathrm{nj}}. \end{cases} \end{gathered} \tag{15}$$

For every global node $\mathbf{x}^{\mathrm{m}}$, $1 \leq \mathrm{m} \leq \mathrm{M}$, we define the set $\mathrm{C}_{\mathrm{m}} \subset \Gamma_{\mathrm{coil}}$ of triangular boundary elements Γ_{n}, $1 \leq \mathrm{n} \leq \mathrm{N}$, adjacent to $\mathbf{x}^{\mathrm{m}}$, i.e.

$$\mathrm{C}_{\mathrm{m}} := \bigcup_{\substack{\mathrm{n}=1 \\ i(\mathrm{m},\mathrm{n}) \neq 0}}^{\mathrm{N}} \Gamma_{\mathrm{n}}, \qquad 1 \leq \mathrm{m} \leq \mathrm{M}. \tag{16}$$

The vector basis function $\mathbf{f}^{\mathrm{m}}$ associated to the global node $\mathbf{x}^{\mathrm{m}}$ is defined by

$$\mathbf{f}^{\mathrm{m}}(\cdot) : \Gamma_{\mathrm{coil}} \longrightarrow \mathbb{R}^3, \qquad \mathbf{f}^{\mathrm{m}}(\mathbf{x}) = \begin{cases} \mathbf{v}^{\mathrm{n},i(\mathrm{m},\mathrm{n})}(\mathbf{x}) & \text{if } \mathbf{x} \in \mathrm{C}_{\mathrm{m}} \\ \mathbf{0} & \text{if } \mathbf{x} \notin \mathrm{C}_{\mathrm{m}} \end{cases} \tag{17}$$

and, clearly, its support is a subset of C_{m}, i.e. $\{\mathbf{x} \in \Gamma_{\mathrm{coil}}\, |\mathbf{f}^{\mathrm{m}}(\mathbf{x}) \neq \mathbf{0}\} \subset \mathrm{C}_{\mathrm{m}}$.

3.3 Surface current density

The current density $\mathbf{J}^{\rm coil}$ on the coil surface $\Gamma_{\rm coil}$ is then approximated by

$$\mathbf{J}^{\rm coil}(\mathbf{x}) \approx \sum_{\rm m=1}^{\rm M} \rm I_m \mathbf{f}^m(\mathbf{x}) = \sum_{\rm m=1}^{\rm M} \rm I_m \sum_{\substack{\rm n=1 \\ i(\rm m,n)\neq 0}}^{\rm N} \mathbf{v}^{\mathrm{n},i(\mathrm{m,n})}(\mathbf{x}), \quad \mathbf{x} \in \Gamma_{\rm coil}, \tag{18}$$

where $\rm I_m \in \mathbb{R}$, $1 \leq \rm m \leq M$, are unknown coefficients that correspond to the stream function intensities. For direct problems, the stream function intensities are determined from appropriate boundary conditions, while in the case of inverse problems, they are obtained by solving a minimisation problem.

It should be noted that the degree of the approximation (18) for the surface current density $\mathbf{J}^{\rm coil}$ is one degree less than the degree of the triangular boundary elements $\Gamma_{\rm n}$, $1 \leq \rm n \leq N$, since the vectors $\mathbf{v}^{\rm ni}(\xi,\eta)$, $1 \leq \rm i \leq 3$, are related to the derivatives of the geometrical shape functions $\rm N_i(\xi,\eta)$, $1 \leq \rm i \leq N_e$, associated with the triangular boundary element $\Gamma_{\rm n}$, see eqns. (9) – (13). More precisely, linear, quadratic and cubic triangular boundary elements provide constant, linear and quadratic approximations for the surface current density, respectively. From eqns. (12) and (13) it follows that for every triangular boundary element $\Gamma_{\rm n}$ the vectors $\mathbf{v}^{\rm ni}(\xi,\eta)$, $1 \leq \rm i \leq 3$, and the outward unit normal vector $\boldsymbol{\nu}^{\rm n}(\xi,\eta)$ are orthogonal and hence expression (18) enforces the approximated current density $\mathbf{J}^{\rm coil}$ to lie in the plane tangent to the coil surface $\Gamma_{\rm coil}$, i.e. condition (3_2) is satisfied. Furthermore, the interpolation given by eqn. (18) is divergence-free pointwise, i.e. condition (3_1) is satisfied, since $\nabla \cdot \frac{\partial \mathbf{x}}{\partial \xi} = \frac{\partial}{\partial \xi}(\nabla \cdot \mathbf{x}) = 0$ and $\nabla \cdot \frac{\partial \mathbf{x}}{\partial \eta} = \frac{\partial}{\partial \eta}(\nabla \cdot \mathbf{x}) = 0$.

3.4 Magnetic vector potential and magnetic flux density

According to eqns. (6), (7) and (18), the magnetic vector potential $\mathbf{A}$ and magnetic flux density $\mathbf{B}$ are approximated by

$$\mathbf{A}(\mathbf{x}) \approx \frac{\mu_0}{4\pi} \sum_{\rm m=1}^{\rm M} \rm I_m \sum_{\substack{\rm n=1 \\ i(\rm m,n)\neq 0}}^{\rm N} \int_{\Gamma_{\rm n}} \frac{\mathbf{v}^{\mathrm{n},i(\mathrm{m,n})}(\mathbf{x}')}{|\mathbf{x}-\mathbf{x}'|}\, \mathrm{d}\Gamma(\mathbf{x}'), \qquad \mathbf{x} \in \mathbb{R}^3 \tag{19}$$

and

$$\mathbf{B}(\mathbf{x}) \approx \frac{\mu_0}{4\pi} \sum_{\rm m=1}^{\rm M} \rm I_m \sum_{\substack{\rm n=1 \\ i(\rm m,n)\neq 0}}^{\rm N} \int_{\Gamma_{\rm n}} \frac{-(\mathbf{x}-\mathbf{x}') \times \mathbf{v}^{\mathrm{n},i(\mathrm{m,n})}(\mathbf{x}')}{|\mathbf{x}-\mathbf{x}'|^3}\, \mathrm{d}\Gamma(\mathbf{x}'), \qquad \mathbf{x} \in \mathbb{R}^3. \tag{20}$$

4 Description of the algorithm

If the z-component of the magnetic flux density $\mathbf{B}$ is known at L points in the region of interest Ω then the BEM discretisation of the inverse problem (8) yields the following system of linear algebraic equations

$$\mathbf{H}\,\mathbf{I} = \widetilde{\mathbf{B}}_z. \tag{21}$$

Here $\mathbf{H} \in \mathbb{R}^{L\times M}$ is the BEM matrix used for computing the z-component of the magnetic flux density $\mathbf{B}$ given by eqn. (20) calculated at L points in the region of interest Ω, $\widetilde{\mathbf{B}}_z = (\widetilde{B}_z^1, \ldots, \widetilde{B}_z^L)^T \in \mathbb{R}^L$ is a vector containing the z-component of the magnetic flux density at L points in the region of interest Ω and $\mathbf{I} \in \mathbb{R}^M$ is a vector containing the unknown values of the stream function I_m, $1 \le m \le M$, at the global nodes.

The system of linear algebraic equations (21) cannot be solved by direct methods, such as the least-squares method, since such an approach would produce an inaccurate and/or physically meaningless solution due to the large value of the condition number of the system matrix $\mathbf{H}$ which increases dramatically as the BEM mesh is refined. Several regularization procedures have been developed to solve such ill-conditioned systems [6, 7]. In the sequel, we only consider the Tikhonov regularization method and for further details on this method, we refer the reader to [6].

4.1 Magnetic energy and regularization

The magnetic energy W defined by

$$W = \frac{1}{2}\int_{\Gamma_{\text{coil}}} \mathbf{J}^{\text{coil}}(\mathbf{x}) \cdot \mathbf{A}(\mathbf{x})\, d\Gamma(\mathbf{x}) \tag{22}$$

is approximated, according to eqns. (18) and (19), as

$$W \approx \frac{1}{2}\sum_{m=1}^{M}\sum_{n=1}^{M} L_{mn}\, I_n\, I_m, \tag{23}$$

where the components of the inductance matrix $\mathbf{L} = [L_{mn}] \in \mathbb{R}^{M\times M}$ are given by

$$L_{mn} := \frac{\mu_0}{4\pi} \sum_{\substack{m', n' = 1 \\ i(m,m') \neq 0 \\ i(n,n') \neq 0}}^{N} \int_{\Gamma_{m'}} \int_{\Gamma_{n'}} \frac{\mathbf{v}^{m', i(m,m')}(\mathbf{x}) \cdot \mathbf{v}^{n', i(n,n')}(\mathbf{x}')}{|\mathbf{x} - \mathbf{x}'|}\, d\Gamma(\mathbf{x}')\, d\Gamma(\mathbf{x}). \tag{24}$$

The approximated magnetic energy W given by eqn. (23) is a quadratic and positive definite form which induces the following discrete energy norm:

$$\|\mathbf{I}\|_W^2 := \|\widetilde{\mathbf{L}}\mathbf{I}\|^2 = \sum_{m=1}^{M}\sum_{n=1}^{M} L_{mn}\, I_n\, I_m = 2W, \tag{25}$$

where $\widetilde{\mathbf{L}} \in \mathbb{R}^{M\times M}$ such that $\widetilde{\mathbf{L}}^T = \widetilde{\mathbf{L}}$ and $\mathbf{L} = \widetilde{\mathbf{L}}^T\widetilde{\mathbf{L}}$.

The Tikhonov regularized solution $\mathbf{I}_\lambda$ to the inverse problem (8) is sought as [6]

$$\mathbf{I}_\lambda \in \mathbb{R}^M : \quad \mathcal{F}_\lambda(\mathbf{I}_\lambda) = \min_{\mathbf{I}\in\mathbb{R}^M} \mathcal{F}_\lambda(\mathbf{I}), \tag{26}$$

where $\mathcal{F}_\lambda$ is the Tikhonov functional given by

$$\mathcal{F}_\lambda(\cdot) : \mathbb{R}^M \longrightarrow [0,\infty), \quad \mathcal{F}_\lambda(\mathbf{I}) = \frac{1}{2}\|\mathbf{H}\,\mathbf{I} - \widetilde{\mathbf{B}}_z\|^2 + \frac{1}{2}\lambda\|\mathbf{I}\|^2_W, \tag{27}$$

with $\lambda > 0$ the regularization parameter to be chosen. Formally, the Tikhonov regularized solution $\mathbf{I}_\lambda$ of the minimisation problem (26) is given by the solution of the regularized normal equation [6]

$$\left(\mathbf{H}^T\mathbf{H} + \lambda\widetilde{\mathbf{L}}^T\widetilde{\mathbf{L}}\right)\mathbf{I}_\lambda = \mathbf{H}^T\,\widetilde{\mathbf{B}}_z. \tag{28}$$

5 Numerical results

In order to present the performance of the proposed method, we solve the inverse problem (8) for a hemispherical coil $\Gamma_{coil} = \partial B(\mathbf{0}, R) \cap \{z \geq 0\}$, where $R = 0.175\,\mathrm{m}$, whilst the region of interest is a sphere of radius $r = 0.065\,\mathrm{m}$ and centered at $\mathbf{x}^c = (0, 0, 0.081)$, i.e. $\Omega = B(\mathbf{x}^c, r)$. Since the geometry of the coil considered in this paper is symmetrical with respect to the z-axis, it is sufficient to investigate only the design of x- and z-gradients, i.e. $\widetilde{B}_z(\mathbf{x}) = G_x\,x$, $\mathbf{x} \in \Omega$ and $\widetilde{B}_z(\mathbf{x}) = G_z\,z$, $\mathbf{x} \in \Omega$, where $G_x = G_z = 1.0\,\mathrm{T\,m^{-1}}$.

The choice of the regularization parameter λ in the minimisation process of the Tikhonov functional (27) is crucial for obtaining a stable, accurate and physically correct numerical solution $\mathbf{I}_\lambda$. The optimal value λ_{opt} of the regularization parameter λ should be chosen such that a trade-off between the two quantities $\|\mathbf{H}\,\mathbf{I} - \widetilde{\mathbf{B}}_z\|$ and $\|\mathbf{I}\|_W = \|\widetilde{\mathbf{L}}\mathbf{I}\|$ involved in the minimisation of the functional (27) is attained. To do so, we introduce a global measure for error that relates the computed and desired z-components of the magnetic flux density in the region of interest Ω, namely the maximum relative percentage error

$$\mathrm{Err}\,(B_z; \lambda) = \max_{\mathbf{x}\in\Omega} \frac{|B_z^\lambda(\mathbf{x}) - \widetilde{B}_z(\mathbf{x})|}{|\widetilde{B}_z(\mathbf{x})|} \times 100 \tag{29}$$

where $B_z^\lambda(\mathbf{x})$ is the numerical z-component of the magnetic flux density calculated at the point $\mathbf{x}$ in the region of interest Ω, for a given regularization parameter λ, by employing the BEM-based algorithm described in Section 4. On assuming that a deviation $\epsilon > 0$ from the desired z-component of the magnetic flux density $\widetilde{B}_z$ is admissible in Ω, such that

$$\widetilde{B}_z^\epsilon(\mathbf{x}) := \widetilde{B}_z(\mathbf{x})\,(1 \pm \epsilon)\,, \qquad \mathbf{x} \in \Omega, \tag{30}$$

then the choice of the optimal regularization parameter λ_{opt} is made by employing the maximum relative percentage error given by eqn. (29) and the admissible level of noise in $B_z|_\Omega$ defined by relation (30), namely

$$\lambda_{\text{opt}} = \max\left\{\lambda > 0 \middle| \text{Err}\,(B_z;\lambda) \leq \epsilon\right\}. \tag{31}$$

The numerical solution $\mathbf{I}_\lambda$ of the regularized system of normal equations (28), with $\lambda = \lambda_{\text{opt}}$ given by eqn. (31), provides only a discrete distribution of the stream function at the global nodes of the BEM mesh employed. However, these discrete values should be extended to a continuous distribution of the numerical stream function over the entire coil surface Γ_{coil} and this is achieved by employing the contours of the stream function using its discrete distribution and the Matlab (The Mathworks, Inc., Natick, MD, USA) contouring function. Hence, in the sequel, the numerically retrieved solutions of the inverse problem given by eqn. (8) are presented in terms of the contours of the stream function as described above.

Figures 1(a) and (b) present the contours of the stream function in the $\theta - \cos\phi$ plane corresponding to the hemispherical x- and z-gradient coils, respectively, obtained using the optimal regularization parameter λ_{opt} given by eqn. (31), L = 351 internal points in the region of interest and N = 2840 linear, quadratic and cubic triangular boundary elements. It should be noted that, the so-called Lambert cylindrical equal-area projection, i.e. the $\theta - \cos\phi$ plane, has been used to represent the 2D contours of the stream function. From these figures it can be seen that, for the examples investigated in this study, the numerical results retrieved using linear boundary elements are more inaccurate than those obtained by employing higher-order boundary elements, with the mention that there are no major quantitative differences between the contours of the stream function corresponding to quadratic and cubic triangular elements.

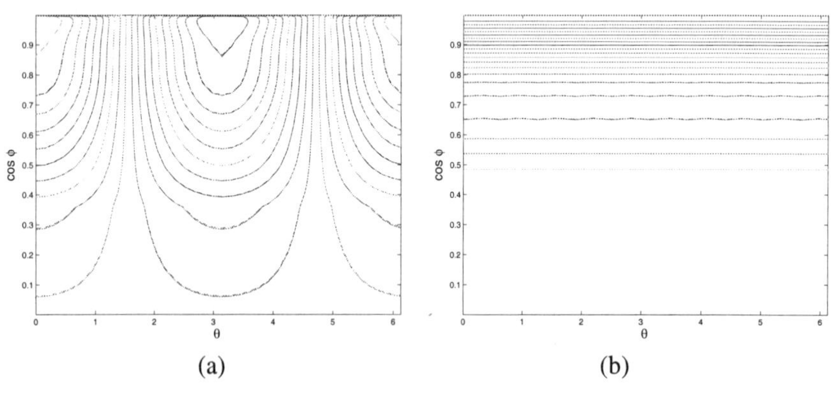

(a) (b)

Figure 1: The contours of the stream function corresponding to the hemispherical (a) x-, and (b) z-gradient coils, obtained using $\lambda = \lambda_{\text{opt}}$, L = 351 internal points in Ω and N = 2840 linear (—), quadratic (– –) and cubic (···) triangular boundary elements.

The convergence of the proposed numerical method with respect to refining the BEM mesh size is illustrated in Figures 2(a) and (b) which present the contours of the stream function corresponding to the hemispherical x- and z-gradient coils, respectively, obtained using the optimal regularization parameter λ_{opt} chosen according to eqn. (31), $L = 351$ internal points in the region of interest and various numbers of quadratic triangular boundary elements ($N_e = 6$), namely $N \in \{1128, 1888, 2840\}$. Although an analytical solution for the contours of the stream function is not available, we can conclude from these figures that the Tikhonov regularization method described in Section 4, in conjunction with the divergence-free BEM presented in Section 3, is convergent with respect to increasing the number of boundary elements used to discretise the coil surface Γ_{coil}. Furthermore, the finer the BEM mesh size is then the smoother contours of the stream function corresponding to the hemispherical x- and z-gradient coils.

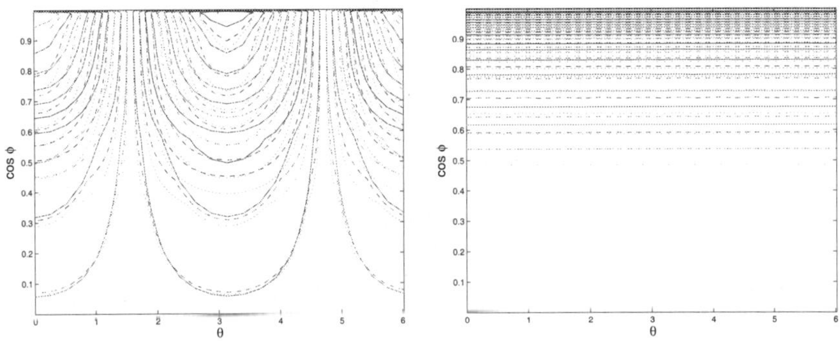

(a) (b)

Figure 2: The contours of the stream function corresponding to the hemispherical (a) x-, and (b) z-gradient coils, obtained using $\lambda = \lambda_{\text{opt}}$, $L = 351$ internal points in Ω and various numbers of quadratic triangular boundary elements, i.e. $N_e = 6$, namely $N = 1128$ (—), $N = 1888$ (– –) and $N = 2840$ ($\cdots$).

6 Conclusions

In this paper, we have investigated the design of hemispherical gradient coils for MRI by considering the reconstruction of a divergence-free surface current distribution from knowledge of the magnetic flux density in a prescribed region of interest. This inverse problem was formulated in the framework of static electromagnetism using its corresponding integral representation according to potential theory. In order to retrieve an accurate and physically correct numerical solution of this inverse problem, a minimisation problem for the Tikhonov functional was solved, in conjunction with a novel higher-order BEM which satisfies the continuity equation for the current density. The numerical solutions were presented in terms of the contours of the stream function and using various types of boundary elements. For the examples analysed, it was proved the efficiency of the proposed method, as well as an improvement in the accuracy of the numerical solutions in the case of higher-order elements. However, there are no major quantitative differences between the contours of the stream function corresponding to quadratic and cubic triangular elements.

References

[1] Turner, R. Gradient coil design: A review of methods. *Magnetic Resonance Imaging*, **11**, pp. 903–920, 1993.

[2] Leggett, J., Crosier, S., Blackband, S. & Bowtell, R.W. Multilayer transverse gradient coil design. *Concepts in Magnetic Resonance B: Magnetic Resonance Engineering*, **16**, pp. 38–46, 2003.

[3] Green, D., Leggett, J. & Bowtell, R.W. Hemispherical gradient coils for magnetic resonance imaging. *Magnetic Resonance in Medicine*, **54**, pp. 656–668, 2005.

[4] Jackson, J.D. *Classical Electrodynamics*, John Wiley & Sons: New York and London, 1962.

[5] Brebbia, C.A., Telles, J.F.C. & Wrobel, L.C. *Boundary Element Techniques*, Springer-Verlag: Berlin and New York, 1984.

[6] Tiknonov, A.N., & Arsenin, V.Y. *Methods for Solving Ill-Posed Problems*, Nauka: Moscow, 1986.

[7] Hansen, P.C. *Rank-Deficient and Discrete Ill-Posed Problems: Numerical Aspects of Inversion*, SIAM: Philadelphia, 1998.

Author Index